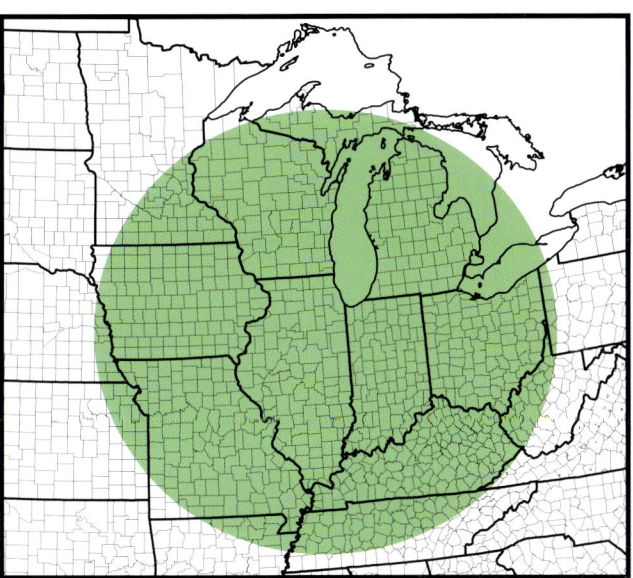

The large majority of all shrubs and woody vines that occur in the
highlighted area are found in this book.

Shrubs and Woody Vines of Indiana and the Midwest

Shrubs and Woody Vines of Indiana and the Midwest

Identification, Wildlife Values, and Landscaping Use

Sally S. Weeks

Harmon P. Weeks, Jr.

with a Foreword by Michael A. Homoya

West Lafayette, Indiana / Purdue University Press

Range maps by Michael A. Black, Rita Blythe, Suzannah Armstrong Rogers, and Amy Wetzel.

All images by Sally S. Weeks.

Drawings in the glossary section are taken from C. S. Sargent, *Silva of North America* (Boston, New York, Houghton, Mifflin and Co., 1890–1902).

Library of Congress Cataloging-in-Publication Data

Weeks, Sally S., 1956-
  Shrubs and woody vines of Indiana and the Midwest : identification, wildlife values, and landscaping use / Sally S. Weeks and Harmon P. Weeks, Jr. ; with a foreword by Michael Homoya.
       p. cm.
  Includes bibliographical references and index.
  ISBN 978-1-55753-610-5 (pbk. : alk. paper) -- ISBN 978-1-61249-144-8 (epdf) -- ISBN 978-1-61249-145-5 (epub) 1. Shrubs--Indiana--Identification. 2. Shrubs--Middle West--Identification. 3. Woody plants--Indiana--Identification. 4. Woody plants--Middle West--Identification. I. Weeks, Harmon Patrick, 1944- II. Title.
  QK159.W38 2012
  582.1609772--dc23
                              2011033215

Contents

Foreword

As a field botanist I am always looking for good reference material regarding our midwestern flora. The early twentieth century works by Charles C. Deam, including his monumental *Flora of Indiana*, have been invaluable and continue to be so, but over the years new information has been garnered, including changes in taxonomy and nomenclature. Until recently our best source for shrubs and woody vines of the state has been the 1932 edition of Deam's *Shrubs of Indiana*. Now 80 years later I am pleased to report that we have a much needed update on that group in this, *Shrubs and Woody Vines of Indiana and the Midwest*, by Sally S. Weeks and Harmon P. Weeks, Jr.

This book is a treasure trove of information about shrubs that botanists and novices alike will find useful. It is replete with excellent photos of multiple observable features of each shrub species, presented in a systematic way to allow quick and efficient comparison between them. The text includes considerable information not only on how to identify the plant in hand, but also details on its habitat, wildlife uses, and possible use in the home landscape. Regarding the latter, such information is particularly pertinent today given the negative consequences and economic impacts that are occurring from those exotic landscaping plants that are invasive and/or vectors of disease and pests.

I have had the pleasure of knowing and working with Sally and Harmon (Mick) for many years and am aware of the considerable expertise and effort that they bring to this publication. They have traveled extensively within Indiana and the Midwest in their hunt to take photos and study the plants in their native habitat. They have also grown many of them on their property and have carefully documented their growth and development. All of this information and much more are included in this attractive and useful book. It is clearly a must-have book for anyone interested in learning about our native shrubs.

Sally credits Charles Deam and his knowledge and enthusiasm for plants as inspiration to produce this important work. Charlie was indeed a dedicated student of our native flora, and he appreciated those people who felt similarly. With this book I can certainly imagine that he would be well pleased. I think you will be too.

Michael A. Homoya
State Botanist for the Indiana
Division of Nature Preserves

Preface

For the past quarter century I have worn out all available midwestern field guides if they had any reference to native shrubs, and I have traveled extensively in search of these woody gems. Some of the classic books that are my favorites include *Flora of Indiana*, by Charles C. Deam; *Michigan Flora*, by Edward G. Voss; *The Woody Plants of Ohio*, by E. Lucy Braun; *Trees and Shrubs of Kentucky*, by Mary E. Wharton and Roger W. Barbour; *Plants of the Chicago Region*, by Floyd Swink and Gerould Wilhelm; *Shrubs and Woody Vines of Missouri*, by Don Kurz; and *Shrubs of Ontario*, by James H. Soper and Margaret L. Heimburger. Publications by Robert H. Mohlenbrock from Illinois are a must as well.

For nearly the same length of time, I have envisioned a field guide focusing on shrubs and woody vines with color plates that would be a more powerful tool to use in educating people about the plethora of natives we might encounter and *could* be using (or at least attempt to use) in our landscapes. In 2008, Welby R. Smith, a botanist with the Minnesota Department of Natural Resources, published *Trees and Shrubs of Minnesota*—a nearly perfect version of my "dream book"—which certainly helped spur me on to write my own. So it is with great excitement (and relief) that this project has come to fruition.

There are several important things my husband, Harmon P. Weeks, Jr., and I want to point out regarding how and why we wrote what we wrote. Our backgrounds are in wildlife and forestry, not horticulture and landscaping. Because of these backgrounds, our ideas of using natives are different from those of someone who has been trained horticulturally. The great botanist Charles C. Deam of Indiana said to know the plants you have to live with them, and so we have. Over the past twenty years, we have propagated and purchased as many native shrubs and woody vines as possible and watched them grow (or not). Much of our experience, and many of our comments, stem from this venture. Our personal landscaping goals have always revolved around attracting wildlife to the area and improving the landscape from an aesthetic as well as a diversity standpoint.

We do not claim that *only* natives or that *all* natives should be planted. There are many desirable introduced species that have proven themselves "well-behaved," which have no invasive tendencies. Many add texture and color that are often not available with our native species. It also would be difficult to plant exclusively and extensively with native shrubs simply because so many are not currently commercially available. Additionally, there are some natives that are adapted to very unique microhabitats that are simply not reproducible in the average landscape.

But the Midwest has long been a stronghold for promoting natives. Jens Jensen and O.C. Simonds created the "prairie style" landscape design in and around Chicago beginning in the early 1900s, and they are credited with helping create a strong midwestern native plant ethic. This style was a version of earlier British natural landscaping, but incorporated native plants from the region. Their style was very popular and utilized extensively in the Chicago area. "Weeds" such as sumacs and dogwoods were incorporated in an effort to capture the essence of nature in an artistic arrangement. It is our hope that our publication will continue that native plant ethic and bring awareness to those who are interested, thereby triggering a supply-and-demand scenario. The more people know about our natives, the more they should request them from commercial nurseries, which will hopefully increase the attention of suppliers to the need for meeting the demand.

One word of caution: this book is not designed to promote illegal harvesting of any kind. When purchasing natives, be sure to inquire as to the propagation methods used by the nursery. Do not support disreputable companies. Many states require permits even for those collecting seeds or cuttings on public lands, which should be the standard practice for propagation of natives.

It has been an absolute joy to work on this project, although there was much frustration with updating information along the way. My co-author and husband has been the best person imaginable with whom to work, and is the brain behind all things wildlife–related. His ability to write all the keys, some of which have never been attempted, has been truly remarkable; making keys understandable and easy to use was a special challenge.

I have to mention my baby brother, Matthew, an engineer, who opened my eyes to seeing things a little differently. Upon a personal request, he set out for a willow patch near his home (and several hours from mine) to see if any were flowering. He called me and said that all he saw were large, caterpillar-like things (the catkins). From that point on, I tried to look at these plants with a bit less of a trained eye in order to describe characteristics in a less technical manner, because he made me realize that many people *probably do* see caterpillars!

Many people have helped over the years, from those located in Tennessee to the Upper Peninsula of Michigan and Ontario, Canada, and many names have not been adequately recorded, to my great embarrassment. But to everyone— thank you. Special thanks go to several folks—

Charles Watkinson, Bryan Shaffer, Katherine Purple, and the late Margaret Hunt at Purdue University Press, who have been very helpful and have supported us through several projects; Michael Homoya, Lee Casabere, Ron Hellmich, Roger Hedge, and Rich Dunbar of the Indiana Division of Nature Preserves, who provided important information for many rare species; Barbara Plampin, Myrna Newgent, Noel Pavlovic, and Dave Hamilla, who all live in and love the Indiana Dunes area and shared their great expertise. Others who have helped along the way include Steve Olson, George Parker, Walter Beineke, Kay and George Yatskievych, George Argus, and Dan Mason.

Range maps were created with much skill by Amy Wetzel, Rita Blythe, and Suzannah Armstrong Rogers. Maps were generated from many sources, including the experience of the authors, but heavy reliance was on the USDA Plants Database (plants.usda.gov) and the Biota of North America Program (Kartesz 2011).

Last but not least, I must mention Charles C. Deam, whose knowledge of and enthusiasm for all plants impressed me many decades ago, and inspired me to attempt this project. I have tried to carry on the tradition.

Sally S. Weeks
July 14, 2011

Introduction

Identification of woody plants is challenging, yet is a fun avocation for some, and a necessary component of the job for others. Regardless, acquiring skills in identification requires dedication and good resources. We hope this book, along with our *Native Trees of the Midwest*, will satisfy the latter and enthuse you sufficiently to achieve the former. Attention to detail and repetition is the key to learning; the more you see plants, particularly in their natural environs with all of their variability, the more adept you become at their identification. Remember one important thing: most species can be identified with just a few crucial distinguishing characteristics. We have supplied both summer and winter keys that use those characteristics to direct you to a specific group or to a particular species. Although we have attempted to use a minimal amount of technical terminology, there are some terms that you will need to know. The illustrations in the next section (Illustrated Glossary) demonstrate specific features and characteristics of plants that should be studied to better understand plant identification.

What is a Shrub?

Technically, a shrub is a woody plant with multiple stems from a common base that grows no taller than 20 feet or so. Of course, there are always exceptions to this "rule," and some species such as *Viburnum prunifolium*, blackhaw, will grow with a single stem, appearing as a small tree. Several species included in this book are considered "sub-shrubs," and they are only woody at the base. Shrubs are much smaller than trees, and shorter-lived when considering individual stems. Most, however, sucker freely from either roots or the base, and can live for many decades in this manner.

Since there are more shrubby species, identification can be tricky; however, unlike the larger trees, there is rarely any difficulty accessing a specimen.

Nomenclature

Nomenclature, or the naming of plant species, is an important part of their study. Scientific (Latin) names and common names are listed for all species; the Latin name is italicized and is recognized worldwide. The common names often vary from region to region and are frequently colloquial (there should be only one recognized and accepted scientific name). These common, or vernacular, names usually result from the way a species is utilized by people in a given region of the country. It is always helpful to learn both common and scientific names.

The scientific names in this book generally follow the taxonomy used in *Flora of North America*, if available. Otherwise, the USDA Plants Database or several regional guides such as *Michigan Flora* and *Plants of the Chicago Region* were consulted. There are sometimes changes in the scientific names, usually as a result of a historically accepted name taking precedence over a current one. Prior names are often mentioned in the text.

Classification

The classification, or grouping, of woody plants into categories is a hierarchy, with each level becoming more unique as they are separated from the previous. The major groups dealt with in plant identification are the family, genus, and species. As an example, pussy willow (the common name) has the specific Latin name *Salix discolor*. The genus is *Salix*; the species name (the specific epithet) is *discolor*; and it is in the family Salicaceae. There are many members of the Salicaceae (willow) family, and there are quite a few members of the genus *Salix;* however, there is only one specific species named *Salix discolor,* because it has features unlike any other shrub species, and a few characteristics unlike any other willow. So the species is a very specific level in the hierarchy. The genus is a broader group covering many similar shrubs, in this case, all the willows. The family is an even broader group that encompasses many species with similar floristic characteristics, including trees such as other willows, cottonwoods, and aspen.

Range Maps

Range or distribution maps have been provided as a general key to the ranges of all our native species. Exact designations of distributions are difficult, and we have incorporated data from many sources, including the Internet, which may not be supported by voucher specimens (actual collected specimens housed in herbaria). Additionally, many species may not have been reported for particular counties in which they occur because of a paucity of survey effort. Thus, while occurrences are indicated on a county-by-county basis, ranges may best be assessed by observing trends in the distribution as opposed to specifics. Ranges of plants are dynamic, being impacted by a species' adaptability and movement by humans, but those of many rare species with specific site requirements have remained virtually unchanged for a century.

Wildlife Values

We have included rather detailed accounts of wildlife values in the text for each species. This information comes principally from the observations of the authors, and, unlike information that is generally available, covers not only food-production characteristics, but also value as cover. The classic work in this area is *American Wildlife*

and Plants (Martin, Zim, and Nelson 1951), which has been used as a major source of information by many authors. We used this reference as well, but tempered our interpretations with personal observations, since the original authors discussed values of plant genera rather than of individual species. Species within the same genus may have strikingly different values to wildlife, and generalizations, thus, are occasionally incorrect.

We have avoided enumeration of wildlife species that use various plant species, since such numbers are largely subjective and biased. However, we have often given examples of wildlife species from the Midwest that frequently use a given species in a particular way, although such lists should by no means be considered all-inclusive. For additional information, the reader is directed to other books that we referenced and found useful. These include Harlow (1942), Halls (1977), Kurz (1997), Leopold, McComb, and Muller (1998), Miller and Miller (1999), and De-Graaf (2002).

Landscaping Value

The notion of using native "woodies" as ornamentals seems to be foreign to many, and nurseries supply what is in demand. Unfortunately, the demand often seems to be for something new, different, and usually exotic. However, in recent years, some nurseries have begun to recognize the importance of planting native species and are offering a limited number of shrubs and vines. Many of the natives listed in this volume are logical choices when selecting ornamentals for landscaping. They are well-adapted to the midwestern climate and tend to be more disease- and insect-resistant than exotics. There is virtually something for everyone—they come in all shapes and sizes. The information provided for each species will give the reader an idea of its requirements and what it will provide in return. Fall color, shade qualities, and longevity are some details discussed. For more information, suggested readings include Hightshoe (1988) and Dirr (1998).

Several terms that are commonly used need

to be discussed here. **Cultivars** (derived from "cultivated variety") and **varieties** are basically the same, and are a subdivision of a species. They are horticultural variants that are usually developed from a wild shrub and breed true, but are maintained under cultivation by sexual or asexual means. **Form** is also a variant of a species that has a subtle difference from the true species, such as a different color flower, but does not breed true. They are usually derived from a wild plant and can be maintained in cultivation by asexual reproduction.

Similar Species Distinctions

Boxes are provided for each species. They offer those species most closely resembling, or those most often confused with, the species and provide clues for distinguishing among them.

Factors Influencing the Distribution of Midwestern Shrub and Vine Species

This field guide concentrates on those shrubs and woody vine species found in the region generally known as the Midwest, with its epicenter roughly located in the northwestern corner of Indiana; however, most of the species included are widely distributed throughout the geographic areas known as the Central Hardwood and Lake States regions. These regions include over 400 million acres, extending from the prairies in eastern Nebraska to the foothills of the Appalachian Mountains in eastern Ohio and from northern Tennessee to northern Michigan.

The general distribution of shrubs and woody vines within the region is influenced by east-west moisture gradients and north-south temperature gradients. These same environmental factors influence tree distributions, and most "forest types" are classified by these overstory trees, not by the smaller understory woody plants. Nevertheless, they are all part of the floral community and for the most part vary together. The oak-hickory forests in the drier western part of our region may have understories

with viburnums and greenbrier, even with blueberries on the dry ridges. In the more mesophytic northern and eastern parts of our region, understories may be more likely to have leatherwood and spicebush. In deciduous forests in the southern part of our region, overstory trees with southern affinities (e.g., southern red oak and yellow buckeye) appear in the overstory, and similar "southern" shrubs and vines (e.g., raccoon grape and swamp dogwood) begin to populate the understory. A quick look at the distribution map for each species will indicate which species occur widely throughout the Midwest and which are limited to the northern or southern extremes of our region.

With shrubs and woody vines, much more so than with trees, climatic influences on distribution are tempered by factors such as light, moisture, soil factors, landscape position, and aspect. Many shrubs and vines occur in forest understories, but a far greater number occupy or do best in open situations, where they do not have to compete with tall canopy trees for sunlight. These oases of light can come in a variety of forms, each with a unique woody plant community: an open bog with few or no trees, an abandoned agricultural field or pasture; a dry, sandy pineland that burns frequently; or a savannah with only scattered overstory trees. Even in forested areas, shrub diversity is greatest near edges where light is more available, and vines often fight their way into the treetops for that precious resource.

Moisture is second to light availability as a factor influencing occurrence. Shrubs seem especially adept at evolving ways to occupy habitats replete in or devoid of water. A northern bog may have one or two tree species but a dozen or more shrubs that seem to thrive in that waterlogged world. A dry, sandy plain may have a similarly depauperate tree diversity and many low shrub species that seem oblivious to the perpetual drought.

Soil type and origin is important in woody plant distributions. Some species seem to compete best in certain soil types (although many are bimodal)—from deep sands to heavy organic soils. Probably most pervasive is the dependence of many species on soils of a certain pH—some requiring acid soils, others doing best (or only occurring) on soils of limestone origin. We have tried to point out such requirements when we discuss the distribution and landscape values of individual species.

Other factors that influence distribution—slope, landscape position, aspect—are all actually functions of the former three. Ridges in our region tend to be dry with acid soils; riparian bottomlands tend to be moist with alkaline soils. North-facing slopes are more moist (and cooler); south-facing slopes drier (and warmer). Steep slopes tend to be drier toward the top and more moist toward the bottom, often with very wet seeps near the base. Woody plant communities ebb and flow with all of these variables; a set of conditions may produce a rather scattered distribution or poorer plant condition for one species, while it may lead to a total absence of another.

Disturbance from human activities has been widespread across the region since the 1400s, when Native Americans were most abundant in the region. Activities such as land clearing, drainage, grazing, and fire, common from the mid-1800s to the 1940s as Europeans colonized the area, have had the greatest influence on the composition of forests and other habitats present today. In the southern part of the region, ridges, stream valleys, and many slopes were cleared and farmed. Grazing on remaining lands left forests with little understory. By the 1930s many of these lands were abandoned because of severe soil erosion, and large areas were returned to public ownership to form state and federal parks and forests. Subsequently, abandoned agricultural lands have succeeded into brushlands and then seral woodlands.

The more level northern parts of the Central Hardwood Region continued to be cleared for intensive agriculture, so that remaining native habitats exist as small woodlands or narrow riparian strips along streams. These habitats cover about 6 to 10 percent of the land but are rich in species. Very diverse shrub and vine communities also dot the landscape in patches that have been minimally impacted by humans, largely because they could not be cleared for farming. These systems include bogs, fens, some swamps and marshes, and sand ridges and dunes.

Increased protection of remaining forests from fire and grazing since the 1940s in the Central Hardwood Region has allowed forest understories to redevelop, albeit somewhat diminished in shrub and vine diversity as a result of the previous uses. The overstory is likely to be oak/hickory, but the understory may be quite variable, depending on the type and duration of former disturbances. These understories not only contain shrubs and vines, but also regenerating young trees that will move into the canopy as the forest ages. In much of the Central Hardwood Region, dominant understory trees are shade-tolerant sugar maples, American beech, and elm, signaling a future condition that will likely impact the understory shrub and vine communities as well.

Much of the land in the Upper Lake States Region has returned to forest cover following the extensive logging and clearing for farming in the late 1800s. Historical fires following logging greatly increased the widespread occurrence of quaking aspen and paper birch across the region. This type also has a rich and unique shrub community. Fortunately, much of this area has been transferred to public ownership and is thus accessible for citizen's enjoyment.

Illustrated Glossary

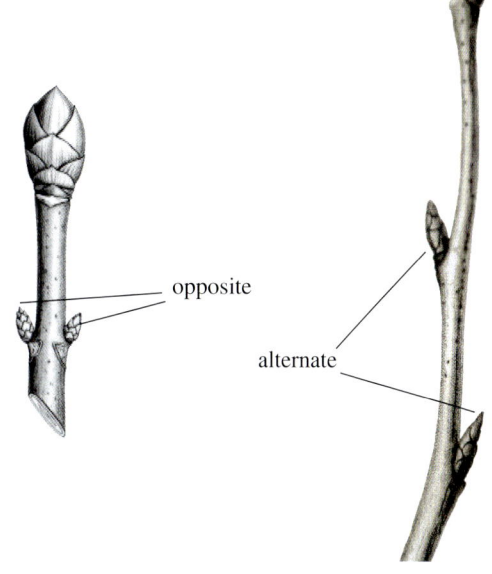

There are many features used to determine the species of a given shrub or vine. First and foremost, you need to determine whether the leaves and branches are arranged in an **opposite** or an **alternate** fashion. This is an important feature to recognize, because if your specimen is opposite, you have ruled out roughly 75% of all the possibilities.

opposite

alternate

Leaves, if available, are highly variable but are usually a useful feature for proper identification; however, they are only available during the growing season (unless the species is evergreen!). There are many shapes, sizes, and textures of leaves. Leaves are categorized by whether they are **simple** or **compound**. Simple leaves are attached directly to the twig, either beside or on top of a bud, whereas compound leaves have various numbers of **leaflets** attached to a stalk (**rachis**) that is attached to the twig where a bud occurs. The stalk that attaches a simple leaf to a twig is called a **petiole**. Most shrubs and vines have simple leaves. In our native shrubs and vines, compound leaves are either **pinnately** compound or **palmately** compound. To determine whether your specimen is a simple leaf or a leaflet from a compound leaf, look for a bud at the base of the petiole or rachis.

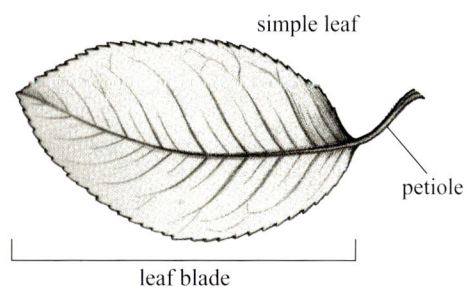

simple leaf

petiole

leaf blade

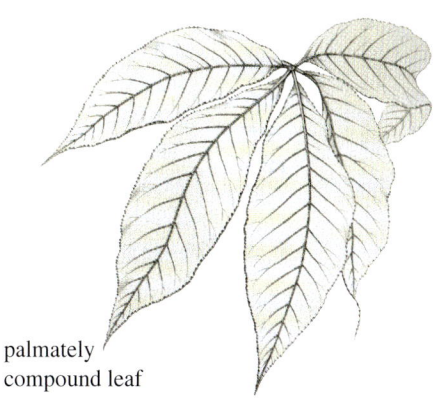

palmately compound leaf

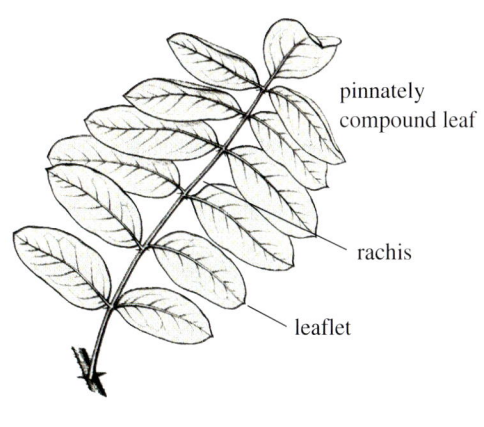

pinnately compound leaf

rachis

leaflet

5

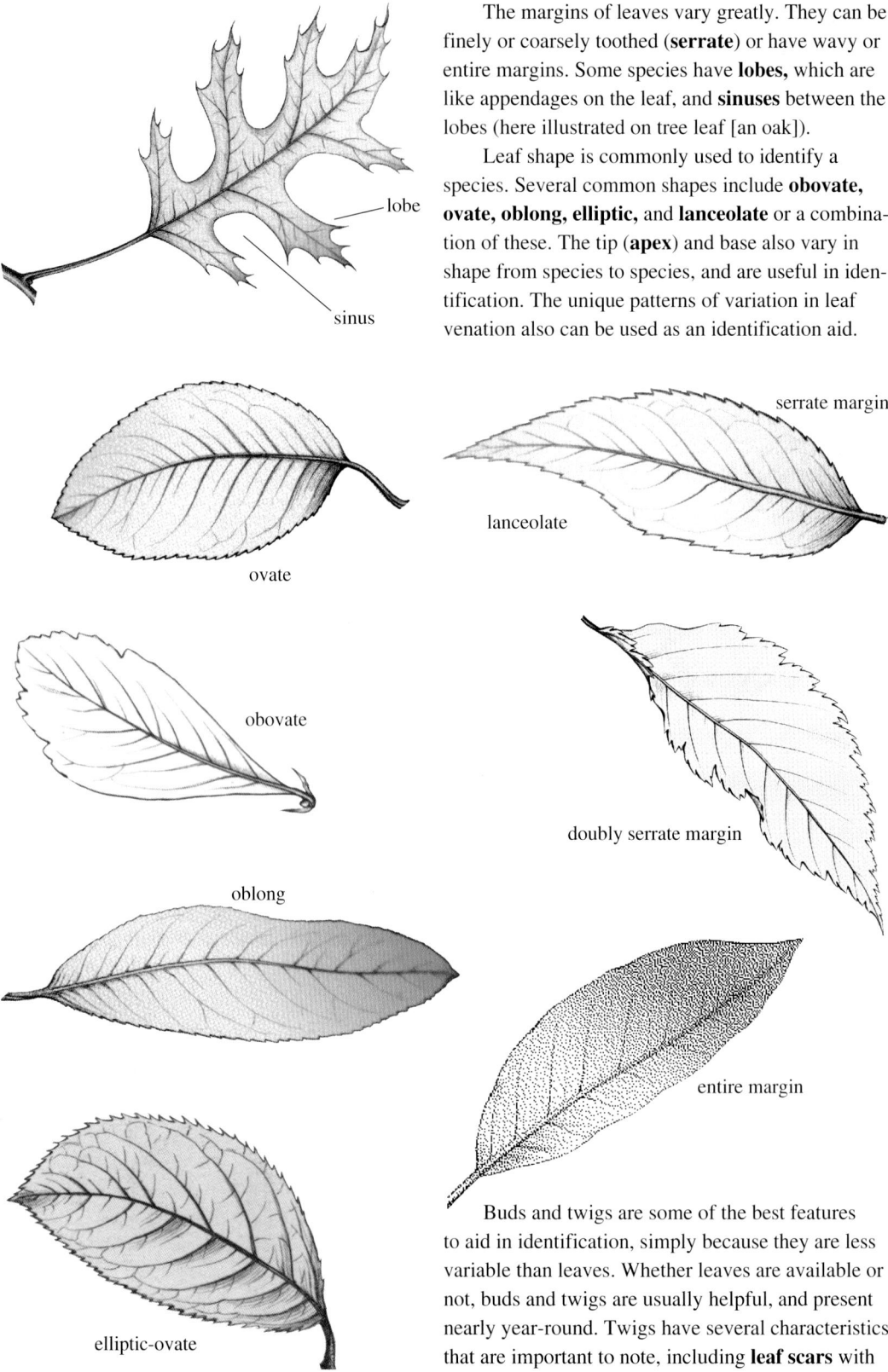

The margins of leaves vary greatly. They can be finely or coarsely toothed (**serrate**) or have wavy or entire margins. Some species have **lobes,** which are like appendages on the leaf, and **sinuses** between the lobes (here illustrated on tree leaf [an oak]).

Leaf shape is commonly used to identify a species. Several common shapes include **obovate, ovate, oblong, elliptic,** and **lanceolate** or a combination of these. The tip (**apex**) and base also vary in shape from species to species, and are useful in identification. The unique patterns of variation in leaf venation also can be used as an identification aid.

lobe

sinus

serrate margin

lanceolate

ovate

obovate

doubly serrate margin

oblong

entire margin

elliptic-ovate

Buds and twigs are some of the best features to aid in identification, simply because they are less variable than leaves. Whether leaves are available or not, buds and twigs are usually helpful, and present nearly year-round. Twigs have several characteristics that are important to note, including **leaf scars** with their differing number and patterns of **bundle scars**, **lenticels**, **stipule scars, thorn-like structures**, and **pith**. These characters vary from genus to genus, and even among species within a genus. While they are usually unique to a species, they are often similar enough to other closely related species to group it with a given taxon, such as the sumacs.

    Leaf scars are found on a twig where a leaf was attached. They are best observed after a leaf has fallen naturally. Within the scar are 1 to many **bundle scars**, which mark the spots where nutrients and water were passed to the leaf. **Lenticels** are raised slits, usually pale in color, that are "breathing" sites for the twig. **Stipules** are small, leaf-like appendages that are at the bases of leaves and usually shed with those leaves, leaving a scar near the top of the **nodes**. Some twigs have thorns, **prickles,** or **spines**. They can be scattered along the twig or paired at the nodes. **Pith** is the center or core of the twig. It is often substantial and colored; and it may be solid, **diaphragmed** or **chambered**.

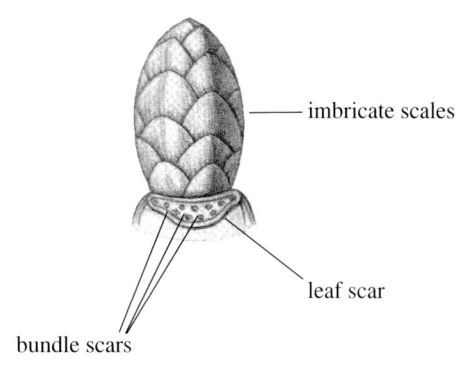

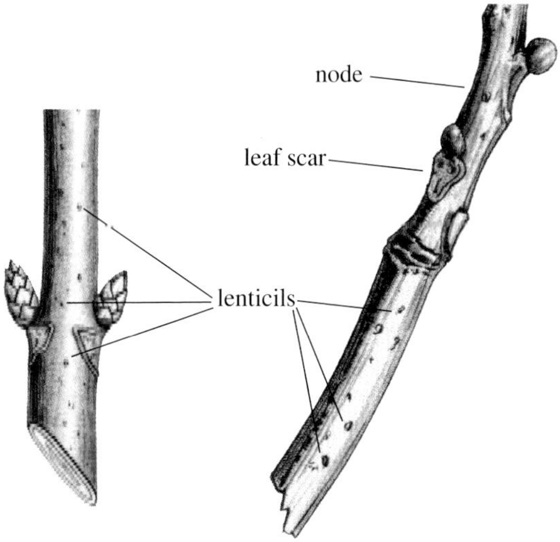

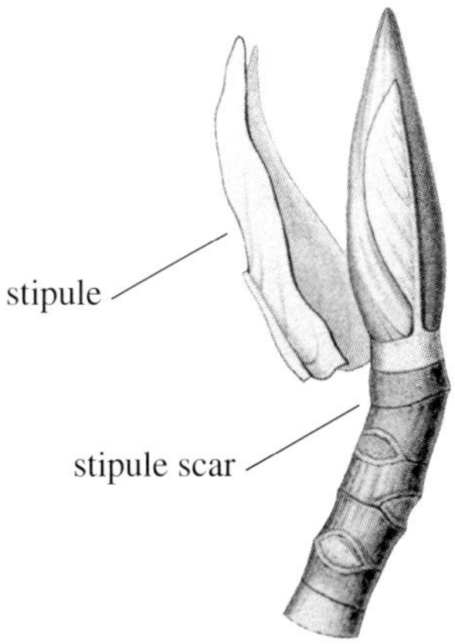

imbricate scales

leaf scar

bundle scars

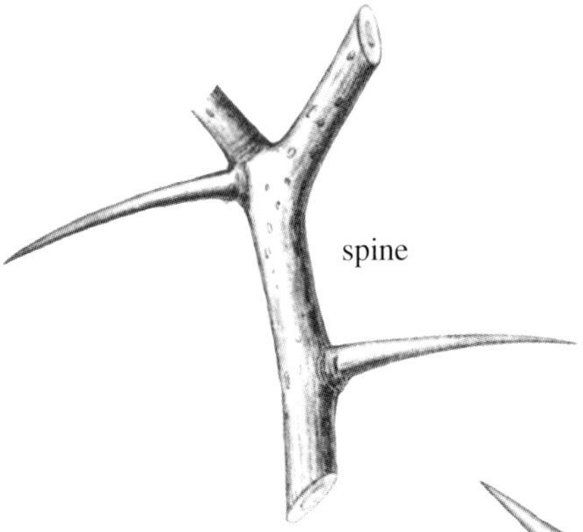

node

leaf scar

spine

lenticils

prickle

naked bud

stipule

stipule scar

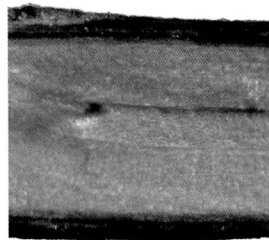

solid pith (willow)

chambered pith
(black walnut)

diaphragmed pith
(blackgum)

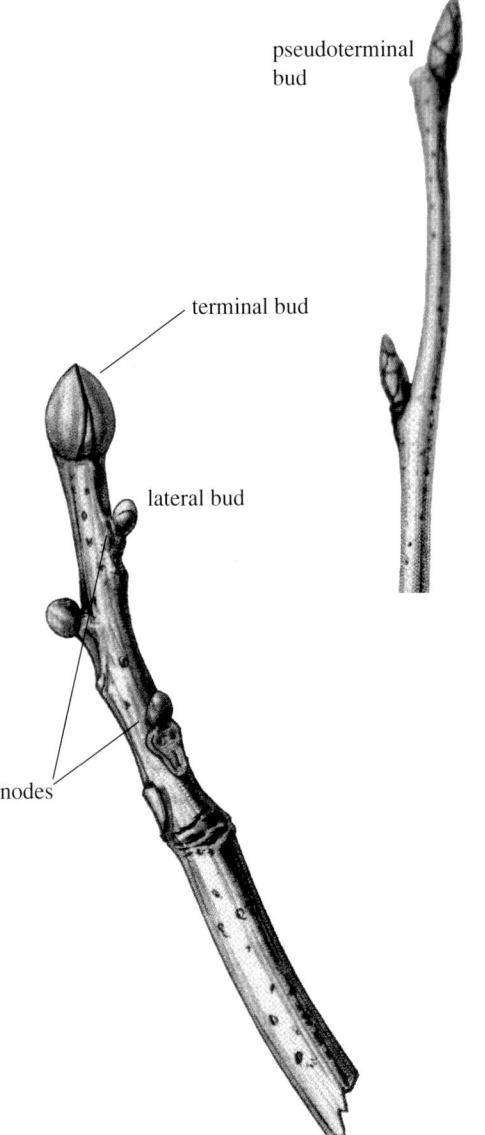

pseudoterminal
bud

terminal bud

lateral bud

nodes

Buds are usually covered with a varying number of scales. There can be a single scale, as in the willows, or numerous, **imbricate** (overlapping) scales, as in the serviceberries. Some buds, like those of poison-ivy, are **naked** buds, meaning that there are no bud scales. There are various textures and colors of buds and their scales.

Buds at the ends of twigs are classified as being either **terminal** or **pseudoterminal**. A true terminal bud is positioned exactly on the tip of the twig; a pseudoterminal or false terminal bud is a **lateral bud** that has taken the position of a terminal after the branch tip is shed in the fall. You can usually distinguish between the 2 by the presence of a branch scar beside the pseudoterminal bud.

Flowers are relatively easily used as a distinguishing feature on shrubs and vines, largely because they are within easy reach. They are, however, fairly short-lived, so it is necessary to time a visit to an unknown species properly. Within some taxa, there are really no recognizable differences between the species' flowers. For instance, the flowers of most viburnums are not discernable to species, so identification is not possible by this feature alone. A basic flower structure of an angiosperm is illustrated. It represents a perfect flower with functioning male and female parts.

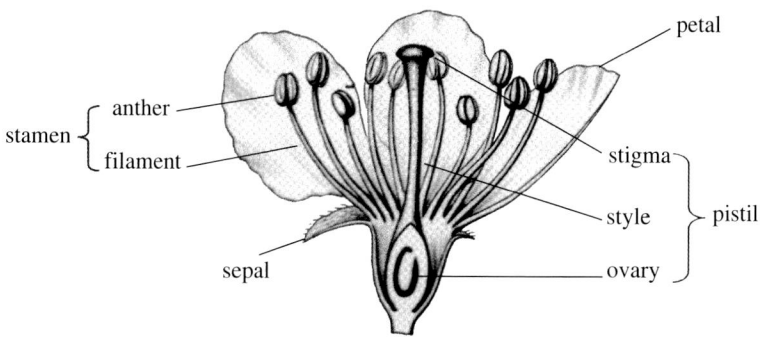

A perfect flower with functioning male and female parts.

Fruit is an excellent characteristic for distinguishing among taxa. Most fruit matures in the late summer and fall. The majority of our native shrubs and vines produce fleshy fruits, and depending on numerous factors, including overall fruit abundance that year and preference by wildlife, they may be available as an identification tool. Some species naturally retain fruit well into the winter, while in others fruits fall or are taken almost immediately by birds or mammals.

Bark is another feature of woody plants useful for identification, but it is rarely as well-developed on shrubs and vines as it is on trees. Life span, especially of individual stems, hinders the development of much bark. Some shrubs, such as bladdernut and round leaf dogwood, have unique coloration or patterns along their stems. Several vines, such as Virginia creeper, are long-lived, and develop distinctive, tree-like bark.

Form is a characteristic often used to identify open-grown trees. Generally, that is not used with shrubs or vines. Size categories such as small, medium, or large might be more appropriate descriptors. There are several species with unique form, such as leatherwood, and the creeping, evergreen groundcovers. But for the most part, the majority of native shrubs are medium- to large-sized, multi-stemmed, and bushy. Occasionally, some species such as blackhaw develop a single, small, tree-like stem. Shade-grown shrubs are typically more spindly. However, if the same species is planted as an ornamental, where it is receiving more light and generally being pampered, it will often look completely different.

Vines are usually climbing up and over other plants, and often there seems to be no limit to how large they can grow. As long as there is a support, most vines will continue climbing up, so size and form descriptions are rather pointless.

Willow identification is a rather complex task. For best results, the specimen in question should be observed during the entire growing season. The two most instructive characteristics are the female flower or fruit (the entire **catkin** [also known as an ament] is best) and the medial leaf. A **medial leaf** is fully developed and found halfway between the top and bottom of a branchlet that was borne in the spring. Several other terms used in the *Salix* key include the **proximal** end of a leaf (toward the base) and the **distal** end (toward the tip).

Each female catkin is comprised of numerous ovaries; unique characteristics of the ovaries and their placement on the ament are keys to identification. In addition, **ovary, stigma,** and **floral bract** (found at the base of each ovary) shape and texture are commonly used in species determination and often mentioned in our keys. Species vary in how densely (or loosely) flowered (or fruited) they are, that is, how tightly packed the ovaries are on the ament. Illustrated is the densely flowered catkin of *Salix discolor.*

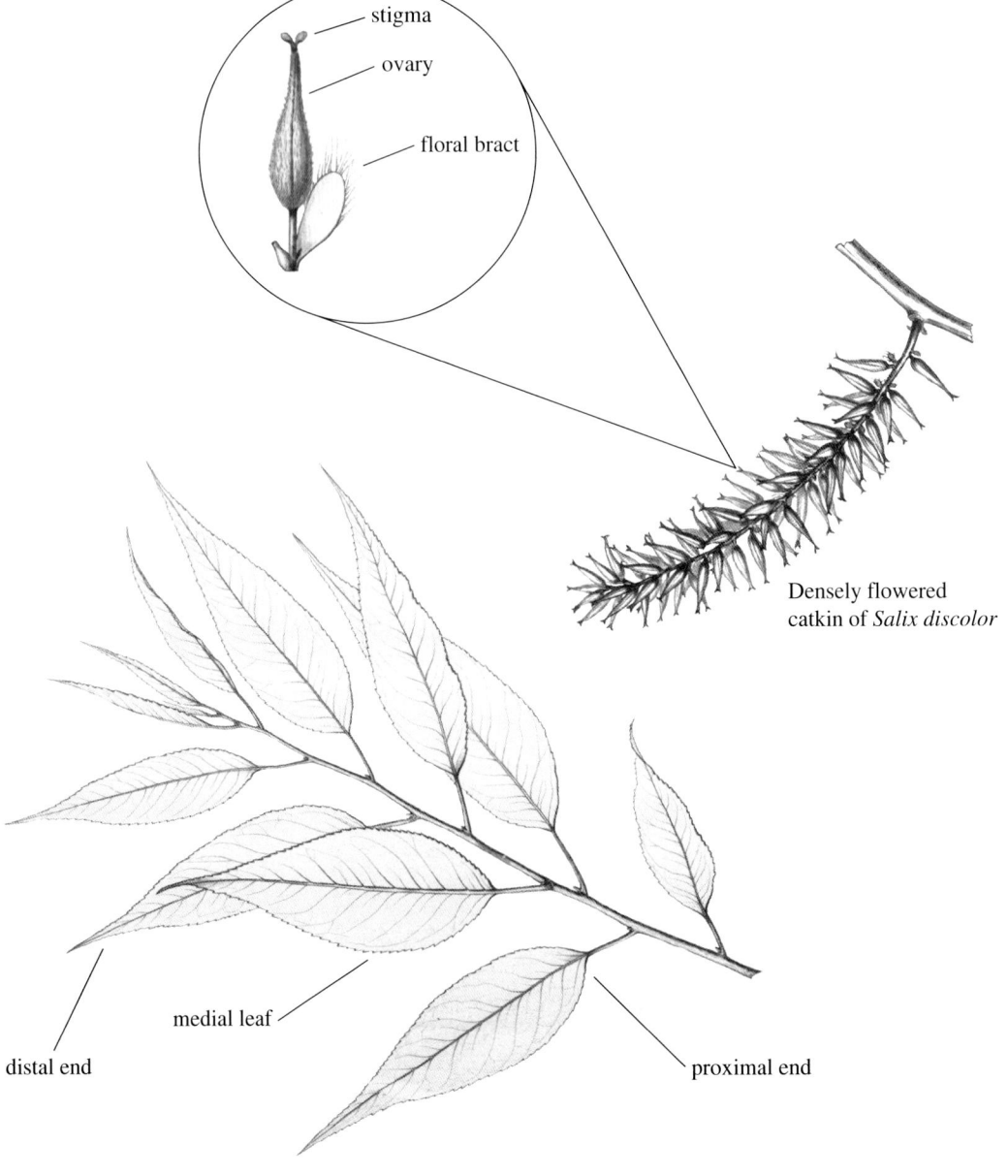

stigma

ovary

floral bract

Densely flowered
catkin of *Salix discolor*

distal end

medial leaf

proximal end

NATIVE SHRUBS

Conifers

common juniper, ground juniper

Juniperus communis L.
Juniperus communis var. *depressa* **Pursh.**
Family: Cupressaceae

Common juniper is 1 of only 3 coniferous shrubs native to the Midwest. Creeping juniper, *J. horizontalis* Moench, is found in the upper Midwest and does just what its name suggests—creeps along exposed bedrock and sandy dunes. Common juniper occurs from coast-to-coast in northern North America, and it is highly variable across its range. It also holds the record for having the largest natural range of any tree or shrub in the world. It is currently listed as state rare in Indiana, threatened in Kentucky and Illinois, and endangered in Ohio.

Form and Size: Common juniper is a low-growing, spreading, multi-stemmed shrub to 4 feet in height and is usually wider than tall. It forms dense, low mats that creep along rocky, dry sites and is rarely over 1 foot tall. This species is reportedly long-lived for a shrub—150 years or more.

Habitat: It is found on stable dunes, growing in full sun along the shores of the Great Lakes. However, other habitats include rocky fields and slopes and abandoned fields. Common juniper requires well-drained soils, but it will tolerate a wide range of soil pH.

Wildlife Uses: Common juniper plays an important cover role in the open landscapes in which it often occurs, supplying thermal cover for wintering birds and nesting and escape cover for birds and small mammals. It has many of the food values for wildlife as its larger congener, *J. virginiana* (Weeks, Weeks, and Parker 2010). It is browsed lightly but consistently by white-tailed deer. Its fruits, however, are highly prized and used heavily by small mammals, game birds (bobwhite, ruffed grouse), many songbirds (notably evening and pine grosbeaks, purple finch, and cedar waxwing), and woodpeckers.

Landscaping Value: Hardy to Zone 2, this species has limited use in the southern part of the Midwest because of intolerance to high humidity. It is adapted to some of the harshest growing environments imaginable, but it prefers well-drained sites. It is slow-growing and long-lived. It makes a nice, low ground cover, but some plants' needles turn a brownish color during winter months. Needles appear blue-green and are sharp-tipped. The bluish fruit overwinter unless taken by wildlife. There are several horticultural varieties that have been developed, which are sporadically found on the market today.

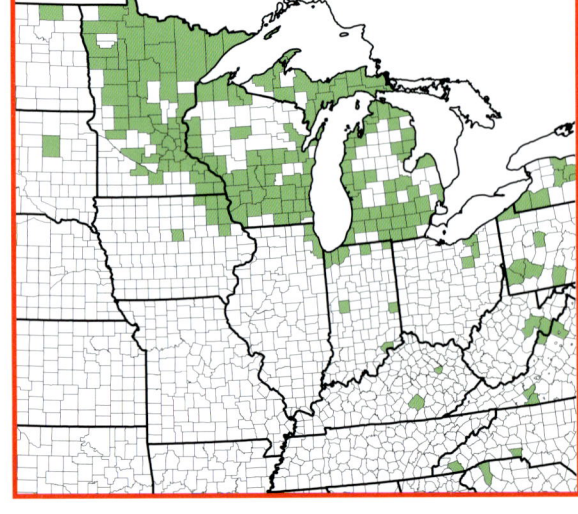

Similar Species Distinctions:
—No other native, shrubby conifer has sharp, awl-shaped needles. Introduced junipers are common in landscapes.
—**Creeping juniper** (*J. horizontalis*) needles are tiny, scalelike and stacked 1 on top of another.

Needles are evergreen, whorled around the twig in 3's, and very sharply pointed. They are up to ¾ inch in length and diverge from the stem. There are no scale-like needles as are found on *J. virginiana*. The upper surface has a wide, central band of stomata, while the lower surface is entirely green. A close look at these photos will reveal the presence of 1 of the juniper scale parasites, *Carulaspis* spp., which seems to be prevalent on plants in the Dunes of northwestern Indiana. As in all junipers, twigs are brown with many grooves and ridges that help to create the appearance of woody, armor-like scales. The needle attachments are slightly raised ridges.

Fruit is a fleshy, berry-like cone that ripens in the fall after 3 growing seasons. They are about ⅓ inch long and dark blue with a whitish coating when ripe. Unlike other junipers, common juniper cones are axillary, not terminal. Each cone generally contains 3 angled seeds.

Flowers are separate sex, usually on separate plants, and appear in late April to early June in northern Indiana. They are tiny, ⅛ inch or less, axillary, and covered with numerous pointed, greenish scales. Males are pictured above left, females above right.

Juniperus horizontalis, creeping juniper

Mature bark is gray-brown and thin, with many shreddy strips that exfoliate with age. The inner bark is an iridescent purple.

Canada yew, American yew

Taxus canadensis **Marsh.**

Family: Taxaceae

Yews are encountered in every city and town in the Midwest. Unfortunately, these are the introduced Japanese yew, *T. cuspidata*, and a hybrid, *T.* x *media*. It is currently listed as threatened in Indiana and Kentucky, and endangered in Tennessee.

Form and Size: It is a somewhat prostrate, spreading shrub with ascending branches that can reach 7 feet in height; a large shrub spreads to cover a sizable area. It is a shade-loving species, but it can be found growing in full sun in more northerly regions. Branches that come in contact with the ground root, and the species can eventually form dense colonies.

Habitat: Yew is found in a variety of habitats across its midwestern range, including coniferous or mixed hardwood forests on steep ravines, gravely slopes and banks, and forested swamps. It prefers acidic soil.

Wildlife Uses: In spite of the fact that Canada yew is toxic to livestock, it is browsed heavily by white-tailed deer, to the degree that it has been essentially eliminated in areas peripheral to its major distribution, occurring only in locations essentially unreachable by deer. It is also a major winter browse for moose; rabbits seem not to prefer it. Its unique red fruits are taken by songbirds and game birds, such as ruffed grouse, but do not seem highly preferred. As 1 of our few coniferous shrubs, it supplies important winter cover for mammals and birds; similarly, it is frequently used for nesting by songbirds, especially early nesters, such as song sparrows.

Landscaping Value: Hardy from Zones 2–6, this species' range is restricted in the Midwest by climate. It does not tolerate hot, humid summer weather. It is never used as an ornamental and is rarely found in native-plant nurseries. Its form is described as straggly when compared to the over-used introduced species mentioned above. Growth rate is slow, but poor soil and drought do not seem to be an issue for this plant. Poorly drained soil is not tolerated. Its spreading nature makes it ideal for a ground cover where deer are not a problem.

Similar Species Distinctions:
—**Hemlock** (*Tsuga canadensis*), a tree, has similar, short needles that have white lines underneath.
—**Balsam fir** (*Abies balsamea*), a tree, has similar needles that also have white lines underneath.
—**Japanese yew** leaves are usually not 2-ranked and sit upright along the twig (creating V-shapes).
—*T.* x *media* has distinctly 2-ranked leaves.

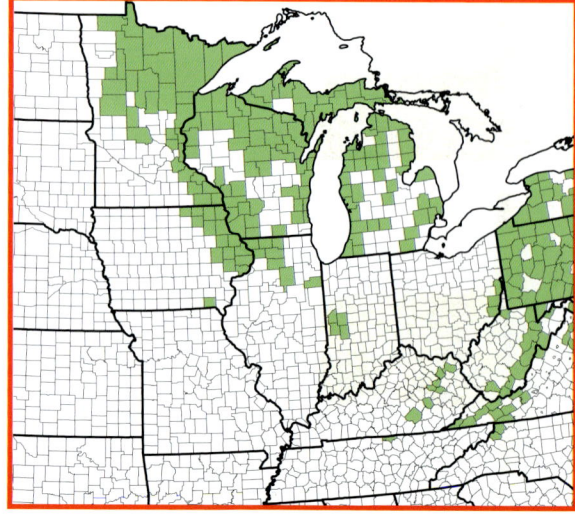

This evergreen conifer has needles that are about ¾ inch long, green on the top, and light green on the bottom. The needle tip is sharp but not painfully so. Needles are usually spirally arranged around the twig, but they are attached to a short stalk (sterigma) that allows them to tilt away from the twig in any direction. This sometimes gives the illusion that the shrub has opposite needles. Fall color is often reddish brown.

Fruits ripen in the fall and are about ⅓ inch in diameter. They have red, jelly-like flesh that surrounds a single, black seed. The fruit is an aril, and the seed is reportedly toxic.

The new twig color is lime green, but with age, it turns brownish and develops cracks running along it. Green, lateral flower buds appear in the early spring, but they soon open and are gone.

The tiny, green flowers appear in April in the leaf axils. Each flower is only ⅛ inch long at most. Male and female flowers are on separate shrubs. The male "cones" (below) appear tan-colored when releasing pollen. Females (left) have green scales and are often tipped with a drop of liquid.

The bark is brown, and it develops slight fissuring and flaking with age.

NATIVE SHRUBS

Broadleaf

partridgeberry, partridge berry

Mitchella repens **L.**

Family: Rubiaceae

This dark green, trailing, evergreen ground cover is a delight in the winter when almost every other plant is brown from the cold. Add in the persistent, red fruit, and you get the feeling of Christmas nearly year-round. Partridgeberry is listed as threatened in Iowa.

Form and Size: Partridgeberry is a prostrate, creeping shrub that forms mats many feet long and wide. It is nearly impossible to determine where a single stem begins and ends, as all stems are intertwined.

Habitat: Its habitat varies from black sand of oak woods (rich in organic matter) and beech and maple deciduous woods and coniferous woods in the northern part of our region, to low, flat, sweetgum and beech woods and crests and slopes of sandstone ridges and outcrops in the southern parts. This species requires a fairly acidic soil, but it makes a good ground cover for shaded locations.

Wildlife Uses: Partridgeberry's prostrate, creeping form supplies little wildlife cover value, except in unusual circumstances. Its fruits are not produced in large quantities, but they are used by game birds (e.g., ruffed grouse and wild turkey) as well as by several mammals (e.g., raccoons, red fox, bear). Leaves are regularly taken by white-tailed deer and ruffed grouse, occasionally forming a substantial portion of the diets.

Landscaping Value: Hardy to Zone 3, this tiny, creeping ground cover grows slowly but has several great attributes. It has dark green, tiny leaves with white veins, an extended flowering period of several weeks, and dark red fruit that persists throughout the winter. It does require well-drained, acidic soil and lots of shade. Stems root at the nodes, making propagation by cuttings easy. A white-fruited form *leucocarpa* is available, but partridgeberry is rarely found in nurseries.

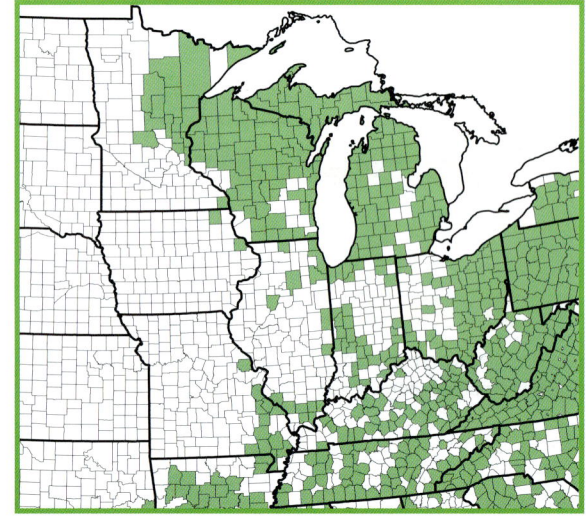

Similar Species Distinctions:
—There is nothing like partridgeberry, either native or introduced, that would create confusion.

Leaves are opposite, fleshy, evergreen, and have short petioles. The terminal pair tends to be cocked to the side of their petioles, while lower leaves have longer petioles and diverge from the twigs. Leaf margins are entire, and the central vein, in particular, is pale-colored when compared to the deep green of the upper leaf surface. The lower leaf surface is somewhat paler.

The white flowers are funnel-shaped and usually paired at the terminal end of the twigs. Each flower is about ½ inch long and has 4 fused, hairy, re-curved petals. Flowers are perfect and appear over a several month period beginning in early May in the southern parts of our region. Swink and Wilhelm (1994) in *Plants of the Chicago Region* have a late-flowering date of September 5.

Buds are only visible at the terminal end of the slender, trailing stems. Flowers and vegetative growth occurs from this point, where buds are compressed between the petioles of the terminal pair of leaves. Buds are pale green and few scaled.

Fruits are berries about ¼ inch in diameter that resemble an inflated pair of shorts because the ovary is shared by paired flowers. They ripen in the fall. Fruits are nearly always scarlet-colored, but they are occasionally white. Each berry contains about 8 seeds. Fruits are persistent, and they are commonly found throughout the winter.

Twigs are slender, greenish or reddish, and covered with short, straight, stiff hairs. Older growth becomes brown and smooth. No lateral buds are visible, but rooting commonly occurs at the nodes of the twigs.

American mistletoe, American Christmas-mistletoe, oak mistletoe

Phoradendron tomentosum **(DC.) Engelm. ex A. Gray**
Phoradendron leucarpum **(Raf.) Reveal & M. C. Johnst.**
Phoradendron serotinum **(Raf.) M. C. Johnston**
Family: Viscaceae

This native woody plant is classified as a shrub, but it lives as a semi-parasite on some of our common trees, including black walnut and American elm. It reaches the northern-most limits of its range along the Ohio River, and it is increasingly common as one heads south from there. It is especially evident in the winter, since it is evergreen. The genus *Phoradendron* is Greek, and it literally means "tree thief." It is listed as extirpated in Pennsylvania.

Form and Size: Mistletoe grows very slowly, commonly adding only a pair of leaves per twig per growing season, so a large, 3-foot specimen is old. It is usually found growing in the upper part of mature tree canopies of a variety of species. In the South, it tends to prefer oak species. Mistletoe sends modified roots (haustoria) into the xylem of trees, where it acquires necessary water and nutrients. Breaking the parasite from the tree branch does not kill the mistletoe plant. Extensive parasitism on a single tree can weaken the tree's immune system, which sets the stage for insect and disease attack.

Habitat: Mistletoe grows exclusively on the branches of trees. In southern Indiana, elm trees (when they were common) were its preferred host. Farther south, oak trees seem to be its target species, particularly willow oak.

Wildlife Uses: This woody parasite grows on older deciduous trees in the southern portion of our region, giving some green context even in the winter. These "bunches" of mistletoe give a dense substrate that songbirds occasionally select as a nest site; although not reported, we would be surprised if birds did not also choose these plants as roost sites in the winter, as they do dead leaf clumps on marcescent trees. The fruits are eaten by numerous songbirds, especially cedar waxwings; shrubs are reportedly browsed occasionally by white-tailed deer, but the clumps are usually high in trees, well out of reach of deer.

Landscaping Value: Probably hardy to Zone 6, this shrub is not recommended, as it is a parasite on trees that deforms branches where it is attached.

> **Similar Species Distinctions:**
> —This is the only mistletoe in the eastern United States. Nothing is similar.

Leaves are opposite, oblong to obovate, thick and leathery, and up to 2 inches long. They generally have 3 major veins that run parallel to the leaf margins. Leaves are yellow-green on both sides, and they may have a few scattered, short hairs. Leaf margins are entire and the tip is rounded. Deam (1932) claimed that the leaves of the female plants are darker than those of the males.

Fruit matures in 1 or more growing seasons in clusters on upright spikes. The berry is whitish, translucent, and about ¼ inch in diameter. Fruit tissue is sticky, which helps seeds to adhere to bark when deposited there by birds. Fruit ripens in November and persists throughout the winter.

Twigs are green, somewhat rough, thick, and brittle. Newer growth has short, pale, scattered hairs. The green twigs have photosynthetic capabilities. Twig branching is opposite, giving the plant a rather angled appearance. Nodes are commonly jointed, a feature not found in the Loranthaceae family, in which mistletoe was once placed. Shown immediately above is a branch deformity caused by the attached mistletoe.

Flowers appear in mid-to-late October over a several week period. They are born on upright spikes that develop from the leaf axils. Flowers are separate sexes, tiny, and yellowish green with 3 tepals (modified petals). Males are seen in top photo; females in bottom photo.

downy bog rosemary

Andromeda polifolia L. var.
glaucophylla (Link) DC.
Andromeda glaucophylla Link.
Family: Ericaceae

Downy bog rosemary is a small, beautiful, evergreen shrub restricted to acidic
bogs in the Lake States. Its extended
flowering period, small size, evergreen
leaves, and beautiful, bell-shaped,
pink flowers seem to make it ideal for
landscape use. Site selection is critical,
however, as this is an acid-loving plant.
Because of habitat loss, it is listed as state rare in Indiana and Pennsylvania, and it has been extirpated
from Ohio. It is often confused with the opposite-branched *Kalmia polifolia* (bog laurel), which occurs in
northern Minnesota, Wisconsin, and Michigan.

Form and Size: Downy bog rosemary is a small, few-stemmed, acid-loving shrub arising from a creeping
stem. Only rarely is it taller than 2 feet.

Habitat: In the wild, it is found in small, scattered clumps on bog mats and hummocks among deep
sphagnum moss, cranberries, and leatherleaf. It prefers moist conditions and full sun.

Wildlife Uses: Its close association with sphagnum in bogs limits its availability to browsers, and no use
has been reported by deer or snowshoe hares. It is a small plant that grows complexed with sphagnum in
hummocks; there it helps supply cover to nests of songbirds such as Nashville warblers and white-throated sparrows, which nest in those hummocks.

Landscaping Value: Hardy from Zones 2–6, this species is limited by both habitat availability and climate. It does make a lovely addition along
a walkway or in a bog garden and can be
grown in a 1:1 ratio of sand and peat. It has
year-round appeal and flowers earlier than
most native shrubs. Pruning old, scraggly
stems will encourage new growth. There
are several horticultural varieties available,
and it is only occasionally offered by native-
plant nurseries.

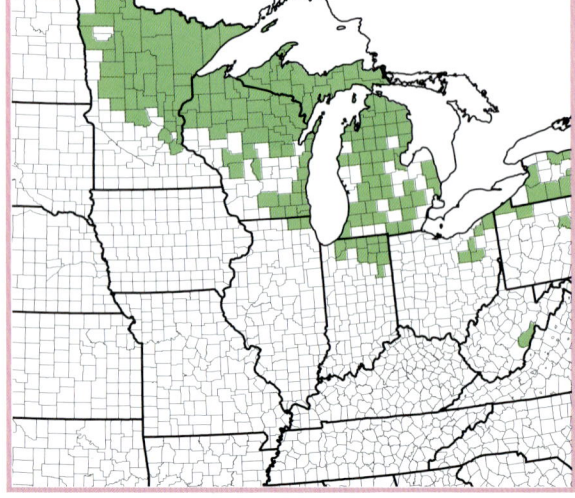

Similar Species Distinctions:
—**Bog laurel** (*Kalmia polifolia*) occurs
alongside downy bog rosemary and
looks very much like it, but its evergreen
leaves are opposite.

The thick, glossy, evergreen leaves are alternate, narrowly linear, dark bluish green above, and white and often lightly hairy beneath. The venation is impressed from above, and there are no petioles. The margins roll under (revolute) and are entire. The leaves average about 2 inches in length; the width is only about ⅛ inch. The oldest leaves are deciduous with time, and they leave a minimal leaf scar.

Fruit is a persistent, turban-shaped, 5-parted capsule. They ripen in late summer and are filled with many tiny brown seeds.

Flowers are in terminal, nodding clusters that first appear in early May; flowering continues into June or later. The flowers are bell-shaped, whitish pink, and about ¼ inch long. There are usually from 5 to 10 flowers per cluster.

Though this species does not gain much girth, it does develop thin, reddish brown, peeling bark with age.

Twigs are pale green, sometimes a bit 3-sided, and slender. There can be a whitish, glaucous coat on the twigs and especially on buds. Overwintering buds are the same pale green color as the twigs, often with the whitish cast; they are broad at the base, become flattened in the middle, and finally are pointed at the tip.

Kalmia polifolia, bog laurel is on the left; downy bog rosemary is on the right.

bearberry, kinnikinnick

Arctostaphylos uva-ursi (L.) **Spreng.**

Arctostaphylos uva-ursi **var.** *coactilis*
Fern and Macbr.

Family: Ericaceae

Over the years, bearberry's habitat has been destroyed, and it is listed as rare, threatened, or endangered in many midwestern states. Indiana lists it as rare, Illinois and Iowa as endangered, and Ohio and Pennsylvania presumed extirpated.

Form and Size: Bearberry is a prostrate, creeping shrub that forms mats where it grows. It carpets bare ground and can be the only plant around for several feet from the main "trunk."

Habitat: It is found along the shores of the Great Lakes on crests and slopes of stable fore-dunes and in black oak savannahs in pure sand. It can also be found in dry, sandy pine forests or even growing from rock crevices. This tough plant often withstands extreme heat and drought in the environment in which it commonly occurs. It grows best in full sun, but it handles partial shade.

Wildlife Uses: The prostrate growth of bearberry yields little in the way of cover except for small mammals that live under it or ground-nesting songbirds that nest within its cover. However, its urn-shaped flowers are used by hummingbirds, and the fruit is eaten by a plethora of species, including some songbirds, but principally small mammals, grouse, turkey, deer, and bear. The fruits are evidently not highly preferred, however, and frequently last through the winter—bears especially feed on them immediately upon emergence from hibernation. There is some disagreement on the palatability of its evergreen leaves for white-tailed deer, but most observers class it as a "fair" food source.

Landscaping Value: Hardy to Zone 2, this plant has an amazing tolerance for heat and drought. It makes a wonderful ground cover, but removing tree leaves and other plants' seedlings is a bit of a chore in the spring, if the bearberry patch is of any size. Its glossy, evergreen leaves are attractive year-round and turn burgundy-red by late fall. This color works well with the deep pink fruits that usually last through the winter. Sandy soil is best, but a mix with various organic matter seems to suit it. Pruning in early spring encourages denser growth. There are several cultivars available, including 'Alaska' and 'Wood's Red,' and bearberry is fairly easy to purchase from more northerly native-plant nurseries.

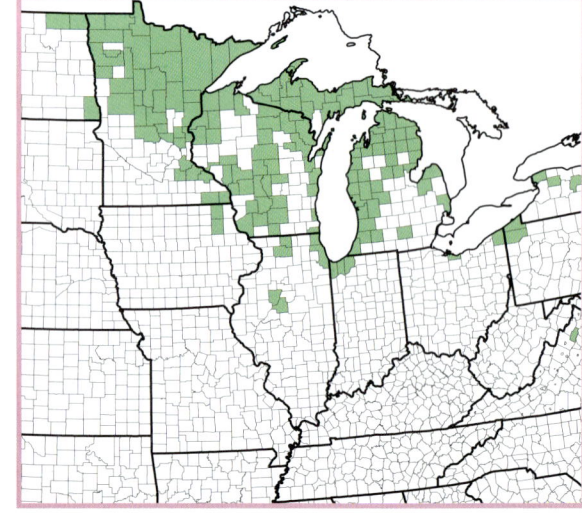

Similar Species Distinctions:
—In the Midwest, there is not much of a chance for mistaken identity, but in extreme northwest Wisconsin, on Isle Royale, Michigan, and in northern Minnesota, the similar *Vaccinium vitis-idaeus* (**mountain cranberry**) occurs. It has smaller, shiny leaves that are mostly hairless.

Leaves are alternate, evergreen, thick, and leathery; they are just over 1 inch long and ½ inch wide. The upper leaf surface is green and shiny; beneath it is pale, commonly with some degree of hairiness. The entire margins are often lined with fine hairs. Leaf tip is rounded while the leaf base is narrow. Petioles are short, reddish (usually), and slightly hairy. Venation, except for the midrib, is unusual; that is, it is without any obvious secondary veins.

Vaccinium vitis-idaeus (mountain cranberry)

Bearberry flowers in May and early June in the lower extremes of its range. Flowers are tiny, ¼ inch long, and hang in terminal (usually) clusters like bells. They are whitish pink or even a rosy pink.

Fruit is a drupe that is a bright reddish color. It is about ⅓ inch wide, dry, astringent, and unpalatable. They ripen in July and August and persist throughout the winter.

Twigs are slender and pinkish red during most of the season. The new growth is covered with thin, long, whitish hairs, most of which fall before wintertime.

Buds are small, round, and covered with several overlapping scales that appear whitish on the margins; they are pinkish red and can have scattered, thin hairs. Winter buds on plants receiving full sun will turn the same dark red color as the leaves.

Bearberry can develop a sizable trunk for such a small shrub. As it matures, its bark becomes dark reddish brown, thin, and peely.

leatherleaf

Chamaedaphne calyculata (L.) Moench.

Family: Ericaceae

Leatherleaf is most often found in bogs where it forms dense, sometimes nearly pure mats. Its common name is in reference to its thick, evergreen leaves. This shrub prefers the cooler climates of the Midwest's northern regions, where it is most commonly found. Leatherleaf is listed as state rare in Illinois.

Form and Size: Leatherleaf is a small shrub, rarely attaining a height over 3 feet. Each plant has an erect, arching, few-branched stem. It forms dense colonies through underground rhizomes, and it can cover large areas as nearly pure stands, particularly in bogs.

Habitat: It is found in bogs, borders of lakes, and other "peaty" areas. Common bog associates include Labrador tea, downy bog rosemary, and sphagnum moss. It is an acid-loving plant and prefers full sun.

Wildlife Uses: As a common component of bog vegetation that forms dense stands, leatherleaf supplies considerable habitat in these environments. The "closed canopy," albeit 1 to 2 feet tall, protects rodents and shrews from avian predation; this same density provides nest sites for wetlands birds, from red-winged blackbirds and song sparrows to mallards. Although the leaves are browsed some by deer, moose, and snowshoe hares, the species generally is considered low in palatability for herbivores.

Landscaping Value: Hardy from Zones 2–7, it is somewhat limited by climate, but it is surprisingly adaptable. In the wild, life in a bog is harsh, and leatherleaf is accustomed to drought and heat. It has an interesting growth form with its nearly vertical stem that gives way to its nearly horizontal upper half. Leaves are evergreen (although it loses the oldest leaves each fall), and cold weather turns overwintering leaves golden brown. It is one of our earliest flowering native shrubs. The authors have grown leatherleaf in an upland site (with some soil amendment) for 5 years. It is growing well, not spreading, and makes a nice border along a sidewalk. This species is sometimes sold at nurseries specializing in natives.

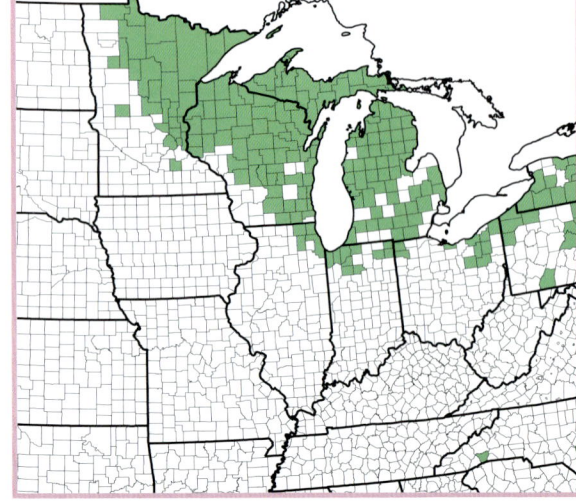

Similar Species Distinctions:
—**Bog willow** (*Salix pedicellaris*) leaves look very much like those of leatherleaf; however, they are not leathery at all and lack the silvery scales.

The oblong leaves are alternate, usually just under 2 inches long and ½ inch wide, thick, stiff, and leathery. The evergreen leaves become light reddish brown as they overwinter. The leaf margin is slightly rolled under and is obscurely, finely toothed. The upper surface is dark green; the underside is paler. Both surfaces are covered with silvery, round scales. In the winter, the silvery scales are especially evident against the reddish brown upper surface. The lower surface turns more silvery. Leaves along the inflorescences are much reduced.

The urn-shaped, whitish flowers have fused petals with upturned margins. They develop in early May and hang along the upper third of the stems from leaf axils like tiny bells. Each flower is ¼ inch long.

The woody fruit is a 5-parted capsule with 5 scurfy, hairy, persistent sepals that together appear star-shaped. Each capsule is about ¼ inch wide. Once the capsules open, they persist on the shrub over the winter. Each capsule contains many tiny seeds. The style becomes woody and remains pointed straight out from the middle of the capsule.

Although leatherleaf never achieves a size where characteristic bark develops, it nevertheless becomes an unusual reddish brown. It usually has thin, gray, shreddy, peeling strips of tissue running the length of the lower stems.

The slender twigs are greenish gray and covered with silvery and rust-colored scales when young. Older stems turn reddish to reddish brown and develop very thin, peeling, shreddy strips of tissue. Leaf scars are somewhat 3-sided. Above each leaf scar is a tiny, reddish bud.

Flower buds are very noticeable on leatherleaf during the dormant season. They are scattered along the upper third of the stem and are ovoid, brownish, and covered with silvery and rust-colored scales. Vegetative buds are along the lower portion of stems above the leaf scars and are tiny, with a few reddish scales.

trailing arbutus, ground laurel, mayflower

Epigaea repens **L.**

Family: Ericaceae

Trailing arbutus is an interesting native ground cover that has been in a state of decline in the lower Midwest and eastern states for some time. It is a northern species, commonly associated with oak, aspen, and pine woodlands. Although tried by many, it is exceedingly difficult to transplant *Epigaea*. Besides having a sizable tap root, it is claimed by some to require a mycorrhizal fungi association for survival. It is particularly beautiful in the early spring when its abundant, fragrant, white flowers appear.

Form and Size: *Epigaea* is a prostrate, evergreen shrub that creeps along the ground, creating a mostly flat ground cover. It is usually less than 3 inches tall, and it becomes quite thick once established in an area.

Habitat: It is usually found growing in partial shade in the understory of woods or along forest edges in moist soil, usually on north-facing slopes. It can also be found growing in nearly pure sand in dunes around the Great Lakes. In more northerly regions, *Epigaea* tends to be common in partially shaded, sandy soils growing among scattered conifers. It tolerates various light intensities, but it is rarely seen growing in full sun. Soil preference is loose and acidic.

Wildlife Uses: Trailing arbutus seems little-used by wildlife as a food source. When it dominates in dry openings, it serves as cover for ground nesting birds, such as hermit thrushes and white-throated sparrows.

Landscaping Value: Hardy to Zone 3, this evergreen ground cover, with its wonderfully fragrant flowers, is a delight for partly shaded areas of a landscape. Unfortunately, it can be hard to establish. Planting it near conifers is probably a good idea (high soil acidity), but removing shed needles is then another job. There are several cultivars, including 'Plena' with double flowers. Trailing arbutus is available through a few nurseries, several of which are from northeastern states, where it is common.

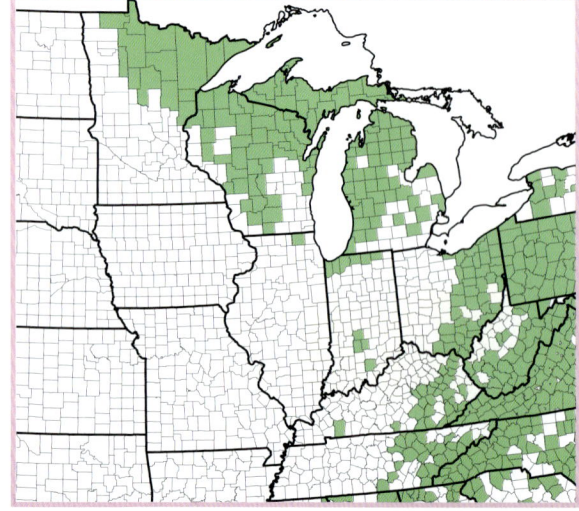

Similar Species Distinctions:
—There is no other native ground cover with evergreen, rust-colored leaves with gland-tipped hairs.

The evergreen, alternate leaves are ovate, oval, or oblong and dark green on both sides. Leaves are 2½ inches long or less. There are varying degrees of bristly, rust-colored hairs, particularly along the lower mid-vein and leaf margins. The margins are entire, and the petiole is usually densely bristly-hairy. Leaves have a fine, netted vein pattern. The hairs on all parts of the plant can be gland-tipped. Hairless leaves can be found.

The white or lightly pink, fragrant flowers appear in early April into May, and they are commonly hidden by the plant's own leaves or by those that fell from trees the previous autumn. They have 5 petals that are hairy near the base. Flowers appear from terminal leaf axils and are single sex on separate plants. With age they can darken to a reddish color. Females are seen above right; male is seen above left.

The newest twigs are greenish and covered with brownish, bristly hairs. Older twigs are brown and become hairless with age. There are no visible lateral buds.

There are no obvious buds during winter months, but as spring progresses, the terminal clusters of green flower buds appear.

The 5-celled, round, dry capsule is about ½ inch in diameter and contains many tiny, black seeds that are dispersed by ants. They ripen in late June and July. Ripening fruits are seen above (right).

wintergreen, checkerberry, teaberry

Gaultheria procumbens **L.**

Family: Ericaceae

To quote Swink and Wilhelm (1994) in *Plants of the Chicago Region*, "The delightful redolence of wintergreen obtained from bruising a leaf of this plant is ineffable during a walk in the woods." Wintergreen leaves were once collected for the extraction of their abundant oil, oil of wintergreen. Most similar oils today, however, are manufactured synthetically. It thrives in the cooler, more northerly climates of the upper United States and Canada. Wintertime leaves will sometimes turn red when exposed to enough sun.

Form and Size: Wintergreen is a low, creeping, evergreen shrub that is never more than 5 inches tall.

Habitat: Its main habitat is moist, sandy, wooded sites dominated by pines in the northern Midwest and oaks in the southern Midwest. It tends to be associated with several species of blueberry. Deam (1932) said it is "adapted only to a moist silicious soil." It also occurs on dry ridges and roadsides. It grows in partial shade to full sun; the more sun it receives, the redder its leaves become in the winter.

Wildlife Uses: This unique species has little cover value for wildlife, but it is used regularly by wildlife as food. The red fruits ripen in the fall, but they remain on the plant into the next summer if not eaten. They are rather dry and not particularly attractive to birds, although grouse and wild turkey take them throughout the year. Mammals are the major users, including a range of species from eastern chipmunk through black bear. Both ruffed grouse and white-tailed deer browse leaves, and in some areas, it is reported as a major winter food for deer.

Landscaping Value: Hardy from Zones 3–6, this tiny plant is probably limited by climate and habitat. It prefers cool summers and loose, acidic soil. It is easily grown given these requirements, and it spreads via underground rhizomes. It flowers over a several month period in the summer, and the fruits can persist nearly a year after maturation. Both contrast nicely with the dark green, shiny leaves. Given the size of the plant itself, flowers and fruits are large and showy. The leaves themselves are attractive. The cultivar 'Macrocarpa' produces many flowers and fruits. Wintergreen is occasionally available through native-plant nurseries.

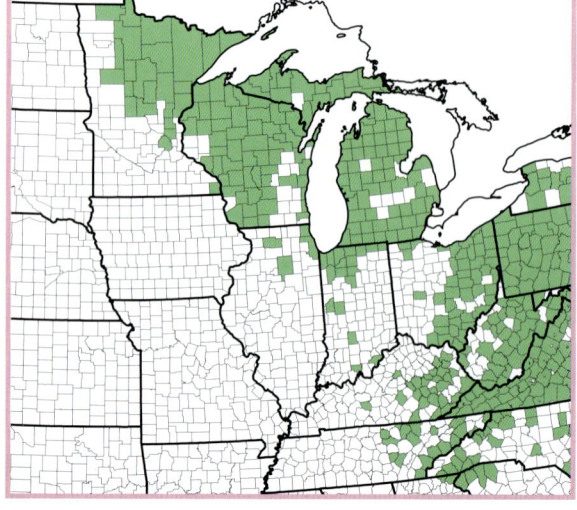

Similar Species Distinctions:
—Once confirmed as being woody, there is nothing else like wintergreen in the wild. Its thick, shiny, fragrant leaves are unique.

The thick, evergreen, alternate leaves are crowded near the tip of the twig. They are oval to elliptic, but they can be somewhat variable in shape. The leaf margin has a few tiny, bristle-tipped teeth that are best seen from beneath. The upper surface is dark green and shiny, with short, appressed hairs along the midrib. The lower surface is pale green. Petioles are short, pink, and hairy. Crushed leaves are very fragrant; young leaves are edible.

Twigs are usually reddish, but the newer ones have a greenish tint. They are slender, and the youngest growth has short, curled, white hairs. Lateral buds are very small and project outward.

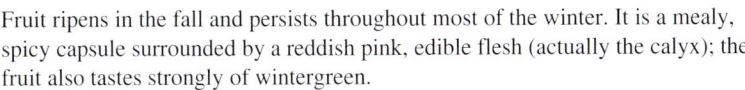

Fruit ripens in the fall and persists throughout most of the winter. It is a mealy, spicy capsule surrounded by a reddish pink, edible flesh (actually the calyx); the fruit also tastes strongly of wintergreen.

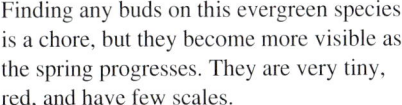

Finding any buds on this evergreen species is a chore, but they become more visible as the spring progresses. They are very tiny, red, and have few scales.

The white (or pinkish), urn-shaped flowers appear from July into August. They usually appear as single axillary flowers on short stalks. Flowers are perfect and can be hidden by the foliage that surrounds them; their presence can be hinted at by the actions of bumblebees, aggressive pollinators of wintergreen.

mountain laurel, mountain-laurel

Kalmia latifolia **L.**

Family: Ericaceae

This beautiful, evergreen shrub of the southern Midwest is much more common in the eastern states and throughout the Appalachian Mountains than it is here. In these areas, it has become so common, especially in forest edges, as to be considered a nuisance.

Form and Size: Mountain laurel is a relatively small shrub in the lower Midwest, rarely reaching 9 feet in height. It is usually much shorter. It grows in woods on rocky, dry, sandy, acidic, thin soils, and it is often associated with sandstone outcrops. In this habitat, it commonly has a gnarled appearance.

Habitat: Although it is almost always found in the understory of woods, it can be found growing in full sun along forest edges. It can become much larger in warmer climates, and it has been recorded up to 35 feet tall. It forms dense thickets in southern parts of its range, but it is only seen as individual specimens in the Midwest.

Wildlife Uses: The evergreen nature and frequent dense growth of mountain laurel results in the shrub being an important cover species in many areas for deer, bear, snowshoe hares, cottontails, and many birds. Songbirds that nest in the shrub layer of mature forests, such as hooded and black-throated blue warblers, frequently choose mountain laurel, as do black bears that occasionally hibernate in "ground nests" in very dense thickets. Although foliage is toxic to livestock, deer browse on it frequently, and in some areas it appears to be a preferred food; other species such as cottontails and ruffed grouse eat leaves occasionally.

Landscaping Value: Hardy to Zone 4, this is one of the most spectacular native shrubs in the eastern United States when in flower. It is slow-growing and requires strongly acidic soil. It handles the full spectrum of light, from full sun to full shade, but the more well-lit the better. It is often difficult to establish, and soil amendments of pine bark and peat are recommended. There are numerous, beautiful, horticultural varieties, many created by Richard Jaynes, which he sells at his nursery, Broken Arrow in Hamden, Connecticut, at www.brokenarrownursery.com.

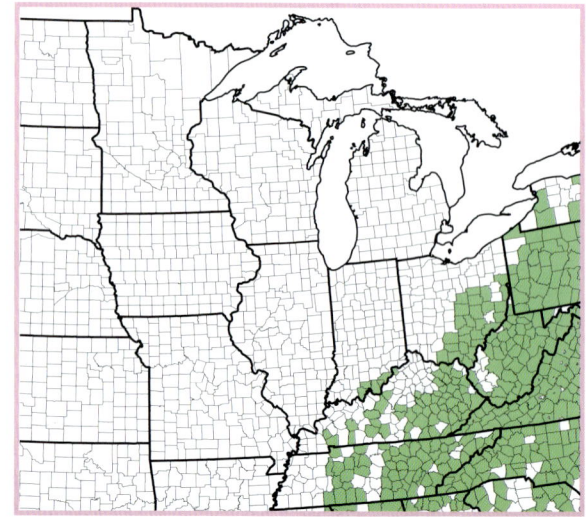

Similar Species Distinctions:
—Other southern evergreens, such as rhododendrons, could be confused with mountain laurel, but no midwestern natives should cause problems.

Leaves are alternate, thick, evergreen, shiny, smooth, and 3 to 4 inches in length. They are dark green above, lime green below, and smooth all over. The common shape is elliptic to oval, and the margins are entire. Petioles are about ½ inch in length.

Fruit is a 5-parted capsule that is somewhat woody when mature in the fall. Each capsule contains many tiny seeds. The capsules are in terminal clusters, and they are persistent though the winter months.

Flowers appear in early May until mid-June on pedicels that are 1½ inches long. The terminal clusters are large and showy, with each pinkish flower almost 1 inch across. Individual flowers resemble little satellite dishes. Inside each flower are 10 stamens, the filament tips of which are held under tension in tiny pockets in the petals. They spring upward when an insect makes contact.

Twigs are light green to yellowish when new, and they turn medium brown when mature. New growth can have dense brown hairs that are shed during the growing season. There are no lateral buds visible along the twigs.

Overwintering flower buds sit atop plants in slender, finger-like, upright clusters.

The only visible winter buds are next spring's flower buds that form in late summer and fall in spike-like, terminal, green clusters. The upright spikes have scattered, single, or paired bracts. Each bract hides a green flower bud. The spikes are covered with very short hairs.

Mature bark is dark brown, thin, and fibrous. The inner bark is reddish orange and often obvious as the thin outer bark exfoliates as the stem ages.

buttonbush

Cephalanthus occidentalis **L.**

Family: Rubiaceae

This ubiquitous wetland shrub probably occurs in every wetland in the Midwest. It is never found far away from water in the wild, but it can handle a somewhat drier site if planted there. Its abundant, large, nectar-laden flowers attract a large variety of butterflies, moths, and bees. Additionally, it exhibits an extended flowering period.

Form and Size: Buttonbush is a sizable shrub that can grow to about 9 feet in height. It is a spreading, multi-branched species that commonly forms impenetrable buttonbush swamps in the wetlands they inhabit.

Habitat: Buttonbush is so content in wetlands that it can handle "wet feet" every day of the year—a very rare feature for woody plants. It prefers full sun, but it can handle light shade. For a shrub, buttonbush has a fairly long life span.

Wildlife Uses: Buttonbush is 1 of the few woody plants in our region that can grow in water year-round. In many wetlands habitats, it is a common woody component, providing important cover for duck broods and roosting ducks. Many wetland birds (e.g., green herons, kingbirds, least bitterns) nest in buttonbush in standing water, achieving added protection from terrestrial predators. Swamps that have a large quantity of buttonbush are used heavily by wood ducks as roosts, especially in the fall. The ducks roost under the cover of the shrub on the water's surface. The flowers are very attractive to butterflies and hummingbirds; seeds produced typically fall into water and are used by dabbling ducks. Leaves have some toxic qualities and are little-used by deer.

Landscaping Value: Hardy to Zone 5, this is the best native shrub for a wet, poorly drained site. It has shiny leaves, large, showy flower clusters that are produced over a month or more in the heat of summer, and pretty, red, ripening fruit. It can be grown on an upland site, but it is susceptible to dry weather and must be watered until established. Buttonbush is easily purchased from nurseries specializing in natives and habitat restoration.

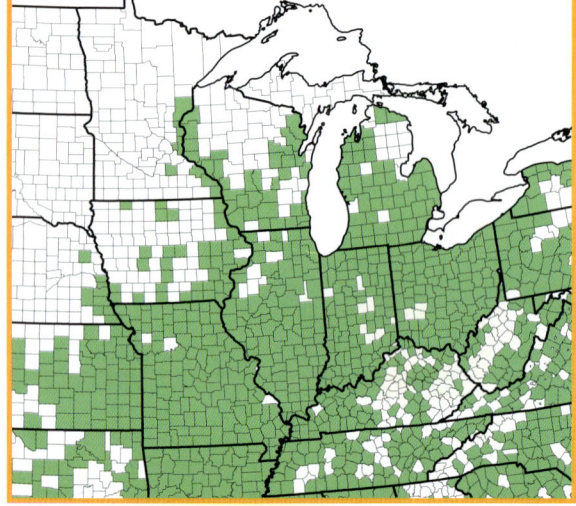

Similar Species Distinctions:
—The aquatic nature and mostly whorled branching of this shrub help rule out any other native (or exotic) shrub.

Leaves are usually whorled (sometimes opposite); they average about 4 inches long and 2 inches wide but can be much larger. The overall shape is ovate or more elongated, especially near the tip. They are dark green above and pale beneath, especially the major veins that can appear nearly white. Leaf margins are entire and often a bit wavy.

The fragrant, white flowers are in rounded balls that are on long stalks. Each "ball" is about 1 inch in diameter. The perfect flowers produce a great deal of nectar and are attractive to many insects. Flowering will begin in June and extend into the fall when weather conditions permit.

The newest growth is a pinkish tan color with large, obvious, raised, white lenticels. Older twigs become silvery brown. The number of leaf scars at a node varies, but they are most often in a whorl of 3. There can be as few as 2 (opposite leaf scars), or as many as 4 in a whorl. The scars are nearly circular with raised rims, and they appear deeply sunken. The bundle scars create a picture of a mustached man. There are tiny, barely protruding buds visible, occurring a slight distance above each leaf scar and are almost completely sunken into the twig.

Bark on younger plants is gray and somewhat peeling. With age, plants develop furrowed, interlacing, brownish bark that somewhat resembles that of pignut hickory.

Fruit matures in the fall, and many of the reddish brown heads persist into the winter. Each head is composed of numerous, hard, brown nutlets. The fruiting heads shatter when the plant is dormant, scattering seeds all over the ground or into the water in which the plants often grow. The receptacle (what seeds were attached to) remains on the shrub even after its seeds have all dispersed.

Dogwoods

Genus: *Cornus*

The Midwest is a great place to live if you enjoy dogwoods. There are 9 native species, which include the small-tree-sized *Cornus florida* and the petite sub-shrub *Cornus canadensis*. Most dogwoods have year-round appeal, beginning with the springtime profusion of white flowers, followed by the beautiful fall leaf color combined with the very patriotic red, white, or blue fruit. Added to this is the fact that most species of dogwood have red or green twigs that add winter interest.

Only *C. canadensis* is particular about where it grows. The other species handle shade well, but the more sun they receive, the more they give— more flowers, more fruit, brighter fall color. Pests and diseases are generally not a problem with shrubby dogwoods. If they are planted, however, one will probably have to deal with "suckering" from the stoloniferous root system. This is a common form of reproduction for shrubs that gives rise to patches or even thickets of dogwoods. One may be left having to make the decision of just how many new plants are desirable. Dividing or simply mowing over new stems are both options.

For identification purposes, learning pith color of both first and second year's twigs helps deter-mine species, and it is best done during the dormant season. Most often, pith will be the same color in both aged twigs, but sometimes not, as can be the case with *C. drummondii*.

There are 8 species of shrub dogwoods in the Midwest that contrast rather strikingly with flowering dogwood (*Cornus florida*) (Weeks, Weeks, and Parker 2010) relative to growth form, fruit characteristics, and wildlife values. Because such characteristics vary among the shrub dogwoods as well, and shrub dogwoods occupy a range of habitats, the wildlife relationships are also diverse. It seems that shrub dogwoods, in general, are not used for nesting by songbirds to the degree one would expect, given their abundance and, thus, availability. The authors suspect this is because the opposite branching and substantial internodel distance limits the locations where a bird can find multiple (i.e., greater than 2) limbs to support a nest. In contrast, the fruits of dogwoods are generally highly preferred and quickly disappear in the late summer and early fall; however, there is variability in palatability among species. Some species are browsed rather heavily by white-tailed deer and other herbivores, although use is unpredictably variable.

Pith color of first and second year twigs is usually a good tool for identification. Some dogwoods look similar from a leaf standpoint, so one should use the pith in combination with other characteristics to confirm identification. Silky dogwood, shown here, has brown pith throughout (first year twig on the left).

Leaves of dogwoods have an unusual venation technically known as arcuate, which means that the major veins form an arc that curves with the leaf shape so that they never grow toward the sides, as do those of most species. Fall color is usually dark burgundy. Shown here is alternate-leaf dogwood, the only native dogwood with alternate branching.

Dogwoods, especially when given a fair amount of sun, flower profusely. All our native dogwoods have white flowers. Seen here is gray dogwood.

bunchberry, dwarf cornel

Cornus canadensis **L.**

Family: Cornaceae

This small, herbaceous-looking plant is actually classified as a sub-shrub. It has a woody, underground rhizome from which short, upright woody stems arise. It is common in the northern regions of the Midwest, but its southern range is restricted, in part, by its inability to survive in soil temperatures greater than 65 degrees. Habitat loss has played a role in its demise in northern Illinois and Indiana. It is currently listed as state endangered in Illinois, Indiana, and Iowa, and state threatened in Ohio.

Form and Size: Bunchberry is a small, colonial sub-shrub that reaches a maximum height of 8 inches. It can cover the ground in large colonies as it spreads by underground rhizomes. Growth rate is slow, but individual stems are long-lived; annual growth rings totaling 35 years have been documented.

Habitat: Bunchberry is commonly found in moist, well-drained, shady boreal forests, but it also grows in open bogs and drier forest types, particularly where acidic soils are present. It is found growing in full shade and sometimes nearly full sun.

Wildlife Uses: Bunchberry's diminutive size makes it unique among dogwoods, not only morphologically, but also in its use by wildlife. This species' red fruits are readily eaten by songbirds (e.g., veery, vireos) and game birds (e.g., wild turkey, ruffed grouse), and it is regularly browsed by deer. As one moves north and west from our region, it becomes even more of an important browse species for ungulates. In mature, cool forests, clones become rather large, forming a ground cover in which ground-nesting songbirds like the veery conceal their nests.

Landscaping Value: Hardy from Zones 2–6, this species requires a cool climate to survive. It also needs acidic soils with a high organic component. The combination fire engine red fruit clusters with the bright green leaves is delightful. It has beautiful red fall color, especially in cooler climates. Bunchberry is available through a limited number of nurseries, mostly from eastern states.

Similar Species Distinctions:
—No other woody shrub looks anything like this. Sometimes, flowering dogwood (*Cornus florida*) seedlings can be found with bunchberry in forest settings, and their leaves can look very similar. Juvenile bunchberry plants can have 2 or 3 leaves, which is when they most resemble flowering dogwood.

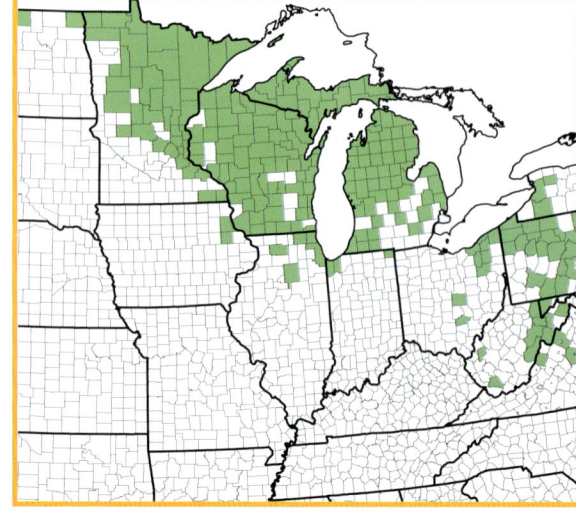

Leaves are in a single, terminal whorl. Fertile stems usually have 6 leaves; sterile stems have 4. They are lanceolate, oval to obovate, and up to 3 inches long. They are shiny and smooth above and slightly paler beneath. Petioles are short, and margins are entire. Leaf venation is of the type typical of all dogwoods—arcuate. Leaves tend to be semi-evergreen.

Flowers appear in May and June. While they seem to be solitary, in actuality, flowers are tiny, greenish, and clustered in the center of 4 whitish bracts. Each combination of flowers and bracts is terminal—usually 1 per stem and 1 inch across. They are a smaller version of those found on *Cornus florida*, flowering dogwood.

The bright red fruit (a drupe) ripens in the fall in terminal clusters. They are rounded, ¼ inch in diameter, and on long stalks (peduncle). If not taken by wildlife, they will persist into early winter.

Twigs are slender, greenish or reddish, non-woody, and somewhat angular. They have 1 or 2 pairs of bracts along the stem below the terminal leaves. New spring growth is shown in the photo to the left.

swamp dogwood, southern swamp dogwood, stiff dogwood

Cornus foemina **Mill.**

Cornus stricta **Lamarck**

Family: Cornaceae

This native dogwood is very similar in appearance to gray dogwood, *C. racemosa*. Even though their ranges do overlap, swamp dogwood is just that— a species of poorly drained woodlands of the Lower Wabash, Ohio, and Mississippi rivers. Although its leaves and flowers look much like those of gray dogwood, the fruit color is a major difference—it is blue.

Form and Size: Swamp dogwood can grow to a height of 9 feet and is usually found in the understory of swampy woods, which produces shrubs that are tall and somewhat spindly.

Habitat: This is a species found in wet woods and along the edges of swampy woods. It sometimes occurs along roadsides in flood-prone areas as well. It tolerates poorly drained soils and heavy shade, but it grows well with more sunlight.

Wildlife Uses: Swamp dogwood is found in wet woodlands in the southern parts of our region. Its cover value is unexceptional, but it has good wildlife food value. It is browsed by white-tailed deer, cottontails, and beaver. Its pale blue to dark blue fruits are eaten by many birds, including ducks, especially wood ducks. Squirrels, chipmunks, and raccoons also regularly take the fruits, although fruit production is often sparse under the shade of bottomland forest canopy.

Landscaping Value: Hardy to Zone 5, it is a shrub of poorly-drained bottomland forests, but it would probably grow well in more upland sites. It has slender twigs that are colored similarly to those of flowering dogwood—the side facing the sky is red, while the lower side is green. Fall color is deep red, which makes for an interesting color combination with its pale blue fruit. The authors have never seen this species offered at any nursery, and there are no known cultivars.

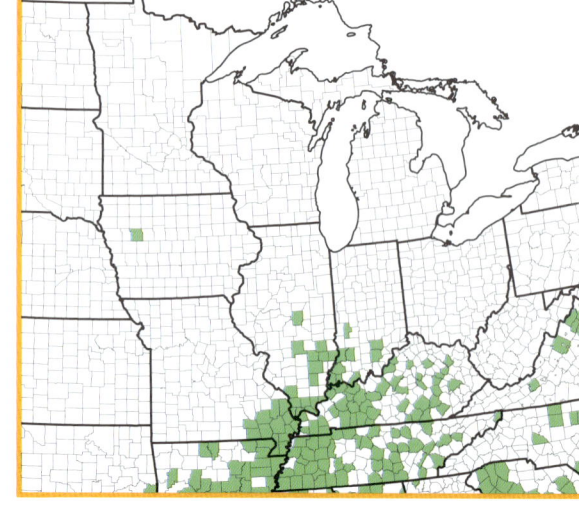

Similar Species Distinctions:
—**Gray dogwood** (*C. racemosa*) is closely related, and the leaves are similar. Fruit color is white. Newest twig color is red.

Leaves are opposite, about 4 inches long and ovate to lanceolate, with 3 to 4 pairs of lateral veins per leaf. They are mostly hairless at maturity, pale green above, and somewhat whitened (glaucous) beneath. The leaf margins are entire and somewhat wavy.

Twigs are slender and usually 2-toned, especially on the youngest twigs. One side of the twig (away from the sun) is greenish; the other side (toward the sun) is reddish. Sometimes the young twigs are just 1 of these colors, and with age they all turn gray. The pith is white throughout. Lateral buds are similar to the terminals but usually shorter.

The white, convex-shaped flower clusters appear terminally in late May and early June.

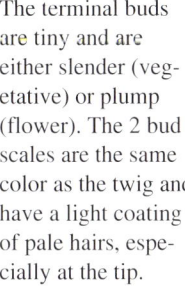

The terminal buds are tiny and are either slender (vegetative) or plump (flower). The 2 bud scales are the same color as the twig and have a light coating of pale hairs, especially at the tip.

Fruit ripens in late summer and is usually a pale blue-violet (it can be white). Each fruit is about ¼ inch in diameter.

Mature bark is brown, fissured, and develops a larger, less scaly pattern than that of gray or rough leaf dogwoods. This species is always found in flood-prone areas, and its bark is commonly dirty from mud residue.

rough leaf dogwood

Cornus drummondii C. A. Meyer
Cornus asperifolia Michaux.
Family: Cornaceae

Rough leaf is 1 of our largest native shrubby dogwoods, commonly growing to 12 to 15 feet in height or more. It is a common roadside shrub that, because of their many similar characteristics, is often confused with *Cornus racemosa*, gray dogwood. Rough leaf is named because of the somewhat scabrous hairs that cover the lower leaf surface. However, as the summer progresses, leaves gradually lose them, and by fall there may be none at all. This does not help the novice with correct identification!

Form and Size: Rough leaf is a large shrub that is less colonial than gray dogwood. It develops into a broad, dense specimen with outer limbs that tend to be horizontal or even drooping. It freely suckers and spreads rather quickly when given room.

Habitat: It is commonly found along wet roadsides, stream banks, the borders of lakes, ponds, and woods, and fencerows. It tolerates a great deal of shade; in some of these habitats, however, it is just as much at home in full sun.

Wildlife Uses: Rough leaf dogwood is very similar to gray dogwood, morphologically and in its value to wildlife. It produces a fruit that is small and white when ripe that is relished by mammals and both resident and migrant birds, including warblers, vireos, thrushes, and eastern phoebes. The species is quite common and occurs as individuals along forest edges and in dense thickets; in the latter instance, they are readily used as nest sites by cardinals, catbirds, Bell's vireos, and other shrub nesting species.

Landscaping Value: Hardy to Zone 4, this is another easily-grown native shrub, but it needs room to grow. Unless suckers are controlled (usually by mowing), the plant will spread many feet outward from the original stem. It tolerates most soil types and moisture regimes, but it grows best on moist, rich sites. Fall color is deep red, which blends well with the white fruits that are attached to orange-red stalks. Rough leaf is often sold, mistakenly, for gray dogwood, or vice-versa, but either way is available from native-plant nurseries.

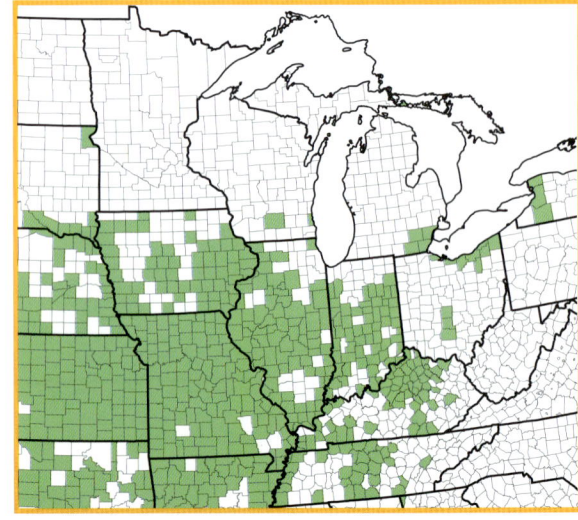

Similar Species Distinctions:
—Gray dogwood (*C. racemosa*) is similar; rough leaf usually has a distinctly roughened leaf texture and larger leaves, compared to the smooth, small leaves of gray. The number of pairs of leaf veins is less in gray dogwood.

Leaves are opposite and up to 5 inches long and 2 inches wide. The upper surface is dark green and lightly hairy, especially when young. The lower surface is lighter in color, mainly because of the many hairs. The leaves feel rough most of the growing season. Leaves usually have 5 or 6 pairs of veins per leaf. The margins are entire but wavy.

First year twigs are reddish brown, straight, and usually rough from flattened hairs. By the second year twigs have begun to turn gray and most hairs are gone. The pith color is variable on this species of dogwood. First year twig pith color can be white or tan; second year twig pith color is usually dark tan to brown. Lateral buds are somewhat appressed, brown, and hairy. Leaf scars are V-shaped.

Flowers appear in early to mid-June in upright, terminal clusters. Each cluster can be 2 inches wide or more and convex in shape. The white, somewhat ill-scented flowers are perfect. Flowering is usually extended over several weeks in a given shrub.

The white, fleshy drupes ripen in late summer and are in great contrast to the bright orange-red, hairy pedicel. The fruit fully ripens about the time of fall leaf color change, when rough leaf dogwood puts on quite a show. Fruit is about ¼ inch across.

Terminal buds are covered with dark brown hairs and few scales. Flower buds are plump at the base; vegetative buds are flattened. As seen here, they can appear to be stalked.

Rough leaf dogwood bark is thin, scaly, and gray. Inner bark is a dark reddish brown.

silky dogwood, pale dogwood

Cornus obliqua Raf.
Cornus amomum Mill.
Family: Cornaceae

This is another native dogwood with reddish twigs that can be confused with others. The brown pith throughout and hairy twigs help distinguish it. A unique feature of this dogwood is the brown streaking on the lower, larger diameter reddish stems that becomes a distinguishing characteristic of the bark. The correct botanical name for this species is a conundrum; in addition to the above, *C. amomum* Mill. ssp. *obliqua* (Raf.) J. S. Wilson is 1 suggested name; *C. amomum* var. *schuetzeana* (C. A. Mey.) Rickett is another.

Form and Size: Silky dogwood can be a large shrub, up to 15 feet in height, but is usually 6 to 8 feet tall. Its growth form is quite bushy. The more sun it receives, the fuller its growth. It suckers like most dogwoods, and individual plants develop large clumps of stems.

Habitat: Silky dogwood is usually associated with water and is commonly found along ponds, lakes, streams, and swamps; however, it is also quite common on the fore-dunes near Lake Michigan. This is a sun-loving shrub that struggles in even partial shade.

Wildlife Uses: Silky dogwood generally occurs in open, moist sites, but it grows well in uplands. It occasionally is limby enough that birds like catbirds and cardinals find sufficient purchase to place a nest. In open situations, it prolifically produces blue fruits that are large enough so that they are not favored by smaller birds, but regularly used by many larger ones, such as wood ducks, robins and other thrushes, waxwings, and catbirds, as well as by several mammals. Fruits often remain on shrubs for longer periods than in those dogwood species with smaller, white fruit. Silky dogwood is not a favorite browse species for most herbivores.

Landscaping Value: Hardy to Zone 4, silky dogwood is an easy-to-grow species that makes a nice specimen plant or backyard plant for a more naturalized setting. It has year-round interest with its copious white flowers, unusual blue fruit, beautiful fall color, and wine-red twigs and branches. It is not particular about soil type or moisture regime. Silky dogwood is 1 of our more commonly found natives at nurseries throughout the Midwest.

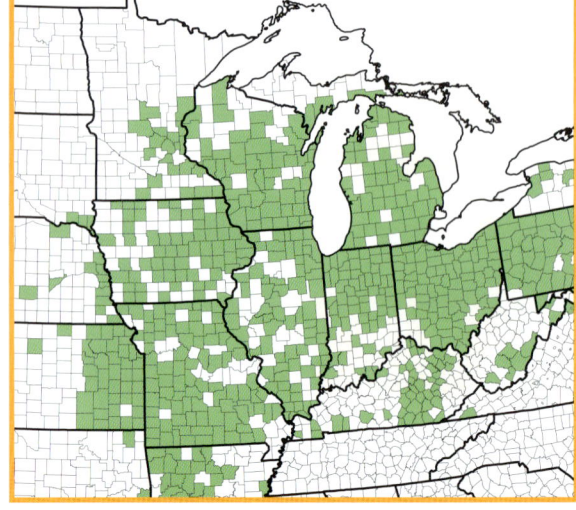

Similar Species Distinctions:
—Dogwoods can be tricky to separate, so the more characteristics you have, the better. Twig pith is usually 1 of the best.
—**Red-osier dogwood** (*C. sericea*), a common associate, has bright, shiny stems that are red nearly to the ground and solid white pith.

Leaves are opposite, ovate, elliptic, or lanceolate in shape and up to 3½ inches long and 2 inches wide. The upper surface is dark green or yellow-green and hairy to slightly hairy. The lower surface is pale and covered with long, appressed hairs that are usually whitish in color, although some are rust-colored, particularly those along the veins. The margins are wavy but entire.

Buds appear 2-scaled, are reddish, and are covered with many brown and grayish hairs. They are pinched at the tip and broader at the base.

The fruit ripens in mid-to late summer and is an unusual shade of gray-blue. Each fruit is about ¼ inch in diameter and contains a single, strongly ridged seed.

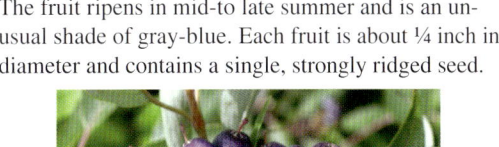

The creamy-white flowers are in upright, terminal clusters that appear in May and June or later. The flowers are perfect, and clusters are more than 2 inches across.

Twigs are reddish with flattened, brown or grayish hairs, particularly on current year's growth. Leaf scars are V-shaped, raised, and have 3 bundle scars. The lateral buds are mainly appressed against the twig and similar to the terminal bud. Pith color is brown throughout the twigs and branches.

Bark is thin and reddish gray with vertical, light brown, shallow fissures and lenticels all along the older stems. It is the only native dogwood that develops these fissures.

gray dogwood, northern swamp dogwood

Cornus racemosa Lamarck

Family: Cornaceae

Gray dogwood, 1 of the Midwest's most petite *Cornus* species, is usually only 4 to 5 feet tall. It is 1 of our most common dogwoods in the northern parts of our region. A near look-alike to gray dogwood is its southern equivalent—*Cornus foemina*, swamp dogwood, and Voss (1985) classes Michigan's gray dogwood as *Cornus foemina* spp. *racemosa*.

Form and Size: Gray dogwood is often only chest high, but it can grow to a taller height. In full sun it remains compact, erect (limbs tend to reach toward the sky), and very leafy. Colonies in the open are often dome-shaped, with the older, taller specimens in the center giving rise to an ever-increasing circle of younger stems.

Habitat: It can be found growing in both wet and dry locations, and it is a common fencerow species. Occasionally it is found growing along marsh edges; in Indiana, Deam (1932) said its preferred habitat is a drained marsh.

Wildlife Uses: Gray dogwood is similar in many respects to rough leaf dogwood. Its small, white fruits are equally relished by songbirds and eaten by wild turkey, ruffed grouse, and bobwhite quail as well. It frequently forms dense clones, especially evident when pioneering old-fields and utility/highway rights-of-way, because they are dome-shaped, with older, taller stems in the middle. These dense clones are good cover for a variety of species, even including white-tailed deer, which use it extensively as a browse species as well. This species has very upright stems that form good attachment sites for songbird nests (compared to most shrub dogwoods), such as those of willow flycatchers, catbirds, and red-winged blackbirds.

Landscaping Value: Hardy to Zone 3, this rather petite shrub dogwood is compact and has slender twigs with burgundy-red new growth, which contrast nicely with the older gray branches. Like all dogwoods, fall color is deep red, which is a nice backdrop for the white fruits on their red stalks. It is adaptable to most soils including dry sites, but it prefers more moisture. Like most dogwoods, insect and disease problems are minimal. There are several cultivars on the market, and gray dogwood can be found without much difficulty in nurseries specializing in natives.

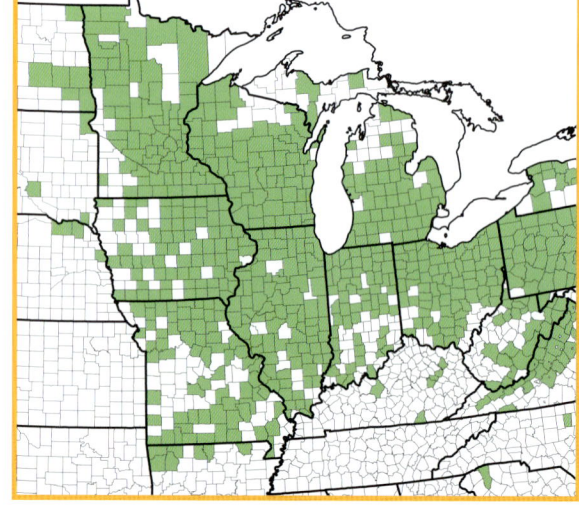

> **Similar Species Distinctions:**
> —See **rough leaf dogwood** for discussion.
> —**Swamp dogwood** (*Cornus foemina*) is a shrub of the swamps. It is a larger shrub with 2-toned twigs and blue fruit. Leaves are very similar in appearance.

Leaves are opposite, up to 4 inches long, and of variable shapes, from ovate (similar to rough leaf) to lanceolate (like swamp dogwood), with either 3 or 4 pairs of lateral veins. They are dull green above with short, appressed hairs, especially when young. The lower surface is paler (glaucous) with few, short, appressed hairs. The leaves feel smooth to the touch. Leaf margins are entire. Pictured on the opposite page are rough leaf leaves (top) and gray leaves (bottom) for contrast.

The youngest twigs are slender, light reddish brown, and usually hairless. Pith color in these twigs is white. Older twigs become light gray and contain tan-colored pith. Lateral buds are similar to the terminal bud; older twigs can have what appears to be stalked buds (these are just developing side branches).

The white, fleshy berries ripen in late summer and stand out against their reddish pedicels. Each fruit is about ¼ inch in diameter and contains a single, shallowly grooved seed. The pedicels are hairless, unlike those of rough leaf dogwood, which are covered with numerous, appressed hairs. Once ripe, the fruit does not last long; it is highly preferred by songbirds.

Terminal buds are brown, small, and have what appears to be 2 lightly hairy scales. The tip of the bud is pinched compared to the base.

Bark is very much like that of several of the Midwest's other native dogwoods, namely rough leaf and swamp dogwood. Mature bark is gray and thin, with a narrow, scaly pattern. The inner bark is dark brown. The trunk diameter of this small dogwood is rarely over 1 inch.

The terminal, upright clusters of white flowers are different from those of other dogwoods, tending to be more tiered or layered than the flat-topped clusters of others species. They begin flowering in late May to early June and have a several week flowering period. The rather ill-scented clusters are over 2 inches across.

round leaf dogwood

Cornus rugosa **Lamarck.**

Family: Cornaceae

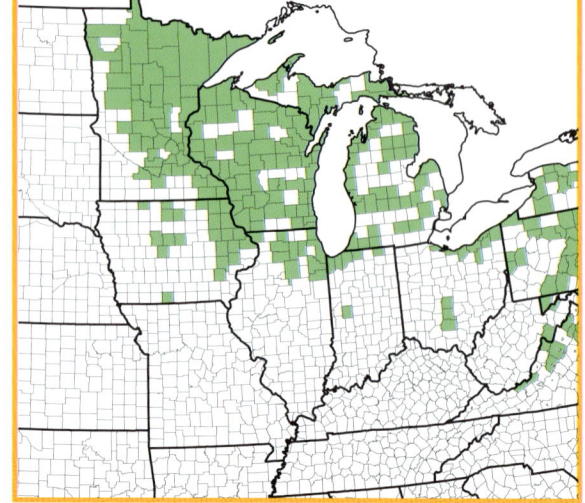

This striking dogwood stands out any time of the year. In particular, its unusual pinkish green, purple, or yellow twigs can be seen in any season, but they are particularly enjoyable to view in the winter. It is a colonial shrub, which can be found inhabiting woods where its common associates include *Viburnum rafinesqianum* and *Viburnum acerifolium*. In Indiana, round leaf dogwood is listed as state rare.

Form and Size: Round leaf dogwood is usually an understory shrub growing in filtered light. Size is commonly 5 to 6 feet tall with a somewhat open form. It suckers from the root crown and creates small clumps. Over time, it can spread to create small thickets.

Habitat: Round leaf dogwood is a rather northerly species in the Midwest. It is found on cooler, north-facing or east-facing, wooded slopes growing with mixed oak species. Soils are usually sandy or rocky, and it seems to tolerate dry conditions. It suckers freely and is shade tolerant, but it seems to be sensitive to extreme summer heat and humidity.

Wildlife Uses: This is a dogwood that is common the upper Midwest under forest canopies. It is a favorite browse species for white-tailed deer and cottontails, which has led to its decline in the few populations that occur in northern Indiana. It produces a light blue (or white) fruit that is very attractive to squirrels, raccoons, ruffed grouse, wild turkey, and songbirds; fruits, however, are often sparse on shrubs under a forest canopy.

Landscaping Value: Hardy from Zones 3–5, its range is limited largely by climate. It truly has year-round appeal from its round leaves, large white flower clusters, reddish fall color, white or blue fruit, and unusual-colored twigs and branches. It is often found growing in poor, dry, acidic soils, but it grows well on more nutrient-rich, moist sites. Heat and humidity cause problems for this species; the authors do not recommend it as an ornamental south of its native range (see range map). Round leaf is found occasionally in nurseries that grow natives.

Similar Species Distinctions:
—Flowering dogwood (*C. florida*) leaves are very similar, but they lack the soft hairs beneath. Twigs are never green or pink like round leaf. Fruits are red on flowering dogwood.

Round leaf has broadly ovate leaves that are very rounded at the base and have an abruptly pointed tip. Like most dogwoods, they are opposite each other. Leaves can be confused with *Cornus florida*, flowering dogwood, as they are similarly round. They can be up to 4½ inches long and 4 inches wide. There are usually 7 to 9 pairs of veins per leaf that run from base to tip. The upper surface is green and lightly hairy, while the lower surface is pale and quite hairy. The margin is entire and somewhat wavy.

Twig colors vary, being pinkish green or yellowish green or purplish. They are dotted with purplish "blemishes" that elongate as the twig becomes thicker. There are also raised, scattered brown lenticels and varying degrees of hairiness. Pith is consistently white throughout. Leaf scars are raised, V-shaped, and contain a single row of 3 bundle scars. Lateral buds are similar to the terminal bud but often appear stalked.

The fleshy drupes are white when ripe in the fall. The pedicels are a bright, pinkish red, which makes for a striking color contrast to the fruit. The fruit is ⅓ inch in length.

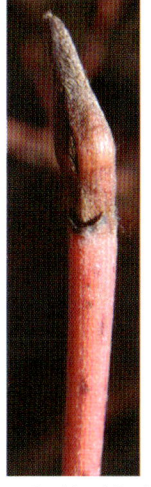

The white flowers appear in terminal clusters during the end of May and into June. Clusters are somewhat flat-topped, and the flowers are perfect.

The terminal bud is slender, few-scaled, and the same color as the twig. It is more elongate than those of most of our native dogwoods. The tips of the scales have tufts of light-colored hairs.

Round leaf dogwood bark becomes rough with age, mainly from the enlarged, raised lenticels that are ever-present. Some fissuring occurs near the base of a large shrub as well. Bark on various aged stems is seen here.

red-osier dogwood, red twig dogwood, American dogwood

Cornus sericea L.

Cornus stolonifera Michx.

Family: Cornaceae

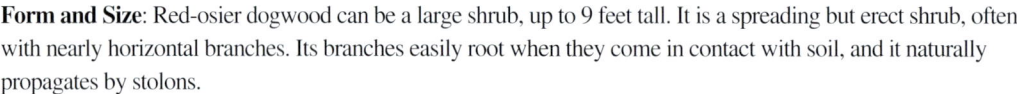

Red-osier is named for the fact that its pliable branches were once used by Native Americans in basketry, similar to how some willows were. It has 1 of the largest ranges of any of our native shrubs, and across this extent, there are several recognized varieties including *C. sericea* L. var. *baileyi* (J. M. Coult. and W. H. Evans) Drescher, which occurs in the Midwest and is distinguished by the fairly dense, partially spreading hairs on the underside of the leaf. The branches, both young and old, are bright red and make it a dramatic winter horticultural specimen.

Form and Size: Red-osier dogwood can be a large shrub, up to 9 feet tall. It is a spreading but erect shrub, often with nearly horizontal branches. Its branches easily root when they come in contact with soil, and it naturally propagates by stolons.

Habitat: This species is common in wet or boggy places including marshes, wet meadows, and streamsides. An exception to this is its occurrence on the sandy fore-dunes close to the Great Lakes. This is a sun-loving shrub, but it will tolerate some shade.

Wildlife Uses: This widely distributed species has tremendous wildlife value, particularly when it grows in thickets in wetland areas—which is its normal habit. In these instances, it serves as cover for everything from songbirds to moose. Wetland-associated songbirds seem to favor it for nesting, likely because frequent heavy browsing by white-tailed deer and snowshoe hares yield numerous sprouts that supply many attachment sites. It produces white fruits, very similar to those of gray and roughleaf dogwoods, which are favorites of songbirds, woodpeckers, and game birds. Shrubs along streams are regularly used as food and construction materials by beaver.

Landscaping Value: Hardy to Zone 2, this is an especially adaptable species, and it is the most widely available native shrub dogwood in midwestern nurseries. Its bright red branches are showy year-round. Fall color is shades of red, which contrasts nicely with the white fruit. It is a large shrub, but it is easy to grow and maintain. There are numerous cultivars including 'Baileyi' and 'Flaviramea,' the latter of which has yellow stems.

Similar Species Distinctions:

—Several dogwood species can be confused with red-osier, mainly because their new twig color is reddish.

—**Silky dogwood** (*C. obliqua*) comes the closest; it has long, red sections of branches. However, toward the base it develops brown streaks in the maturing stem. Red-osier does not.

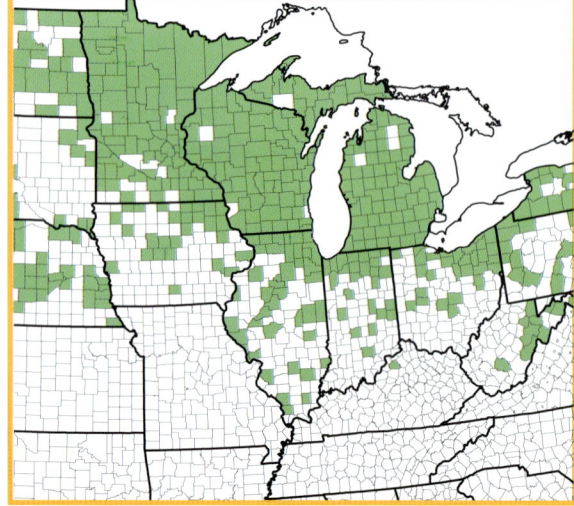

Leaves are opposite and oval, ovate to somewhat ovate-lanceolate; they are about 3½ inches long. They are medium to dark green above and pale beneath. The lower surface is whitish (glaucous) and can have appressed hairs. The petiole is longer than those of most dogwoods, often several inches in length. The margin is entire.

The white fleshy fruit matures in mid-to-late summer on slightly hairy, dark reddish purple pedicels. These are highly preferred by birds and do not last long.

The white, flat-topped flower clusters appear terminally earlier than other midwestern dogwoods. It can begin flowering by early May and continues sporadically throughout the summer. The clusters are 2 inches across.

Buds have 2 scales that are the same red color as the twig; scales have scattered whitish hairs plus dense, brown hairs at the tips. Terminal buds can be the plump flower buds or thinner vegetative buds.

Twigs are wine-red when young with scattered, light-colored, appressed hairs only near the tip. Older twigs are smooth, bright red, with scattered, raised, brown lenticels. Leaf scars are V-shaped and outlined with black, which extends around the twig to meet that of the opposite scar. Lateral buds are similar to the terminal bud, but they are slender and densely covered with light brown hairs. Buds appear to be stalked. The pith color of all aged-twigs is white, and pith is thicker than that of other dogwoods. Branches on red-osier remain red until they are 4 or 5 years of age.

Bark on older specimens turns from red to greenish and finally to grayish brown and develops some fissuring. It is very rough from the numerous lenticels that persist and enlarge with age. In the shade, older branches are green rather than red.

bush honeysuckle, dwarf honeysuckle

Diervilla lonicera **Mill.**

Family: Caprifoliaceae

Under cultivation, bush honeysuckle grows quickly and it is especially attractive when in flower, which occurs over a several week (or more) period. In some parts of the Midwest, it is rather difficult to locate in the wild because of habitat destruction and too many white-tailed deer, which have taken their toll on the populations. It is listed as rare in Indiana and threatened in Tennessee.

Form and Size: Bush honeysuckle is a low, spreading shrub that does not grow over 3 feet tall. It spreads vigorously by root suckers and responds quickly to pruning and fire damage.

Habitat: Bush honeysuckle is found in dry, open woods. It also occurs on rocky, wooded bluffs. It tolerates at least half-day sun, but it is almost always found in partially to heavily shaded, wooded locations. In the Upper Lake States, bush honeysuckle is very common.

Wildlife Uses: Although bush honeysuckle produces seeds in a capsule that are likely used by birds and mammals, its major value is supplying low cover. Its tendency to form clones through root suckering provides protection for small mammals and rabbits and nesting cover for songbirds and wild turkey. Ground nesting species (such as white-throated sparrows) benefit from its low overhead cover, and shrub nesters (such as song sparrows) use the dense shrubs themselves. White-tailed deer browse its leaves, and moose use it as a winter browse when snow depth allows.

Landscaping Value: Hardy to Zone 3, bush honeysuckle is an attractive shrub during the growing season because of its reddish, young stems that contrast with its dark green foliage. Its yellow flowers are borne from the twig tips, easy to see, and continue to be produced until frost on domesticated plants. It is a good choice for mass plantings, but some form of containment should be possible as bush honeysuckle can overtake an area with its prolific root suckers. 'Copper' is a cultivar with copper-colored new leaves. Bush honeysuckle is available through nurseries specializing in natives.

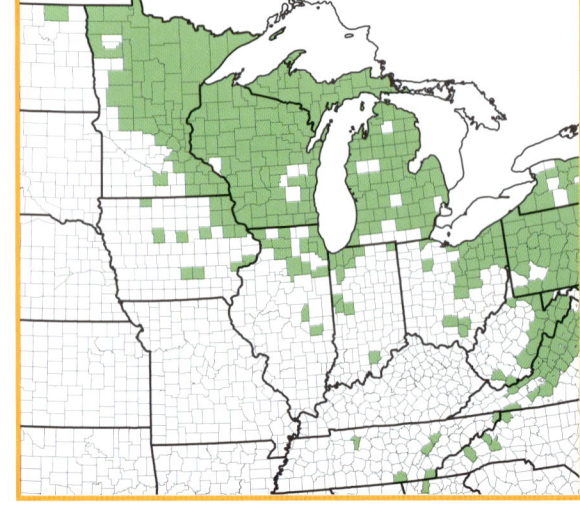

Similar Species Distinctions:
—No other honeysuckle produces woody fruit capsules or has somewhat square stems. It should not be confused with any other native shrub.

Leaves are opposite and may be up to 6 inches long, although generally about 4 inches. They are ovate to ovate-lanceolate, and they gradually taper to an elongate tip. The margins are finely toothed and somewhat hairy. They are dark green and hairless above; the lower surface is paler green and hairy along the midrib. The petioles are very short. Fall color is a pretty red.

The terminal bud is light to dark tan and covered with several scales that appear to be loose. The scales are tipped with whitish hairs.

Flowers are tubular, pale yellow, perfect, and in terminal or axillary clusters. There are usually 3 flowers grouped on long, hanging peduncles. With age they develop a reddish hue. Flowering begins in mid-May and can last for several months.

Fruit is an elongated capsule with flared, bristle-like, slender tips. When ripe it is brown and persists throughout the winter. They ripen in late summer and contain many tiny, unwinged seeds. Each capsule is about ½ inch long.

Twigs are light brown or tan and lined with strong, vertical ridges running along them. New twig growth is especially 4-sided. Leaf scars are raised and nearly V-shaped with 3 bundle scars. Older twigs and buds are gray. Pith is white, and it occupies a sizable proportion of the twig diameter. Lateral buds are like the terminal bud and usually curve toward the twig.

The stem diameter of this shrub is never more than 1 inch, but it does develop gray-brown, thin, peeling outer bark that pulls off in strips.

Wahoos

Genus: *Euonymus*

The genus *Euonymus* contains 3 species native to the Midwest, but they are overshadowed by several aggressive, exotic species, particularly *E. alatus*, winged wahoo, and *E. fortunei,* wintercreeper. These exotics are discussed in the "Introduced Species" section of this book. Another exotic of concern is *E. europaeus*, spindletree, which has escaped in some areas of the Midwest and in the eastern United States.

Wahoos or burning bushes, as they are often called, have 1 obvious feature in common. They have green branches and twigs with some degree of "ridging" along them, which can give them a 4-sided look. Identification, especially of young plants, can be difficult, a problem when one is trying to distinguish between native and exotic. For example, small plants of *E. alatus* often do not develop the exaggerated stem wings, especially when shade-grown, and can easily be confused with *E. atropurpureus*.

One pest that is particularly detrimental to all *Euonymus* but the exotic *E. alatus* is *Unaspis euonymi*, euonymus scale, which can destroy a plant. The insect can be controlled with a fine oil spray in May through August, but timing of application is critical.

Landscaping with *Euonymus*, especially eastern burning bush, is rewarding as there is a 4-season payback. Given at least half-day sun, fall color is gorgeous.

All *Euonymus* species are similar in wildlife value. The most striking characteristic of the group is the prolifically produced unique fruits. These fruits, however, are used sparingly by songbirds, ruffed grouse, and wild turkey, and they often remain unused on shrubs. All species are highly preferred browse of white-tailed deer, which may limit occurrence of some species when deer densities are high. Under forest overstories, plants tend to be spindly and supply little cover, but when they get sufficient sunlight, they become dense enough to supply songbird nest sites.

Fruit is a fleshy aril that is enclosed in a rather spongy capsule. There are claims that some or all parts of the fruit are poisonous, so it is best to avoid eating them.

All members of this genus have opposite branching, and often lateral branches are so strongly angled that they
appear to create 90 degree angles with
the main stem. Eastern burning bush, *E.
atropurpureus*, is pictured.

Twigs of all *Euonymus*
species are green, which is
very unusual for any woody
plant (most are brownish).
Growing along the length
of the twigs are tannish-colored "lines" (especially on
young, fast-growing twigs)
that give them a 4-sided
appearance. Seen here is
E. americanus, strawber-
rybush.

Euonymus scale is
commonly found on
eastern burning bush
and can slowly kill a
plant.

strawberry bush, brook euonymus

Euonymus americanus L.

Euonymus americana L.

Family: Celastraceae

Strawberry bush is a rare shrub of midwestern swamps, but it becomes more common as one proceeds south. An over-abundance of white-tailed deer has taken a toll on this species, as well as another native *Euonymus*, *E. obovatus*. Both have green twigs and are preferred winter browse. The common name "strawberry bush" originates from the ripening fruit that resembles the color and texture of a strawberry fruit. Our common eastern burning bush, *E. atropurpureus*, has similar twigs, but it occurs in almost any habitat. Strawberry bush is listed as endangered in Illinois.

Form and Size: It can become 6 feet tall or more and has an open, angular growth form. Although it is normally found growing in heavy shade, it tolerates nearly full sun and can be planted on a drier, more upland site with success.

Habitat: This species seems to nearly always be associated with low, moist, wooded sites. As an example, in Indiana, Deam (1932) reported it only from "low, flat, beech, red gum (sweetgum), and pin oak woods" in the southern part of the state.

Wildlife Use: This species is uncommon throughout our region, and it occurs most frequently in bottomland forests where shade limits its cover potential. It is, however, a favorite browse species of white-tailed deer; it often shows browse damage at very low deer densities and essentially disappears at high densities. The fruits, while aesthetically pleasing, are used only sparingly by wild turkeys and various songbirds.

Landscaping Value: Hardy to Zone 5, this is a southern species of bottomland forests. It is adapted to wet, slightly acidic soil and partial sun. The authors grow it on an upland site in neutral pH soil in nearly full sun, and it is doing well. Although this is a very attractive shrub year-round, a unique wintertime appeal is its strongly angled, green twigs and branches. Fall color is a burning bush red. *Euonymus* scale can be a problem. Strawberry bush is rarely available from any nursery, but it can be grown from seed or cuttings.

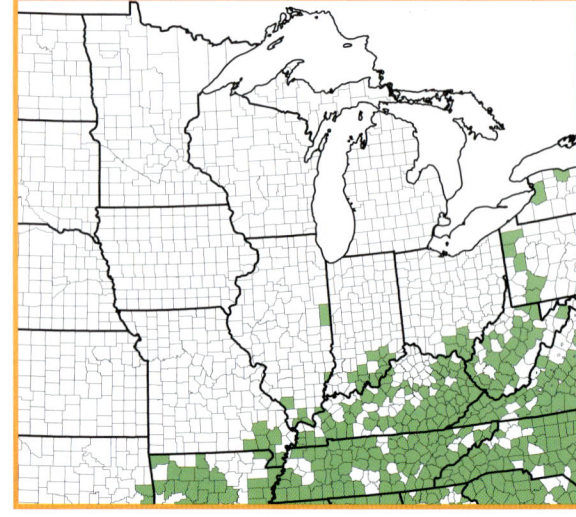

Similar Species Distinctions:
—**Running strawberry bush** (*E. obovatus*) has similar flowers and fruit, but has a somewhat prostrate growth form. Its leaves can be as variable as those of strawberry bush, but they are usually widest above the middle. It is a woodland species of rich, moist forests.

The unusual-looking flowers are 5-petaled, greenish pink, and about ¼ inch across. They are perfect and appear in May in axillary pairs. Each pair of flowers is on a long stalk.

Leaves are opposite, simple, oblong-lanceolate to broadly ovate, and about 3½ inches long. The shape, like other *Euonymus*, can be variable, but each leaf is usually widest at or below the middle. They usually have a short tip and finely toothed margins that are somewhat rolled under. The leaf is grass green and smooth on the upper surface, paler and rather glaucous-white beneath. The petiole is extremely short.

The warty (tuberculate), pink capsule ripens in the fall. When ripe, the pink capsule splits open in 3 to 5 parts to reveal orange-red, fleshy-coated seeds. Each capsule is about ½ inch across.

Bark is green for most of the life of the shrub. With age, it develops tan-colored streaks that run vertically up the trunk. Bark of the oldest shrubs turns gray-brown with a rough surface.

Terminal buds are covered with numerous, loose, pinkish red, pointed scales. There is commonly a pair of smaller, lateral buds beside the terminal.

Twigs are green, especially the 1- to 2-year-old growth. Branches remain green for many years. They usually have tan-colored, lateral lines to varying degrees that create a 4-sided appearance. Lateral buds are like the terminal but smaller. Leaf scars are tannish white, somewhat raised, and have a row of 3 bundle scars.

eastern burning bush, wahoo

Euonymus atropurpureus **Jacq.**

Family: Celastraceae

Wahoo is the Midwest's most common native species of *Euonymus*. It is very adaptable to almost any site and prefers more sun than our other 2. The more sun it receives, the more its leaves will "burn" in the fall. This is the species of burning bush that *should* be planted in the Midwest, instead of the commonly found exotic, *E. alatus*. *Euonymus* scale, *Unaspis euonymi*, tends to be a problem with members of this taxon, both native and exotic. The Latin *atropurpureus* refers to the dark purple flowers.

Form and Size: Wahoo can become a large shrub, up to 15 feet tall, but it is usually smaller. It freely suckers from the roots and forms small colonies over time. It is shade tolerant but can handle full sun.

Habitat: The common habitat is near streams and along woods edges, but it can be found in more upland sites. It grows quickly and is especially attractive in the fall when in full fruit.

Wildlife Use: Wahoo is more abundant and widely distributed in our region than its native congeners. It reproduces via rhizomes as well as seeds and occasionally forms colonies that supply some wildlife cover and nesting sites for songbirds. The fruits, which are produced in great abundance at times, are only minimally used by birds and small mammals. It is browsed by cottontails when growing near forest edges, as well as by white-tailed deer, which seem to prefer it less than they do *E. americanus* and *E. obovatus*.

Landscaping Value: Hardy to Zone 4, this common shrub has the prettiest flowers of any *Euonymus*. It has year-round appeal with its green twigs and branches, purplish red flowers, red fall color, and bright orange-red fruits. It needs more than a half-day sun to get the best fall color, but full sun is a bit much. Fruits are usually retained through late fall, long after leaf drop. This shrub can be planted just about anywhere, and it is worth the effort. *Euonymus* scale can be a problem. Wahoo is available through some nurseries specializing in natives.

> **Similar Species Distinctions:**
> **—Winged wahoo** (*E. alatus*), when shade grown, does not develop the typical stem "wings" as well, and it may be confused with wahoo if one focuses only on the leaves.

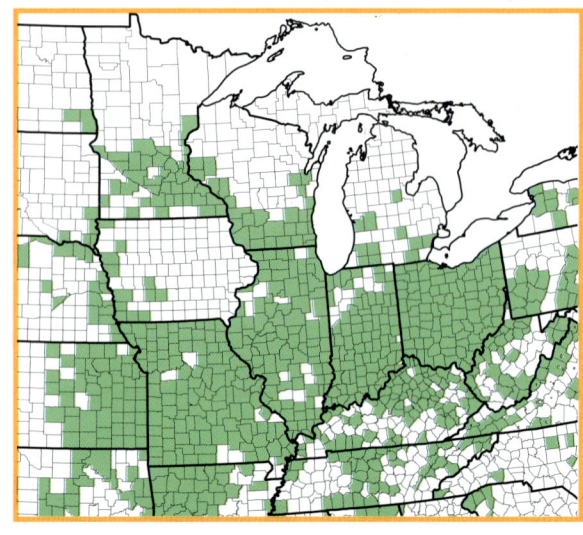

Leaves are opposite and oblong-ovate with an elongated tip—more so (usually) than our other 2 species. They tend to be broadest in the middle of the leaf. The margin is mostly finely toothed, and both leaf tip and base are usually pointed. Size is variable, from 1½ to 6 inches long; petiole length is the longest of our native *Euonymus* species—up to ¾ inch long. Leaf upper surface is dull green and smooth; under surface is slightly paler and somewhat hairy.

The perfect, maroon flowers have 4 petals and are about ½ inch across. This is the only native *Euonymus* with maroon-colored flowers. Flowering begins in late June and continues into July. The flowers arise from the leaf axils on long stalks (peduncles) that are about 2 inches in length.

Fruit matures in the fall and is very obvious, simply because of its bright color. The fleshy, smooth pink capsules split into quarters (usually), revealing the orange-red, flesh-coated seeds (arils) that persist into late fall.

Twigs are usually green but can have a reddish tint. Vigorous growth often has linear, pale-colored ridges that give the twig a 4-sided appearance, a feature common to most *Euonymus*. Lateral buds are similar to the terminal, but smaller. Leaf scars are half-moon shaped with a curved row of 3 bundle scars. When grown in full sun, twigs can develop small wings along the ridges.

Terminal buds are ovoid, with greenish red, sometimes loose-fitting scales. There can be a pair of smaller, lateral buds beside the terminal. Buds are usually sharply pointed.

Bark is mostly smooth and gray until old, at which point it develops some fissuring and roughness.

running wahoo, running euonymus, running strawberry bush

Euonymus obovatus **Nutt.**

Family: Celastraceae

Running wahoo is 1 of the Midwest's native ground covers that gets little recognition, even though it has interesting fruit, pretty fall color, and behaves itself, unlike the commonly planted, exotic *E. fortunei* (wintercreeper). Its only drawback is that white-tailed deer love it. Here it is pictured in early spring mixed with many spring ephemerals. This is a species of special concern in Tennessee.

Form and Size: Running wahoo is a low-growing, mostly prostrate shrub. Some say it has almost vine-like branches. It can "reach upward" to several feet in height, but it usually creeps (runs) along the ground, no more than 1 foot tall. It roots at the nodes and spreads to form loose mats.

Habitat: It is almost always found growing in shady woods in moist, rich soil, but it can be found creeping across large boulders and rock outcrops. It can grow in half-day sun, and with that amount of light, it forms dense mats.

Wildlife Use: This prostrate little shrub is quite different from its congeners in form, but it is similar in wildlife food value. While its fruits are sparingly used by birds and small mammals, it is highly preferred browse for both cottontails and white-tailed deer. In fact, when deer populations are high, this species essentially disappears from the moist slopes and bottomlands where it normally occurs. It provides little cover, because of its decumbent nature, but when it occasionally forms dense mats, it can serve as concealment for ground-nesting songbirds, such as Kentucky warblers.

Landscaping Value: Hardy to Zone 3, this small shrub is often more like an herbaceous plant, and it can be used to naturalize a degraded woodlot or forest edge. If given partial light, fall color is very pretty, and a nice blend with the strawberry-red fruits. It is not particular about site, but it is not tolerant of drought. Pruning will create a dense, compact plant. *Euonymus* scale can impact this species, and deer commonly eliminate it by over-browsing. Running wahoo is available from few native-plant nurseries.

Similar Species Distinctions:
—**Strawberry bush** (*E. americanus*), when small and shade-grown, can be confused with running wahoo because of similar leaves and stems. There is no easy method to separate them. Winter buds are slightly different, as may be seen in the individual species accounts.

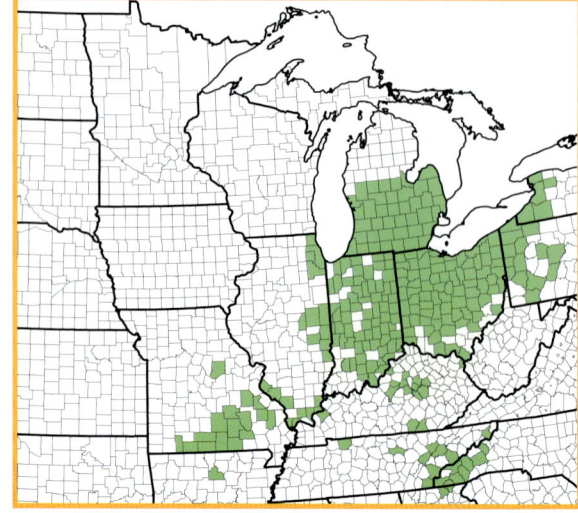

The major leaf difference between this and other native species of *Euonymus* is the fact that this 1 has leaves widest at or above the middle, as the Latin suggests. Leaf shape is variable, however, and they can be elliptic or oblong. They are up to 2½ inches in length, with petioles only about ¼ inch long. The upper leaf surface is dull green and smooth (usually); the lower surface is only slightly paler and smooth. Leaf margins are finely toothed, and the tip is abruptly pointed. The terminal pair of leaves is always the largest, and there are usually only 2 to 4 pairs of leaves per branch.

Fruit ripens in the fall and is similar to strawberry bush with its warty, pink capsules. Running wahoo capsules split into 3 parts that reveal scarlet-orange, flesh-covered seeds. The fruit is about ¾ inch across.

Twigs are green and smooth. Vigorously growing plants will have tan-colored lines running along the twigs that give it a 4-sided appearance. With age, the twigs will turn grayish brown. Leaf scars are pale green and shaped like a half-moon. There is a single, sizable bundle scar in the middle of each leaf scar. Lateral buds are similar to the terminal, but they tend to be green in color.

The unusual-looking flowers are only about ¼ inch across, have 5 greenish pink petals, and appear in clusters of 1 to 3 on long stalks (peduncles). The perfect flowers emerge in May with the leaves.

Terminal buds are ovoid, pointed, and covered with greenish pink scales, usually many fewer than in our other native *Euonymus* species. A pair of lateral buds usually sits beside each terminal.

wild hydrangea, smooth hydrangea, American hydrangea

Hydrangea arborescens **L.**

Hydrangea cinerea **Small**

Family: Hydrangeaceae

Certainly less spectacular than the exotic and commonly planted *H. macrophylla*, which has large pink and blue flowers, the midwestern native wild hydrangea is, nonetheless, quaintly beautiful in a delicate way. In older literature, this species is classified in the Saxifragaceae family. There are still several varieties recognized by some authors, with differences based on the degree of hairiness on the undersides of the leaves.

Form and Size: Hydrangea is a small shrub that commonly attains a height of 3 or 4 feet. On drier sites it will often stand just a couple of feet tall. It spreads from suckers that create a full, rather rounded form. Its sizable leaves usually cover up the fact that there are few branches.

Habitat: It prefers moist soils on shaded wooded slopes of ravines, rock ledges, and wooded stream banks. It does not tolerate full sun. Growth is best in moist soil, but it is often found growing in the crevices of rock outcrops or actually growing on rocks, as long as there is even a little soil buildup and the rock remains moist.

Wildlife Uses: When deer were reintroduced into the Midwest in the 1930-1950 period, their densities rapidly increased in protected areas, and biologists reported that hydrangea was being eliminated by over-browsing. It is a preferred summer browse species and does not withstand heavy browsing by high deer populations. It, however, does persist in the rock outcrops and slopes that are its microenvironment. Seeds are little used. Hydrangea is a favorite nest site for several small songbirds, especially the indigo bunting.

Landscaping Value: Hardy to Zone 4, this species tends to be adaptable to many soil types, as long as it has a source of moisture, as it is sensitive to drought. It does not grow well in full sun, and it makes a wonderful addition to a shaded area of a yard with a fence or building as a backdrop. Pruning after leaf drop promotes a fuller shape but removes the interesting, overwintering, flower-like fruiting heads. It grows slowly, which means little maintenance. The pretty, early summer flowers are similar to the viburnums that produce a ring of sterile flowers around a central cluster of small, fertile flowers. A cultivar 'Annabelle' has large flower heads and barely resembles our native, which is rarely available through native-plant nurseries.

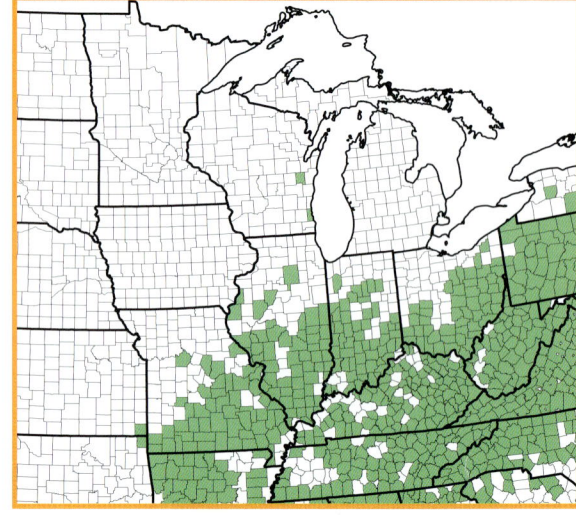

Similar Species Distinctions:
—Small hydrangeas several inches tall can be mistaken for some herbaceous species that have opposite branching and "egg-shaped" leaves.

The leaves are opposite, ovate to orbicular in shape, and range from 2 to 7 inches in length. Petioles are long, sometimes almost as long as the leaf itself. The upper surface is green and somewhat hairy, while the lower surface is pale with hairs running along the veins. Leaf margins are finely toothed, and the leaf tip has a very abrupt, short tip. Fall color is pale yellow, with the veins turning brown last.

The fruit is comprised of a tiny, 2-celled capsule, many of which occur in each disk. Each capsule has many tiny, prominently ribbed seeds. Fruit matures by late summer and persists throughout the winter or longer.

Twigs are smooth, beige to tannish, and very brittle. Inside is almost nothing but pure white pith. Leaf scars are V-shaped with 3 bundle scars inside. Each scar has a distinct ridge along the bottom. Lateral buds are similar to the terminal and usually diverge from the twig.

Terminal buds are difficult to find, except on immature stems and plants, as the twigs are usually terminated by flowers. However, when present, they are ovoid and pointed, with at least 3 pairs of loose-fitting, rust-brown scales. The scale margins usually have at least a few white hairs along them.

Flowers begin to appear in mid-June and continue into July. The perfect, white flowers develop terminally or from upper leaf axils. The inflorescences are disk-shaped and sometimes 4 or 5 inches across. Larger, sterile flowers (¾ inch across) usually surround the disk that is comprised of mostly tiny, fertile flowers.

Hydrangea stems never become larger than ½ inch in diameter and are fairly short-lived. Therefore, there is not much time for bark to develop. The outer tannish bark separates from the stem in large papery sections on the oldest stems. A few scattered, raised lenticels are obvious along stems.

St. John's Worts

Genus: *Hypericum*

There have been several taxonomic changes over the years within the genus *Hypericum*—even the family in which it is included has changed, from its original Guttiferae to Hypericaceae, and finally to its current family, Clusiaceae. This is mainly a tropical family, and there are many more herbaceous than woody members in the Midwest. Within the genus *Hypericum* are several woody species that occur here, the most common of which is *Hypericum prolificum*, shrubby St. John's wort. The rarest, which is only found in a few southern counties in Illinois and several in Kentucky, is *H. lobocarpum* (*H. densiflorum* var. *lobocarpum*), which has deeply lobed fruit capsules.

From a landscape perspective, St. John's wort puts on a show of exploding yellow fireworks (its flowers) in July when nearly all other native shrubs have finished flowering. Their extended flowering period of a month or longer adds to their appeal. Our native shrubs are very hardy, undemanding, and able to grow almost anywhere.

All members of the genus *Hypericum* have opposite leaves with entire margins and yellow flowers. One unusual characteristic of the family is the translucent glands that dot the leaves. A leaf held skyward appears to have tiny holes across the surface.

Perhaps the major wildlife value of this genus is as cover, although seeds are regularly used by birds. The dense growth form of both *H. kalmianum* and *H. prolificum* are very attractive to shrub-nesting birds in the open situations where these species normally occur. *H. hypericoides* is a more diminutive plant that supplies cover to ground nesters. The flowers of all are very attractive to bees and wasps. Deer browse leaves to some degree in spring and summer.

Leaves are always simple with entire leaf margins. *H. prolificum* (seen here) and *H. kalmianum* have pairs of leaflets and bracts at the base of leaves that give a full appearance to the shrub.

St. John's worts are known for their unusual leaf "holes" that are actually translucent glands. A leaf held up to the sky will allow you to see all the dots scattered across it.

Many species of *Hypericum* have so many yellow stamens that the flowers resemble exploding fireworks. Seen here are Kalm's hypericum flowers.

St. Andrew's cross

Hypericum hypericoides **(L.) Crantz**
Ascyrum hypericoides **L.**
Family: Clusiaceae

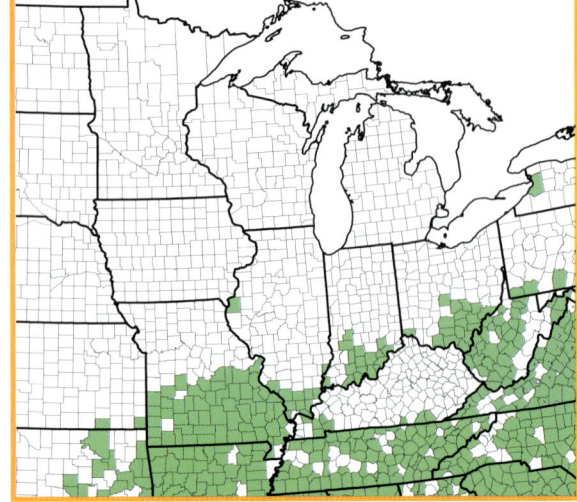

St. Andrew's cross is named after the apostle Andrew who was reportedly crucified on a saltire-shaped (X) cross. He told his executioner that he was not worthy of crucifixion on the same type of cross used for Jesus' execution. The flower of St. Andrew's cross is saltire-shaped and quite distinct. This species is often listed in the Hypericaceae family.

Form and Size: This is a short, spreading shrub that can have straggling, upright branches to 3 feet. Sometimes it hugs the ground and grows no taller than 6 inches or so. Its slender, delicate branches are numerous, and it appears rather spindly, especially when growing in heavy shade. St. Andrew's cross is commonly overlooked during the growing season because it is so small. It has barely woody stems, resembling an herbaceous plant. In the winter when herbaceous vegetation is mostly gone, it stands out because it retains its leaves.

Habitat: St. Andrew's cross is found in dry, rocky, open woods on ridges or slopes in acidic soils. It handles full shade, but it grows best in partial sun. It is a small, multi-stemmed plant that roots along the sprawling stems to form small mats.

Wildlife Uses: St. Andrew's cross is a sprawling shrub that occurs in dry open woodlands, in contrast to our other more openland, upright *Hypericum*s. In these woodland situations, it sometimes produces substantial ground cover that supplies nesting sites for associated birds, especially eastern towhees. Seeds are used in winter by songbirds and gallinaceous game birds; some minor browsing by deer has been reported, but it is not a preferred species.

Landscaping Value: Hardy to Zone 5, St. Andrew's cross is a small, often overlooked shrub of southern portions of the Midwest. It is a wonderful choice for a ground cover along paths or walkways in an urban landscape, and it would grow well in a rock garden. It flowers in the summer when most other shrubs are finished, and it has an extended flowering period of at least a month. Though it is not considered an evergreen, it retains some green leaves throughout most of the winter. Without competition, it grows low to the ground and slowly spreads. It is resistant to heat and drought, but it requires slightly acidic soil.

Similar Species Distinctions:
—**St. Peter's wort** (*Hypericum crux-andreae [H. stans]*) is native to the southern regions of the Midwest and is a threatened plant in southern Kentucky. It is a larger, erect shrub to 3 feet and has distinctly 2-sided twigs with similar, but larger, stalked flowers. Its leaves are larger and elliptic-oblong in shape.

The yellow flowers appear from late June to mid-July (or later) near the tips of the twigs. They are 4-petalled and about ½ inch across. Each flower has 4 sepals of 2 sizes: 2 large and 2 small. Flowers are commonly solitary, but they can appear in small clusters.

Leaves are simple, paired, mostly opposite (the lowest leaves are often alternate), and semi-persistent. They are firm, linear, and up to 1½ inches long. Each pair of leaves can have smaller leaves clustered near its base. Each leaf has a pair of slender, pointed bracteoles near the base that turn brown long before the leaf. The margin is entire, and there is virtually no petiole. Both leaf surfaces are pale green and can be dotted with black glands, which is a characteristic of this family. Fall color is seen in part here—a reddish pink.

Fruit is a small, brown, flattened capsule that splits in half when ripe. Each capsule is sandwiched between 2 sepals, all of which is persistent throughout the winter. The capsules are ⅓ inch long and full of tiny, black seeds.

No terminal buds are visible in the winter, as twigs are tipped with fruit hidden by a pair of dried sepals. Twigs are very thin and reddish brown. They develop shreddy, thin bark that gives the twig an angled or 2-edged appearance. Lateral buds are tiny, brown, few-scaled, and barely visible in the winter.

Kalm's hypericum, Kalm's St. John's wort

Hypericum kalmianum **L.**

Family: Clusiaceae

Kalm's hypericum is a beautiful shrub when in flower, with the many flowers producing an explosion of yellow stamens. Bees are attracted to the copious nectar that is produced. This species has made its way into the horticultural trade, probably because of its ease of propagation, prolific flowers, and extended flowering period. The St. John's worts, because of their early summer flowering period, are named in honor of St. John the Baptist, whose birth date was June 24.

Form and Size: It is usually less than 3 feet in height, with a strong central stem and a bushy appearance that results from its numerous, fine twigs and branches. When found, it is usually in colonies and can completely dominate an area over time, if no other woody competition exists.

Habitat: Kalm's hypericum is found in low depressions and pannes on calcareous, sandy sites near the Great Lakes. In the upper Midwest, it can also be found along riverbanks and in meadows. It prefers full sun, but it tolerates a great deal of shade, and is seemingly adaptable to many soil types.

Wildlife Uses: Kalm's hypericum is essentially a northern version of *H. prolificum* and has almost identical wildlife values to that species.

Landscaping Value: Hardy to Zone 4, Kalm's hypericum is an adaptable shrub that will grow in richer, more organic soil, as long as it is well drained. In full sun, it grows densely and full, with attractive dark green foliage. Flowers are the show-stoppers, and they bloom over a month-long period in the heat of summer. Outer bark is dark brown and exfoliates over time, exposing bronze-colored inner bark. This species is found in nurseries that specialize in natives, and occasionally in other commercial outlets.

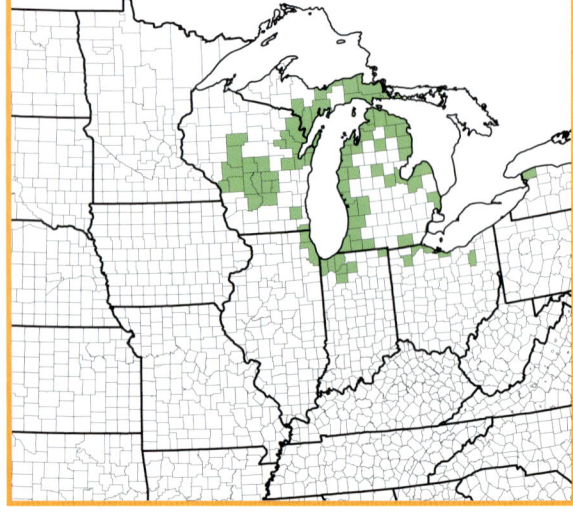

Similar Species Distinctions:
—**Shrubby St. John's wort** (*H. prolificum*) is similar but it is found in different habitats. It has flowers in upright clusters and 3-chambered capsules. This is the only woody St. John's wort with an overlapping range.

The opposite leaves are oblanceolate to linear-oblong, up to 5 inches long, but rarely more than ½ inch wide. There are smaller, paired leaf-like bracts at the base of each leaf. Leaves are grass green and smooth above, whitish beneath. Margins are entire and petioles are nonexistent.

The striking, yellow flowers are about 1 inch across and are so crowded with stamens that they appear pom-pom-like. The perfect flowers are in terminal, flattened panicles and have 5 petals. Swink and Wilhelm (1994) in *Plants of the Chicago Region* report flowers from June 19 to October 3.

Outer bark is thin and brownish colored; it exfoliates in wide pieces, exposing the almost bronze-colored inner bark.

Twigs are very slender, grayish brown with thin, loose bark that gives the twigs a 4-sided appearance. Lateral buds are rarely visible, but they are protected by overwintering, clustered, tiny green leaves. Leaf scars have a definite ridge that is V-shaped.

Fruits ripen in late summer; they are terminal, 4-to-6 chambered, brown capsules about ¼ inch long. Each capsule has numerous, tiny black seeds. Capsules persist throughout the winter.

shrubby St. John's wort

Hypericum prolificum **L.**

Family: Clusiaceae

This is the Midwest's most common and adaptable shrubby *Hypericum*. It will grow in just about any habitat, from wet to dry. It is beautiful when in full bloom, and its seeds are sought after by songbirds as a winter food source, especially when snow cover is deep.

Form and Size: Shrubby St. John's wort is usually 3 to 4 feet tall—a bit taller than Kalm's hypericum. It readily seeds in an area and is usually found in colonies. Growth form is similar to that of Kalm's, but larger.

Habitat: It occurs in many habitats, including abandoned fields, degraded, open, wooded slopes, roadsides, streamsides, and wet woods. It is very adaptable to a wide range of soil types, moisture regimes, and light intensity. It tends to be somewhat short-lived, but it is efficient at replacing itself through seeds.

Wildlife Uses: This is a sun-loving species that is especially fond of open areas with moist soils. In these situations, its dense limb structure is very attractive to nesting shrubland birds. It frequently occurs in rather dense stands that provide excellent cover for wildlife, including bobwhite and cottontails. In winter its seeds, which are prolifically produced, are eaten by bobwhite and many wintering birds, especially tree sparrows and dark-eyed juncos.

Landscaping Value: Hardy to Zone 4, shrubby St. John's wort is very adaptable to a wide range of soils and moisture regimes. It is easy to grow, and it works well in mass plantings. Summer flowers are beautiful against the dark green foliage. Fall leaf color is yellow. Winter is the time to notice the dark exfoliating bark, somewhat leafy stems, and persistent, upright capsules. This species is occasionally sold at nurseries that specialize in natives.

Similar Species Distinctions:
—**Kalm's hypericum** (*H. kalmianum*) has similar characteristics. See comparisons on its species account and above.
—**Golden St. John's wort** (*H. frondosum*) is a native shrub half as tall, which usually has larger, wider leaves and larger (to 1¾ inches), solitary flowers. It is native to Kentucky and southward to s. Georgia on dry, rocky sites.

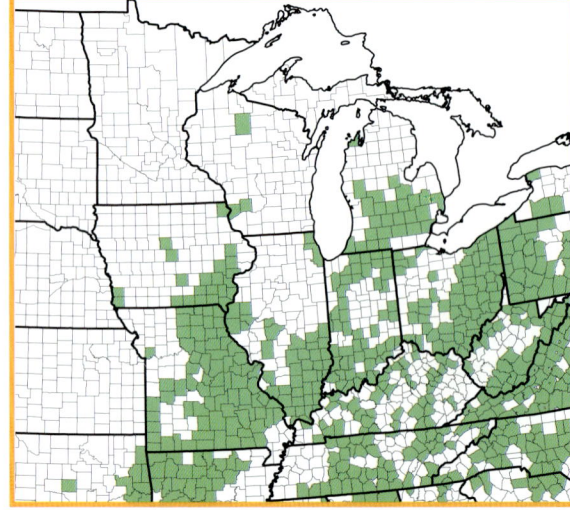

The opposite leaves are linear to oblanceolate and up to 4 inches long. They are usually more than ½ inch wide. The largest leaves are usually twice the width of those of Kalm's hypericum. Margins are entire and the petiole is very short but present. Leaves are bright green, smooth on top, and slightly paler beneath. There are paired leaf-like bracts at the base of the leaves.

Fruit is a 3-chambered, upright, pointed, brown capsule that ripens in the fall. They are about ⅓ inch in length and filled with many tiny, black seeds. The capsule is persistent throughout the winter.

Flowers appear throughout July and into August from leaf axils. The yellow, perfect flowers have 5 petals and a profusion of yellow stamens. They are generally smaller in diameter than those of Kalm's hypericum—only ¾ inch across.

The bark of mature plants is gray or brown, and develops thin, flaking pieces similar to those of a birch. The inner bark is a bronze color.

Twigs are slender, gray, and slightly ridged. The gray coloration is a thin bark that splits to reveal a brownish inner bark. Leaf scars are strongly triangular in shape, with a single, circular bundle scar in the middle. There is a distinct V-shaped ridge along the bottom edge of each leaf scar. Lateral buds are usually "protected" in the winter with a cluster of tiny, green leaves. If visible, the buds are tiny, white, and sunken in the leaf scars, which are often on spur shoots.

American fly honeysuckle, fly honeysuckle

Lonicera canadensis **Marsh.**

Family: Caprifoliaceae

This is 1 of just 3 midwestern shrubby, native honeysuckles. It is a common forest edge species where summer heat and humidity is not too harsh, and it has been described as a "scraggly, running, understory shrub." When grown as a domesticated specimen, it has a full shape. Currently, it is state extirpated in Indiana, and of special concern in Tennessee.

Form and Size: Fly honeysuckle can reach a height of 6 feet, but it is usually smaller. It often grows in the heavy shade of northern woodlands, where it is few-branched and sprawling. It frequently develops loose colonies from low branches that root on contact with the ground.

Habitat: In the upper Midwest, it is common in open woods, forest edges, swamps, and bogs. It is found in wet to moderately dry soils and tolerates partial to full shade. Individual shrubs are short-lived and moderately fast growing.

Wildlife Use: Largely absent in the southern reaches of the Midwest, this species becomes ubiquitous at more northerly latitudes. Its cover value for wildlife is relatively minimal, since it most frequently occurs as individuals under forest canopy where it is rather spindly in growth form. When it does grow in the open, it becomes more full and supplies good nesting cover for songbirds. Regardless of location, it regularly produces fruit, but not in large quantities; fruits are quickly taken by songbirds, ruffed grouse, wild turkey, and small mammals.

Landscaping Value: Hardy to Zone 3, this easy-to-grow species develops into a neat, compact, moderate-sized shrub with less than a half-day sun. It has a shallow root system, but it seems to tolerate drought and heat well. It prefers slightly acidic soil but adapts well to something more neutral. Flowers and fruit appear early in the spring and are not particularly showy. Fall color is pale yellow. Seeds look like sesame seeds, and they quickly germinate once they are spread by birds. No horticultural varieties are available, and this species is yet to be found in any midwestern nurseries.

Similar Species Distinctions:
—**Mountain fly honeysuckle** (*L. villosa*) has similar, dangling flowers, but paired flowers appear to share a single ovary. Leaves are more oblong.
—The introduced honeysuckles are large shrubs with leaves widest at or above the middle. Twig pith of these shrubs is always brown but hollow in the center.

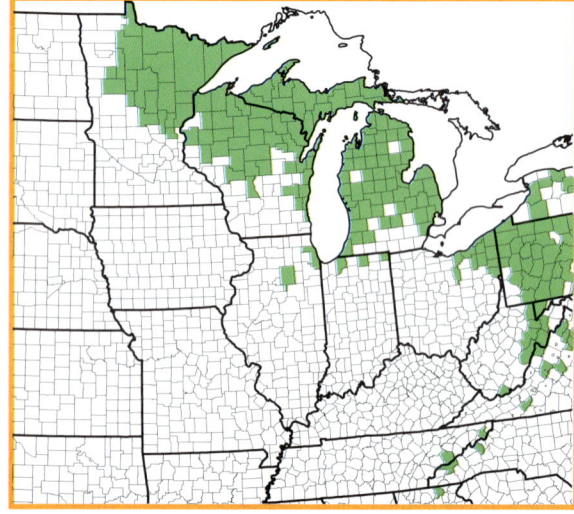

The opposite leaves are about 3 inches long, ovate, and mostly wider below the middle. They are light green and smooth above, slightly paler and smooth beneath. The leaf margin is entire but lined with a row of fine hairs. The short petiole is hairy as well.

This shrub is a harbinger of spring, flowering in early April in the central part of the Midwest. It flowers well into May farther north. Flowers are pale yellow to creamy white, and often tinged with pink. They develop in pairs on short, smooth stalks; they are tubular and nearly 1 inch long.

Fruit ripens in late May through June on 1 inch long fruit stalks (peduncles). Berries are red, fleshy, paired, and about ½ inch long. They are usually difficult to see as they are hidden by the fully developed leaves.

Mature plants never gain much diameter, but the bark is light brown, and with age it becomes shreddy and peels into thin strips.

Buds are ovoid to elongate, with few, smooth, greenish scales. Lateral buds diverge away from the twigs slightly, and beneath them are somewhat diamond-shaped leaf scars. Twigs are smooth and tan-colored. Pith is solid and white.

Russet buffaloberry, Canada buffaloberry, soapberry

Shepherdia canadensis L.

Family: Elaeagnaceae

Russet buffaloberry is a common shrub across much of the United States where cool summer temperatures prevail. It is quite attractive in the fall when the shiny red, ripe fruit is in striking contrast to its dark green foliage. Currently it is state endangered in Illinois and Pennsylvania, and state extirpated in Indiana.

Form and Size: Buffaloberry is a medium-sized shrub that can reach 6 to 9 feet in height, especially in the cooler areas of its range. It has a much-branched, rounded form. It thrives in full sun, but it can be found in partial shade.

Habitat: This species occurs in many habitats across its expansive range. Several in the Midwest include dry, rocky shores, riverbanks, limestone outcroppings, open woods, and dunes. It tends to be colonial—when you see 1, you see numerous plants.

Wildlife Use: This unique little shrub is rare throughout the southern Midwest, so overall impact on wildlife is low. However, locally it may be quite common, supplying fruit to a series of bird and mammal users, including black bears and ruffed grouse in the fall. It is only sparingly browsed by white-tailed deer in summer and snowshoe hares in winter; preference is low. Additionally, its structure is such that it is readily used for nesting by songbirds and is occasionally dense enough to supply protective cover.

Landscaping Value: Hardy from Zones 3–6, buffaloberry's southern range is limited by climate. It is not well suited to the heat and humidity of the southern Midwest. It grows well in dry, alkaline sites, and it is very resistant to drought. It improves soil quality by fixing atmospheric nitrogen. Flowers are unspectacular, but the dark green leaves combined with the bright red fall fruit makes a late season statement. Specimen plants where full sun allows them to develop a full form would be a nice landscape addition. Several cultivars are available including 'Rubra' and 'Xanthocarpa,' which has yellow fruit. Locating buffaloberry in any midwestern nursery has proven impossible to date.

Similar Species Distinctions:
—**Autumn olive** (*Elaeagnus umbellata*), a close relative, is an introduced species escaped throughout much of the Midwest. It has alternate branching and light red, speckled fruit on stalks.

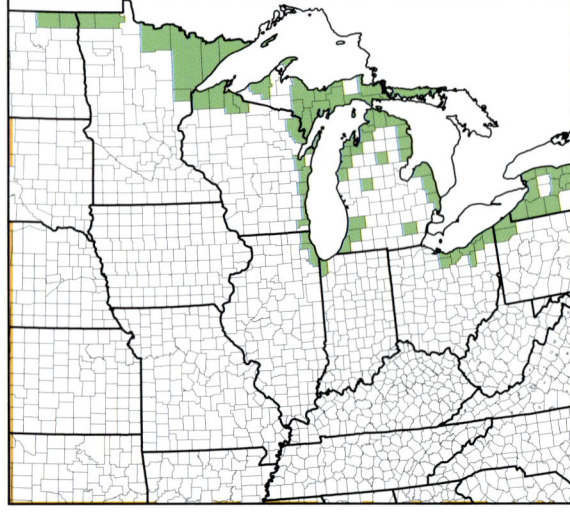

The opposite leaves are up to 2½ inches long and ovate, with rounded base and tip. The margin is entire. The upper surface is dark green with scattered, star-shaped clusters of hairs. The lower surface has silvery hairs and rust-brown scales.

Buffaloberry flowers in May after the leaves have appeared. Flowers are on short spikes on last year's growth. Plants are unisex; both male and female flowers are tiny, with 4 yellowish tepals (combination petals and sepals). Males (left) have 8 stamens; a few unopened flower buds are visible (right).

Pith color is the same as the rust-brown scales. Buds are of 2 kinds. Vegetative buds are valvate with 2 to 4 scales. Bud scales are covered with rust-colored flat scales. The small, rounded flower buds are formed during the summer months and remain clustered near the tips of the twigs. They are dotted with the same rust-colored scales. Twigs are medium brown and covered with brownish, flattened scales.

Mature bark is not particularly unique but is dark brown and slightly scaly.

Fruit ripens in late summer into the fall. They are bright red, fleshy, juicy, red-orange fruits (drupes) that are in tight clusters. Each fruit has pale, scattered scales on the red surface.

coralberry, buckbrush, Indian currant

Symphoricarpos orbiculatus **Moench**

Family: Caprifoliaceae

Coralberry is an interesting small shrub that has its greatest visual appeal in the winter because of the clusters of hot pink fruit that remain on the plant until spring. A planted row of coralberry along a drive is a striking sight in the dead of winter. The lime-green, new leaves on hairy, pinkish stems are very attractive in the spring as well.

Form and Size: It is a small shrub, usually just 2 to 3 feet tall, although it can reach 5 feet in height. The shrub is finely branched, with arching upper twigs that create a rounded crown.

Habitat: Coralberry is common in old-fields, bottomlands, and forest edges in dry or moist soil. It prefers full sun. If growing in shade, it becomes straggly and eventually dies out. It suckers freely from the roots and sends out runners, both of which help the species to develop small colonies in any given area.

Wildlife Uses: This unique shrub, the distribution of which peaks in the southern portion of our region, forms dense clones in open bottomland habitats. These waist-high thickets provide valuable protective cover for small animals and are favorite nest sites for many shrub-nesting songbirds. The species produces prodigious quantities of reddish (coral-colored) fruit that give a colorful glow to patches in the fall. Although there are reports of bird use of these fruits, the authors note little if any bird use, and fruits remain throughout the winter; no doubt some emergency use is made of these persistent fruits. However, fruits, as well as associated leaves and stems, are readily eaten by white-tailed deer and form a substantial part of their fall diets in some areas.

Landscaping Value: Hardy to Zone 2, this small, slender shrub is easily grown and thrives on neglect. With its arching stems, it makes a graceful border along a walk or drive, or even planted along short retaining walls or landscaping timbers. It will grow well in full sun, but half-day will still produce the numerous, hot pink berries. This is 1 hardy plant that will grow in just about any soil. There are several cultivars, and coralberry is occasionally available from nurseries specializing in natives.

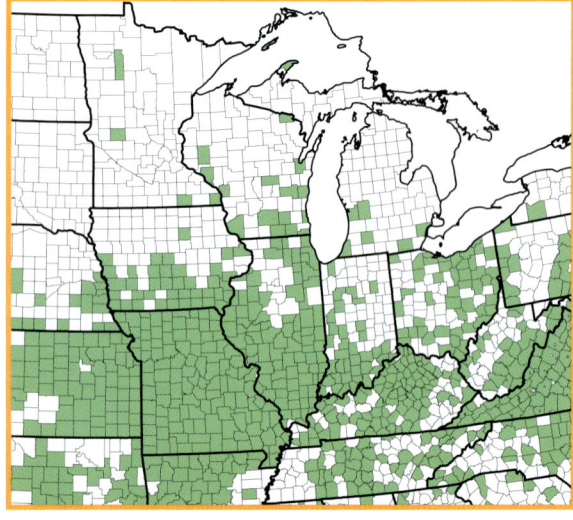

Similar Species Distinctions:
—**Beautyberry** (*Callicarpa americana*) is found in the South and is a much larger shrub, but has hot pink, clustered fruit similar to that of coralberry. It also has opposite branching.
—**Honeysuckle** (*Lonicera spp.*) juvenile shrubs, especially Morrow's, can look similar from a leaf standpoint.

The opposite leaves are no more than 2 inches long and ovate to oval. The upper leaf surface is green and smooth, and the lower surface is paler and hairy. Leaf margins are wavy or entire. Petioles are very short.

Flowers appear from the leaf axils in early July and can be found through August and September. They are only ⅛ inch long and have 5 pinkish green, fused petals that create a bell-shaped flower.

Although coralberry rarely achieves more than an inch in diameter, it does have distinctive thin, brown, peely bark. As it peels, the reddish pink inner bark is revealed.

Terminal buds are rarely seen, since the twig tips die back after leaf-fall. Twigs are very slender, brown, and hairy. Older twigs become smooth. Lateral buds are extremely tiny, and covered with loose, reddish brown scales. Leaf scars are tiny, raised, and V-shaped.

Fruits ripen in the fall and are a berry-like drupe. They are about ⅕ inch in diameter, hot pink in color, and in axillary clusters. They persist throughout the winter unless eaten by white-tailed deer. Each drupe has 2 cream-colored seeds that resemble sesame seeds.

Viburnums

Genus: *Viburnum*

Viburnums and landscaping. There probably is not a single purveyor of woody plants in the United States that does not offer some species of viburnum, whether native or introduced. They are highly desired by many, as evidenced by the number of introduced species available, and an even larger number of their cultivars.

Michael A. Dirr, a professor of horticulture at University of Georgia, is the author of numerous books, including *Viburnums* (2007). He is a huge fan of the taxon, and he once wrote that "a garden without viburnums is like a life without the pleasures of music and art." Anyone interested in more information on native and introduced viburnums should read his book. Flowers and fall color are wonderful on all species, but there is hardly a season they do not shine in a landscape.

Here in the Midwest, we have been rewarded by nature with at least 11 species of these delightful shrubs. Two species are not presented in this book; one is *Viburnum edule*, a boreal species found on Isle Royale in Lake Superior and a few northeastern counties of Minnesota; the other is largely an Appalachian species, *V. alnifolium* (*lantanoides*), found in northeastern Ohio and eastern Kentucky and Tennessee.

Viburnums probably have their major value to wildlife in their fruit production, which can be substantial. With some exceptions (e.g., *V. dentatum*), however, the fruit does not seem to be highly preferred, and it frequently clings to the shrub into the winter. Similarly, the genus is not highly preferred as browse by white-tailed deer, but some are taken regularly by moose. Cover value is variable, depending on species and growth form. Although opposite branching, they have in general greater nesting cover value than *Cornus*, because plants tend to be more "limby" with shorter internodal distances. Individuals provide some protective cover, but several have increased values because of their clonal nature (e.g., *V. acerifolium* and *V. prunifolium*).

All viburnums have opposite branching, which is always a great feature to notice, since it rules out so many of the dominant, alternate-branched species in the decision-making process of species identification. Mapleleaf viburnum, *V. acerifolium*, seen here, like many viburnums, has delicate, red blushing on the new leaves and developing inflorescences.

Many viburnums have both vegetative (right) and floral buds that overwinter. Some can be fairly sizable—over an inch long—and have striking color, like this *V. cassinoides,* which adds a nice touch to the winter landscape.

The very similar-looking nannyberry and blackhaw are distinguished by the leaf petioles. Nannyberry petioles are expanded, as seen on the right, versus the "normal-looking" blackhaw petiole on the left.

Viburnums are well-known for their splendid fall color, most of which are some shade of red. Shown here is *V. rufidulum*, rusty blackhaw, in all its glory.

mapleleaf viburnum, maple-leaved arrowwood, flowering maple, dockmackie

Viburnum acerifolium L.

Family: Adoxaceae (Caprifoliaceae)

Mapleleaf viburnum is 1 of our more common viburnums that is underrated as an ornamental. In the spring, it comes in with a bang with a reddish blush to its leaves and flower buds. In the fall, its subtle pinkish hues, mixed with the fading green leaf color, combined with the shiny, black fruit, are like the big fireworks at the end of the show.

Form and Size: Mapleleaf can reach a height of about 6 feet, but it is usually only 4 feet or so. It has a delicate appearance because of its sparsely-branched nature.

Habitat: Mapleleaf is almost always associated with forested ridges and slopes. It seems to tolerate the driest conditions, but it also does well in moist, mesic habitats. Plants are rarely, if ever, found away from the shade of forests, but it does handle at least a half-day of sun.

Wildlife Uses: Mapleleaf viburnum is 1 of the smaller viburnums and often grows in loose colonies. As such it provides good protective cover for small mammals and birds, as well as nesting cover for birds, both overhead concealment for ground nesters and structure for shrub nesters. Its fruits are similar to those of most viburnums and are not highly preferred, but they are taken by rodents, deer (occasionally), songbirds, and gallinaceous birds, such as wild turkey and ruffed grouse. It is sparingly browsed by deer, moose, and cottontails.

Landscaping Value: Hardy to Zone 3, mapleleaf is adaptable to almost any site, but it grows fuller when given more sun. Conner Shaw of Possibility Place Nursery in Monee, Illinois, grows his mapleleaf in nearly full sun, where they take on the appearance of a completely different species. This species is found in nurseries specializing in natives.

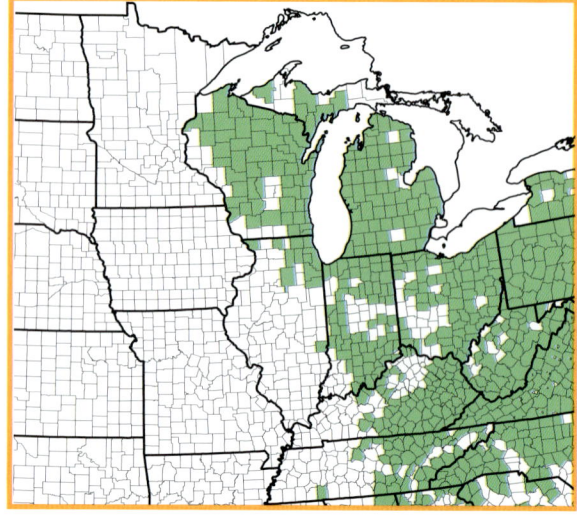

Similar Species Distinctions:
—**American cranberrybush** (*V. tri-lobum*) has fewer hairs on the leaves, essentially smooth with age on upper surface, that are similarly shaped, but usually larger. Its buds are red and hairless. Fruits are red.

The opposite leaves are up to 5 inches long and equally wide, depending on the length of the lobes. There are generally 3 pointed lobes that have margins with sizable teeth. Some leaves may have only a terminal "lobe." The upper leaf surface is green and smooth to hairy at maturity, while the lower surface is somewhat paler and densely hairy and glandular.

Fruits ripen in the fall in upright clusters. Each drupe is about ⅓ inch in diameter and black. It contains purplish flesh and a single seed.

Flowers appear from the middle of May until the middle of June. They are in upright, terminal, flat-topped clusters (cymes) of tiny, white, 5-petaled flowers. The developing flower buds are commonly tinted pinkish red.

Twigs are slender, straight, light brown to reddish brown, and hairy. There are scattered, tan-colored lenticels that are hard to see because of the hairs. Lateral buds are similar to the terminal buds, but they are flattened against the twig with a rounded outer surface. Leaf scars are narrow, V-shaped lines below the nodes that contain 3 bundle scars.

The terminal bud may be of 2 sizes. All buds have a pair of loose-fitting scales at the base. The largest, more plump bud is a flower bud; the smaller bud is vegetative. Both have dark reddish brown, hairy scales that are light brown along the sides, giving the scales a 2-toned appearance.

Mapleleaf does not attain much girth to its stems and, therefore, it remains mostly smooth and brown throughout its lifetime. Lenticels are still visible along older stems.

downy arrowwood, southern arrowwood, arrowwood viburnum

Viburnum dentatum **L.**

Family: Adoxaceae (Caprifoliaceae)

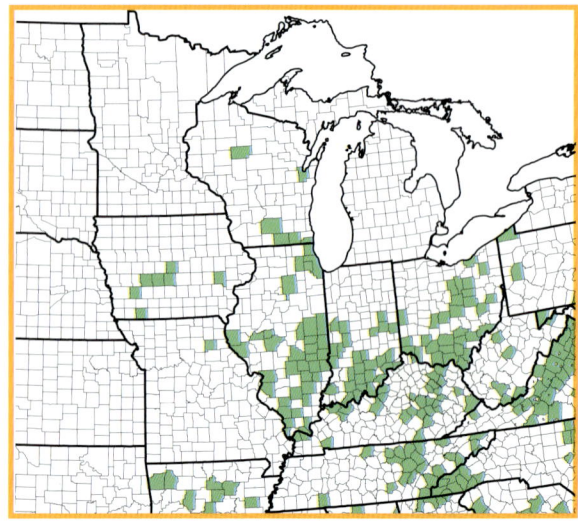

Of all the viburnums, this has to be the 1 that is most confusing from the standpoint of scientific names. Decades ago, taxonomists talked of the fact that *Viburnum dentatum* was a species of the eastern U.S. that did not occur in the Midwest. More recently, its many named varieties have been lumped into 2 species, *V. dentatum* and *V. recognitum*. The authors choose to discuss simply *Viburnum dentatum*. Highlighted locations north of central Indiana and Illinois should probably be referred to as *V. recognitum* or escapes.

Form and Size: It is a tall shrub, reaching a height of at least 8 feet, and has an erect, full form. Shade or partial sun suits this arrowwood fine, but across its range, it can be found growing in full sun.

Habitat: This species occurs in numerous habitats, especially near the base of wooded ravines or along wooded streams, nearly always associated with American beech. The authors have found it at the edge of moist woods. It has been planted often as an ornamental and has escaped cultivation into disturbed habitats, such as roadsides, city parks, and old-fields.

Wildlife Uses: This medium-sized shrub is quite limby compared to most viburnums and has upright limb structure; this combination makes them fairly attractive nest sites for shrub nesters such as catbirds and cardinals. This species generally occurs as individual, bird-dispersed plants that have minimal protective cover value. The fruits of arrowwood tend to be smaller than those of other viburnums and are very preferred by songbirds and game birds such as the ring-necked pheasant. Few last into the winter. It is occasionally browsed by deer and cottontails, but it is not preferred.

Landscaping Value: Hardy to Zone 3, this species has a large, native range, and it tends to be adaptable to almost any site. Its large, dentate teeth on the leaf margins are attractive, and fall color is beautiful. Flowering and fruiting heads are large and showy. It is an easy to grow species and readily available in many nurseries.

Similar Species Distinctions:
—**Missouri viburnum** (*V. rafinesquianum*) has dentate leaf margins but usually has more ovate-shaped leaves.

The opposite leaves are up to 4½ inches long and have from 6 to 10 pairs of impressed veins per leaf. The shape is variable, but it is commonly ovate, orbicular, or nearly round with a sharply-pointed tip. The upper leaf surface is green and smooth or hairy, while the lower surface is lighter green and smooth to hairy. Leaf margins have coarse, pointed teeth.

Fruits ripen in the fall in upright, terminal clusters. The drupes are blue-gray (some say the color of slate) and about ⅓ inch long. Each fruit has a single seed.

Flowers appear in early June in terminal clusters (cymes) that are up to 4 inches wide. Each cluster has many tiny, white, 5-petaled flowers.

Mature bark is not necessarily unique to this species; it is grayish brown, slightly shallowly fissured, and dotted with lenticels.

The terminal bud has at least 7 reddish brown scales that may have scattered hairs. Buds are ovate and quite pointed. Twigs are very straight and tannish, light brown, or grayish. Hairs may be present, especially on the newest growth. Scattered, tan lenticels run along the twigs. Lateral buds are similar to the terminal buds and are mostly appressed against the twig except for their crooked tips. Leaf scars are V-shaped and contain 3 bundle scars.

nannyberry, sheepberry

Viburnum lentago **L.**

Family: Adoxaceae
(Caprifoliaceae)

Nannyberry is a common viburnum
of wet areas. It is easily confused with
another common native, blackhaw
(*V. prunifolium*). Nannyberry leaves
differ in that they are usually wider
and larger, with a more elongated tip.
Its petiole is also flared, sometimes
termed winged by others.

Form and Size: This can be a large
shrub, commonly 15 to 20 feet tall in the Midwest. Michael Dirr (2007) in *Viburnums* reports that the national champion in Oakland County, Michigan, is 50 feet tall by 40 feet wide. It suckers freely to produce
a large, spreading clump, but it can remain single-trunked. The trunk diameter can be substantial—the
largest of our native viburnums. It has strongly angled lateral branches, similar to those of *V. prunifolium*.

Habitat: Nannyberry is found in wet, poorly-drained sites, such as swamp and pond edges or boggy areas. It has been reported from more upland sites as well. In the wild, it is usually found in partially shaded
areas, but it can tolerate a great deal of sun.

Wildlife Uses: Nannyberry often occurs in large clones in open areas or forest edges; in such situations its
wildlife value is very similar to that of blackhaw (*V. prunifolium*). These colonies are so dense that they
shade out ground cover and thus have little cover value for ground-dwelling species. Their fruits are large
and among the least preferred of the viburnums; they often are present into the winter when they are sparingly used by mammals and game birds.

Landscaping Value: Hardy to Zone 3, this
large viburnum would be a nice addition to
any low area of a landscape, as long as there
is plenty of room. Mildew is a problem on
specimen plants the authors have planted.
There are several horticultural varieties
of this species, and a naturally occurring
hybrid between nannyberry and blackhaw
(*V.* x *jackii* Rehd.) is known. Nannyberry is
occasionally available through native-plant
nurseries.

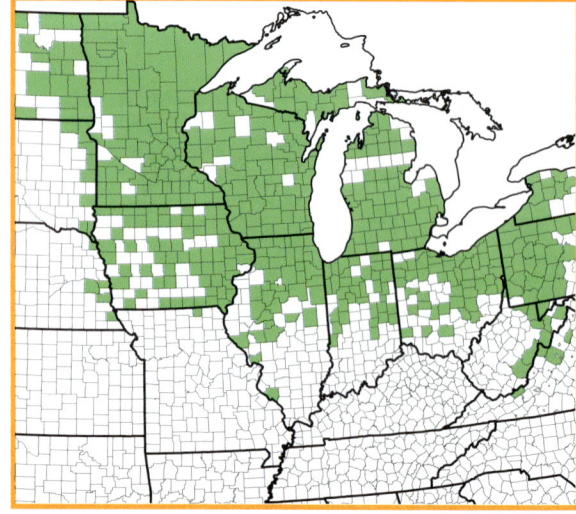

> **Similar Species Distinctions:**
> —**Blackhaw** (*V. prunifolium*) has slender leaf petioles and smaller, narrower
> leaves.

The opposite leaves are mostly ovate but trend toward oval to orbicular. Leaves are up to 3 inches long and have finely toothed margins. The upper leaf surface is green and smooth, while the lower surface is slightly paler and smooth with tiny, reddish scales. The petiole is up to ¾ inch long and expanded to give a winged appearance; it commonly has a pinkish tint and looks ruffled.

Mature bark is brownish gray with a small, scaly pattern. The outer scaly bark sloughs over time to reveal the reddish brown inner bark.

Fruits ripen in the fall in drooping clusters on orange pedicels. The fruits are up to ¾ inch long and black, with a slight glaucous bloom. The fruit, a drupe, is edible and has a single seed.

Flowers appear in mid-May in upright clusters (cymes) that are up to 4½ inches across. The cluster is full of tiny, white, 5-petaled flowers.

The terminal buds are of 2 types. All buds have 2 fleshy, pink, elongated scales that meet at the tip. The larger flower bud is bulbous at the base with an elongated "snout." Vegetative buds are slender. There may be a tiny pair of vegetative buds at the base of each terminal.

Twigs are slender, straight, and pinkish or light brown; they are often partially coated with a grayish cast. There are scattered, tan-colored lenticels along the twigs. Lateral buds are similar to the terminal, except they are smaller and all vegetative; they are appressed against the twig. Leaf scars are V-shaped and contain 3 bundle scars. Lateral buds are often seen on short, strongly angled short twigs.

Kentucky viburnum, softleaf arrowwood

Viburnum molle **Michx.**

Family: Adoxaceae (Caprifoliaceae)

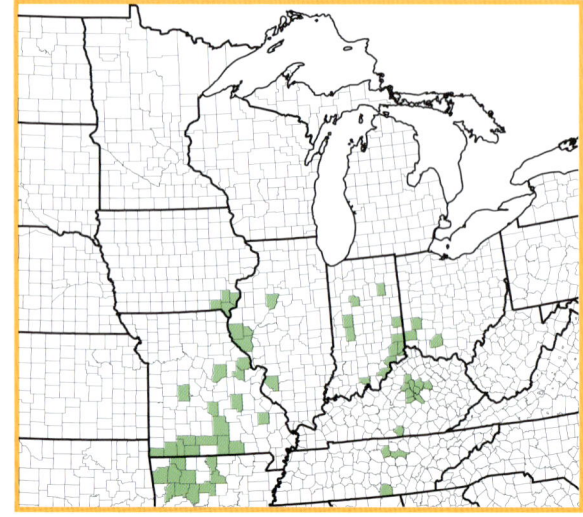

Kentucky viburnum is 1 of those "dentate-toothed" viburnums that look so much alike, based on leaf structure. However, 1 look at the bark of this species rules out all others. It is quite distinctive with its rusty-orange, papery-thin, peely bark. In fact, the bark is very similar to that of paperbark maple, *Acer griseum*. Unfortunately, this is 1 of our rarest native viburnums. It is listed as rare in Indiana, threatened in Illinois and Kentucky, and endangered in Ohio.

Form and Size: This is a suckering species with an erect form and full crown. It can reach a height of 10 feet, but it is usually shorter.

Habitat: Kentucky viburnum is found on gravely, steep, wooded slopes and hills of calcareous origin. It is usually found in partial shade but grows well in full sun. It naturally grows in dry sites but would probably thrive in a rich, moist site.

Wildlife Uses: This species is essentially the southwestern equivalent of *V. dentatum* and *V. rafinesquia-num* in our region. Its value to wildlife more nearly matches the latter. However, it usually occurs as individuals under some degree of forest canopy where it rarely forms clones and is more similar to *V. dentatum* in this respect, limiting its protective cover value.

Landscaping Value: Hardy to Zone 5, this unusual viburnum is our most interesting from a bark standpoint. Very little data are available on the use of this species in the landscape. The authors have been growing it successfully for several years in a rocky, clay loam in half-day sun. However, finding it in a nursery can be difficult. We found it at Johnson's Nursery in Menomonee Falls, Wisconsin.

Similar Species Distinctions:
—**Southern arrowwood** (*V. dentatum*) has similar leaves with large, dentate teeth along the margin, but its buds are pink, and its bark does not peel.

Flowers appear in late May and the first of June in upright clusters (cymes) that are up to 3½ inches across. The clusters are made up of many tiny, white, 5-petaled flowers.

Fruits ripen in the fall in upright clusters. The fruit, a drupe, is blue-black and shiny when ripe, and it contains a single seed.

The opposite leaves are up to 5 inches long and nearly as wide. They have distinctly, coarsely toothed margins and 8 to 10 pairs of veins. Its petioles are the longest of any of our native viburnums—up to 1½ inches. The upper surface is green and smooth (perhaps with some random, flattened hairs), while the lower surface is yellow-green and usually has dense soft hairs, especially along the veins.

The terminal buds have at least 6 reddish brown or grayish brown, 2-toned scales, the outside pair being quite loose. There are often hairs on the outer surface of some scales. The larger flower buds (left) are more teardrop shaped than the vegetative buds, which are shaped like upside-down, pointed ice cream cones.

Twigs are straight, silvery gray, and dotted with tan-colored lenticels. The lateral buds are similar to the terminals, except smaller. They diverge from the twig somewhat. Leaf scars are V-shaped and contain 3 bundle scars.

The bark is thin, reddish orange, and dotted with many lenticels. Over time, the outer bark peels away and falls, leaving smooth bark, except for raised lenticels. The oldest bark becomes brown.

blackhaw, black haw

Viburnum prunifolium **L.**

Family: Adoxaceae (Caprifoliaceae)

Blackhaw is probably the most common viburnum in the Midwest and is found in many habitats; it is not picky about where it grows. This is another beautiful native that has aesthetic value year-round. In the spring, its new leaves are almost copper-colored, and its flowering period lasts several weeks. By fall, the nearly black fruit is in splendid contrast to the brilliant pinkish red foliage. Winter buds are pink, and the strongly angled branching is obvious and unusual.

Form and Size: Blackhaw can be a suckering, colonial shrub, or a single-stemmed, tree-like, large shrub. Lateral branching is at right angles to the main stem, giving the shrub a unique appearance. A large trunk diameter is several inches with a height of 12 feet. Farther south, there are records of specimens that are 33 feet tall.

Habitat: Blackhaw is most common in woods or forest edges. It can also be found along roadside fences and other open areas. It tolerates all but the wettest or driest sites, and it handles full sun to a great deal of shade. However, the more light it receives, the more flowers and fruit it produces, and the more brilliant the fall color it displays. It probably does not grow well in acidic soil.

Wildlife Uses: Blackhaw often grows to large shrub size, especially when it occurs in openings and forest edges; in these locations it typically forms thickets with interlacing branches that supply nest sites for many birds, even those that typically nest higher, such as rose-breasted grosbeaks and cedar waxwings. In the understory, however, they are less limby, and the right angle, opposite limb structure rarely provides nest sites for songbirds. The fruits are readily eaten by mammals (e.g., raccoons and deer), songbirds, turkey, and grouse, but they also occasionally persist into the winter. Leaves are often heavily browsed by deer, especially in the spring.

Landscaping Value: Hardy to Zone 3, this viburnum is a very tough species. It adapts to almost any site and just about any amount of light. Several cultivars and varieties are available commercially, particularly in midwestern nurseries, and it is easily grown.

Similar Species Distinctions:
—**Nannyberry** (*V. lentago*) can be very difficult to distinguish from blackhaw, but it has somewhat expanded petioles and larger, broader leaves.

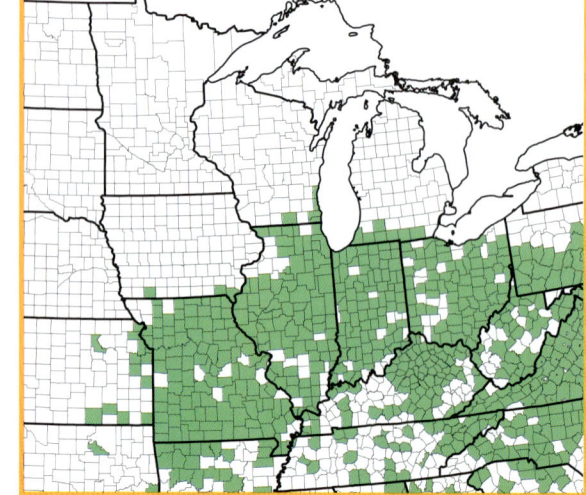

The opposite leaves are up to 3½ inches long and variable in shape. They are ovate to slightly obovate, or narrow-oval to nearly orbicular. The leaf tip is pointed, and the margins are finely toothed. The upper leaf surface is green and smooth, while the lower surface is paler and smooth. The petiole is not winged like that of *V. lentago*.

Flowers appear with the leaves in late April and continue for a several week period. They are in upright, mostly flat-topped clusters (cymes) that are up to 4 inches across. Flowers are white, 5-petaled, and non-fragrant.

Fruits ripen in the fall in dangling clusters. The ripe fruit is black and about ½ inch long. Each fruit, which is a drupe, contains a single, flattened seed.

There are 2 kinds of terminal buds present during winter months. All buds are covered with 2 pinkish, elongated, fleshy scales. The larger terminal buds with a bulbous base are flower buds, while the more slender, snout-like buds are vegetative. There is usually a smaller, flattened pair of vegetative buds at the base of each terminal, both floral and vegetative.

Twigs are slender, straight, and tannish when new. Older twigs turn pinkish and commonly have a silvery coating. Scattered lenticels appear along the twigs. Lateral buds are small, flattened, and pinkish. Leaf scars are broadly V-shaped, slightly raised, and contain 3 bundle scars.

Mature bark is brownish, somewhat scaly, and becomes broken into small, fairly thick, blocky sections with age. The inner bark is dark and reddish brown.

Missouri viburnum, downy arrowwood viburnum, Rafinesque viburnum

Viburnum rafinesquianum/rafinesqueanum **Schultes**

Viburnum affine **Bush**

Viburnum rafinesquianum **var.** *affine* **(Bush) House**

Family: Adoxaceae (Caprifoliaceae)

From the reddish-tinted new leaves in the spring, to the beautiful fall color that can begin in August while the black fruits are present, this is a species worthy of recognition. It has large dentate teeth along the margins, which add appeal to the plant. This species is listed as rare in Kentucky.

Form and Size: Missouri viburnum is a bushy plant that can reach to a height of around 5 feet. It suckers and spreads to form dense colonies.

Habitat: It is usually found in dry, sandy, shaded woods, commonly mixed with *Viburnum acerifolium* in well-drained soil. It is very shade tolerant, but it handles partial sun.

Wildlife Uses: The wildlife value of this species is very similar to that of *V. dentatum*, to which it is similar in size and form. Its fruits are a little larger than those of downy arrowwood and not quite as highly preferred, but they are taken by birds and mammals more regularly than those of other viburnums. When it grows in the open, it can be clonal, and in these circumstances, it provides more protective cover than *V. dentatum*.

Landscaping Value: Hardy to Zone 4, this viburnum has many great ornamental qualities. It is adaptable to full sun and tolerant to heat and drought. Dirr (2007) states in his book *Viburnums* that he discovered a yellow-leafed variety, 'Louise's Sunbeam,' which is slated for release into horticultural trade. This viburnum is occasionally available through midwestern nurseries specializing in natives.

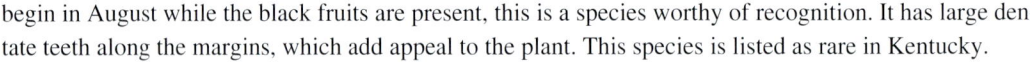

Similar Species Distinctions:

—Southern arrowwood (*V. dentatum*) leaves are sometimes more ovate, which then look very similar to those of this species. However, it still has longer petioles with no stipules.

The opposite leaves are up to about 2½ inches long and 1½ inches wide. Leaf shape is ovate, but the margins are irregular and can have sparse or continuous, large, pointed teeth. The upper leaf surface is dark green and smooth, while the lower surface is paler and hairy along the veins. The leaf tip is usually pointed, and the petiole is as short as those of any of our native viburnums. The petiole is channeled above, with hairy margins; there is a pair of linear stipules at the petiole base.

Fruits ripen in August in upright clusters. The black drupes are shiny and about ⅓ inch long; each 1 contains a single seed.

Flowers appear in May in upright, somewhat round-topped clusters (cymes) that are up to 3 inches in diameter. The individual flowers are white and have 5 petals. They are somewhat malodorous.

The terminal buds are ovoid in shape and are covered with 6 reddish brown scales that have a silvery coating near the tip. The scales have hairs along the margins. There can be a pair of smaller, lateral buds at the base of the terminal.

Twigs are straight, reddish brown, and mostly covered with a silvery-gray coating. There are scattered, rust-colored lenticels along the twigs. Lateral buds are similar to the terminal, only smaller. Leaf scars are broadly V-shaped and contain 3 bundle scars.

Specimens large enough to develop any characteristic bark are usually not found, at least not in our region. The trunks have thin, reddish brown bark that has scattered, rust-colored lenticels.

southern blackhaw, southern black-haw, rusty blackhaw

Viburnum rufidulum **Raf.**

Family: Adoxaceae (Caprifoliaceae)

Rusty blackhaw has so many interesting characteristics that it is hard to know where to start. Some of its most unique features include its dark green, shiny leaves and its dark, rust-colored buds. Fall color is spectacular. It is basically blackhaw (*V. prunifolium*) with rust-colored tomentum covering the bud scales.

Form and Size: This is not a large shrub in the Midwest, but it can reach 10 to 12 feet in height. Farther south, specimens 30 feet tall have been recorded. It is usually a single-trunked, upright shrub growing in partial shade.

Habitat: Rusty blackhaw has several habitat types where it is frequently found. One is on dry, rocky, wooded slopes and forest edges. Occasionally, it can be found along wooded streamsides.

Wildlife Uses: Wildlife values of rusty blackhaw are very much like *V. prunifolium*. It tends to occur more frequently as an individual and, thus, has somewhat less protective cover value.

Landscaping Value: Hardy to Zone 5, this southern blackhaw is simply gorgeous any time of the year. It is extremely drought tolerant and grows well in moist, well-drained sites in full sun. Rusty blackhaw has the best winter characteristics of any native viburnum with its dark, rusty-colored buds and dark, blocky bark. This would make a fine specimen plant in the landscape. This species is rarely offered in native-plant nurseries.

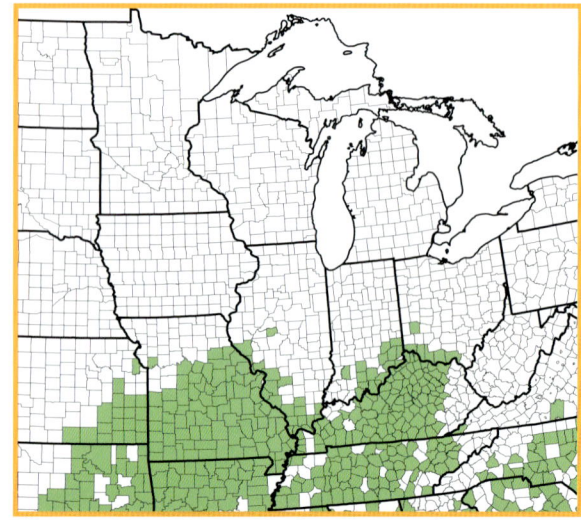

Similar Species Distinctions:
—No other viburnum, or other native shrub, has the combination of characteristics that this species does.

The opposite leaves are up to 4 inches long and are oval, ovate, obovate, or elliptic. They are very thick for a native viburnum. The upper surface is dark green, shiny, and mostly smooth, while the lower surface is paler and mostly smooth, except for some rusty tomentum, especially near the leaf base. The leaf margin is lined with tiny, hooked teeth. Petioles are covered with rusty tomentum and are somewhat expanded or winged.

Flowers open in early June in upright clusters (cymes) that are up to 5 inches in diameter. Each flower is white and has 5 petals.

Terminal buds are covered with 2 fleshy scales that have a rust-colored tomentum that is like no other viburnum. The larger buds that are swollen near the base are flower buds (right), while the more slender buds are vegetative. There is usually a pair of flattened, lateral buds at the base of the terminal.

The fruits ripen in the fall and appear in drooping clusters. The pedicels are orangish or reddish, and the fruits are bluish black with a white bloom. Each fruit, which is a drupe, has a single, flattened seed.

Rusty blackhaw probably has the most distinctive bark of all our native viburnums. Mature bark (right) is grayish brown, thick, and blocky. The inner bark is rusty-brown.

Twigs are pale grayish and mostly smooth, except near the tip of the new growth, where rusty tomentum may be retained over winter. Scattered, tan-colored lenticels dot the twig. Lateral buds are smaller than the terminal and flattened against the twig. Leaf scars are broadly V-shaped and contain 3 bundle scars.

highbush cranberry, American cranberrybush

Viburnum trilobum **Marsh.**

Viburnum opulus **L. var.** *americanum* **Ait.**

Viburnum opulus **ssp.** *trilobum* **(Marsh.) R. T. Clausen**

Family: Adoxaceae (Caprifoliaceae)

Taxonomists have never agreed on whether the North American highbush cranberry is the same species as the European *Viburnum opulus*. Some have named our species a variety of *V. opulus*; others suggest it is distinct enough to warrant species status, namely *V. trilobum*. The main difference used to distinguish between the 2 is the shape of the petiole gland. The North American gland shape is usually short-stalked and flat-topped. The European plants typically have broader, cup-shaped glands that have no stalk. The *V. opulus* is used in landscaping and has naturalized in areas of eastern U.S., compromising the accuracy of our map designations. It is listed as endangered in Indiana, threatened in Ohio, and rare in Pennsylvania.

Form and Size: Our native is a tall shrub, up to 10 feet in height, which has a multi-stemmed, rounded crown.

Habitat: The common habitat of highbush cranberry is swamps, fens, or borders of rivers and streams. It is fairly easy to find an escaped plant, since many nurseries carry what some taxonomists call the European species. Although it naturally grows in moist areas, it grows well in a more upland site. While it is usually found growing in at least partial (if not heavy) shade, it can tolerate full sun.

Wildlife Uses: Highbush occurs in several habitats, but it is most commonly found in quite wet locations. Here it is used occasionally for nesting by songbirds, but its relatively sparse, stout twigs do not usually supply numerous nest support sites. This species and its European cousin, *V. opulus*, are frequently planted, and in open sites they produce large quantities of red fruits that are extremely low in palatability for birds and mammals, often lasting through the winter.

Landscaping Value: Hardy to Zone 3, this is probably the most common native *Viburnum* in midwestern nurseries. There are many cultivars, and it is easy to grow in most soils and moisture regimes. This large shrub is used for specimen plants or for screens, but give it room, for it is often as wide as it is tall.

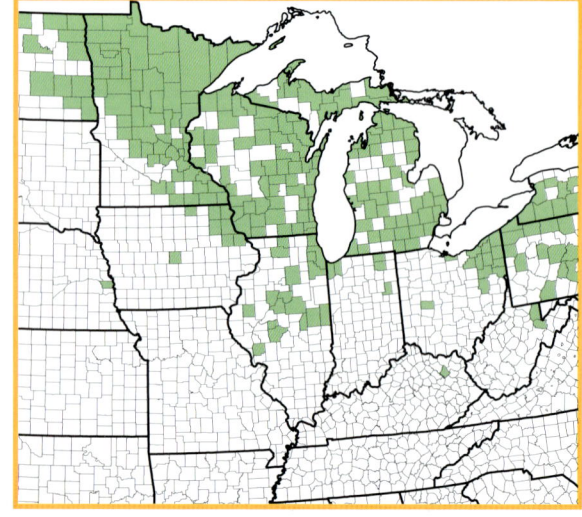

Similar Species Distinctions:
—**Mapleleaf viburnum** (*V. acerifolium*) has 3-lobed leaves that are generally smaller; they are softly hairy beneath and occasionally on the upper surface. Its buds are brownish and hairy as well.

The opposite leaves are 3-lobed and up to 5 inches long. Each lobe is elongated and has coarse, irregular teeth along the margins. The upper surface is green with scattered hairs, becoming smooth with age; the lower surface is light green and hairy (usually), especially along the veins. Petioles can be nearly 1 inch long and have at least 1 pair of flat-topped glands (may have more).

Flowers appear in June after the leaves are mostly developed. They are flat-topped clusters (cymes) that are up to 4½ inches in diameter. The outer, larger, white flowers are actually sterile, but they are the showiest of those in the cyme. The central fertile flowers are small, white, and 5-petaled.

Fruits ripen in the early fall and persist through the entire winter in drooping clusters. Ripe fruit is red. Each fruit is about ⅓ inch long. Pedicels are usually bright orange-red. Each fruit, a drupe, contains a single, somewhat rounded seed.

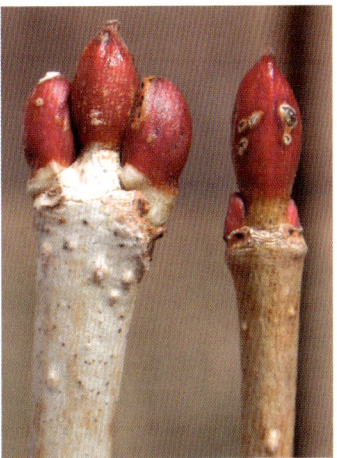

Terminal buds are plump and blunt-tipped with few red, smooth scales covering them. There is usually a pair of lateral buds sitting at the base of each terminal. Twigs are nearly gray or tannish, smooth, and mottled with gray. There are scattered, light-colored lenticels that dot the twigs. Lateral buds are smaller than the terminal and appressed against the twig. Leaf scars are broadly V-shaped and contain 3 bundle scars.

Bark on mature plants is grayish and rough, mainly because of numerous enlarged lenticels that expand with age.

withe rod, witherod, wild raisin

Viburnum cassinoides **L.**

Viburnum nudum **L. var.** *cassinoides*
(L.) Torr & A. Gray

Family: Adoxaceae (Caprifoliaceae)

This is a striking species that has also made
its way into the horticultural trade. Because of
taxonomic difficulties, things get a bit confusing when one tries to determine exact identification. Some
authors name 2 distinct species—*V. cassinoides* and *V. nudum*. Others have determined there is just 1 species—*V. nudum* with 2 varieties: var. *nudum* and var. *cassinoides*. Either way, *V. nudum* var. *cassinoides*
or *V. cassinoides* is the more northerly version. Wild raisin is listed as state endangered in Indiana, Kentucky, and Pennsylvania. This is 1 of the few native viburnums that has fruits that are eaten by humans.

Form and Size: This is a large shrub that can reach up to 15 feet or more. It is multi-stemmed with a
spreading rounded form. Large, sun-grown specimens are as wide as they are tall.

Habitat: This is a species that prefers wet, acidic sites. It is found in low woods and forest edges,
swamps, or edges of bogs, but sometimes it occurs in drier sites with oak or jack pine. It grows well in
partial shade, but it is adaptable to full sun.

Wildlife Uses: The wildlife value of withe rod has similarities to features of nannyberry (*V. lentago*) and
downy arrowwood (*V. dentatum*); it is intermediate to these 2 in size and form, usually occurring as individuals in low, wet areas. In such areas, it serves as nesting cover for associated birds (e.g., red-winged
blackbirds, yellow warblers) in a manner similar to arrowwood. The fruits are somewhat smaller than
nannyberry and use by birds and mammals is higher, but preference levels are not as high as for downy
arrowwood fruit. Winter twigs are browsed occasionally by deer and rabbits.

Landscaping Value: Hardy to Zone 3, withe rod has beautiful, dark green leaves, great fall color, and
produces many flowers when grown in the full sun. Its fruits are particularly pretty as they ripen, succes-
sively changing, as Dirr (1998) states, "from
exquisite pink, rose, robin's-egg blue to
blue-black," with all colors possible on the
same fruiting head. It seems very adaptable
to most sites, as long as its acidity requirement is met. It is becoming more popular
and is occasionally available from larger
nurseries.

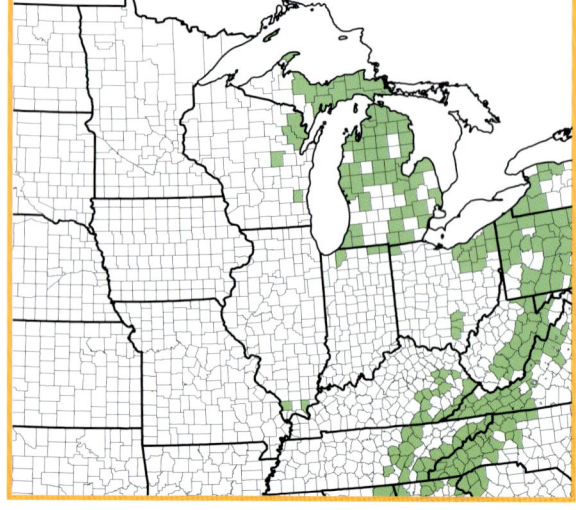

Similar Species Distinctions:
—**Smooth witherod** (*V. nudum*), if truly
a separate species, has waxy, shiny upper leaf surfaces and no teeth along the
leaf margins. Buds are a golden-brown
color. It has a more southerly range.

The opposite leaves are up to 3½ inches long and are elliptic to oval to oblong. Most have a short, pointed tip. The upper leaf surface is green and smooth, with a matte finish, while the lower surface is smooth, but with mealy, rust-colored dots (scurfiness) especially along the midrib and at the leaf base. Margins usually have varying amounts of small, fine teeth.

Fruit ripens in the fall in drooping clusters. The fruits (drupes) are about ⅓ inch long and turn from pink to blue-black, with a glaucous bloom. Each fruit contains a single seed.

Flowers appear in June in upright, flat-topped clusters (cymes) that are up to 5 inches across. Each cluster has numerous tiny, white, 5-petaled flowers.

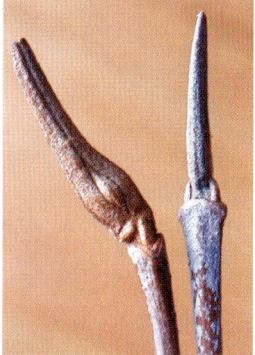

Terminal buds are of several kinds. All are elongate and covered with rust-colored, scurfy scales. The buds have 2 scales but are considered naked—scales are fleshy and wrap the bud inside them. The bulbous buds produce flowers; the slender buds are vegetative. Twigs are slender, straight, and tannish or pinkish. They are dotted with tan-colored lenticels. Lateral buds are similar to the terminal buds, but scales often appear loose, as seen here. Leaf scars are V-shaped and contain 3 bundle scars.

There is no diagnostic bark of withe rod, and it remains brown with light-colored streaking near the base. Raised lenticels are scattered along the trunk.

Elderberries

Genus: *Sambucus*

Only 2 species of elderberry are native to the Midwest. *Sambucus racemosa* is the common northern species, while *S. nigra* ssp. *canadensis* is more of its southern equivalent, although their ranges do overlap.

The elderberries are unusual shrubs because of their combination of opposite branching *and* compound leaves, which is a good thing to remember when looking for them in the summer months. Another unusual characteristic is the substantial Styrofoam-like pith—its color actually determines the species.

Elderberries are more famous for their copious fruit crops than anything else. Although wildlife heavily utilize the small berries, humans have long used them for dyeing, wine-making, and baking. From a landscaping standpoint, they are rarely used, but they certainly should be considered, for when grown in full sun, they develop dense, deep-green foliage and many showy flowers and fruits.

The elderberries are rather unique, too, because they often occur as scattered individuals or clumps in early successional habitats, having been planted through the feces of fruit-eating birds. Both species in our region produce small fruits, often in great abundance, which are highly preferred by birds and mammals, and the supply is quickly exhausted. Red elderberry, the more northerly of the 2, flowers and fruits earlier than common elderberry. So in northern parts of our region where both occur, they complement each other nicely and greatly extend the period of fruit availability for wildlife. From a cover perspective, elderberries offer little; sparse, stout branches do not provide many attachment sites for songbird nests, and their scattered distribution offers poor protective cover.

Fruits are tiny, juicy berries that are important to wildlife. Common elderberry shown here can have fruiting heads as big as a dinner plate.

Leaves of elderberries are large and compound, 2 features not usually seen on shrubs. Red elderberry leaf is shown here; its hairy undersurface easily distinguishes it from the leaf of American elderberry.

The pith of elderberries makes up the largest portion of all stems, which is why they are so easily broken. Pith color is an important identification tool; pictured is *S. nigra* ssp. *canadensis* with its white pith.

common elderberry, American elder

Sambucus nigra L. ssp. *canadensis* (L.) R. Bolli
Sambucus canadensis L.

Family: Caprifoliaceae

Perhaps because this is such a common shrub, common elderberry's ornamental potential has been overlooked. However, a large, robust colony in flower is spectacular. Its fragrant flowers are in clusters the size of a lunch plate (or larger), and flowering occurs in June and July when most other native shrubs have finished. Fruits are usually produced in abundance and are edible by birds and mammals, including humans.

Form and Size: It can attain a height of 10 feet and grows in upright clumps that are produced from root sprouting. Individual stems are short-lived and persist for 3 to 5 years.

Habitat: Elderberry is common throughout the Midwest, and it is most frequent in moist soil along roadsides, ditches, streams, lakes, and in wet woods. Full sun is required for best growth, but it tolerates partial shade.

Wildlife Uses: Common elderberry is a widely distributed early successional shrub in our region. Its major wildlife value is its fruit production; in some years production of clusters of its small fruits is so heavy that the twig end breaks under the weight. Fruits are small enough to be used by the smallest of songbirds, and the list of bird species using the fruit is very long. Mammals such as fox squirrels, raccoons, opossums, and coyotes also heavily use the fruits. The species is little-browsed by white-tailed deer, and the bark is not frequently used by cottontails. Cover value is low, with the thick limbs giving little opportunity for nest placement by birds. There is some wild clonal formation through root-suckering, but rarely does the species supply much protective cover.

Landscaping Value: Hardy to Zone 4, this wide-ranging species gives an informal appearance to an area, and it could be used for naturalizing or to create a border. A backdrop such as a fence brings out the beauty of a patch of elderberry. It is a very hardy plant with few requirements, and no insect or disease problems. It has a very full form during the growing months because of the large leaves. There are several cultivars available, including 'Aurea,' which has red fruit.

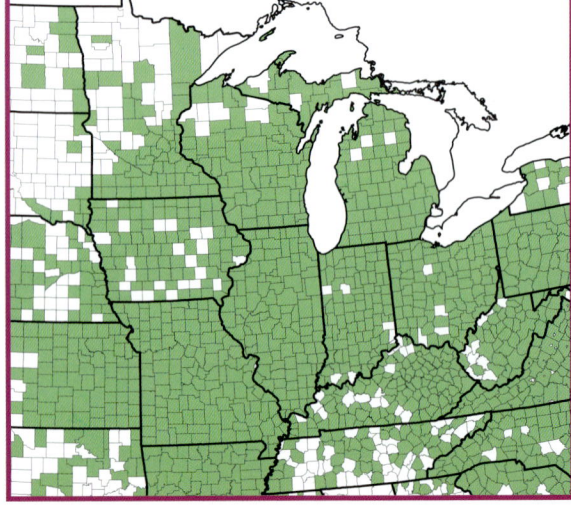

Similar Species Distinctions:
—**Red elderberry** (*S. racemosa*) has hairy leaf backs, flowers earlier, and has red fruit. Its pith color is orange.

The opposite leaves are compound and can be up to 10 inches long. More often, they are 7 or 8 inches long. There may be 5 to 11 leaflets, but usually there are either 5 or 7. Occasionally, the lower leaflets can be divided. Leaves are green and smooth above, while the lower surface is paler with some hairiness, especially along the midrib and primary veins. The leaflets are mostly ovate to oblong, and margins are finely toothed.

Fruits ripen in August and are in large, terminal clusters. Clusters are commonly the size of lunch plates. Ripe fruit is nearly black, juicy, and about ⅛ inch in diameter. The juice is crimson-colored and is used in making jelly and wine.

Flowering begins in mid-June and can continue into August. Numerous tiny, white flowers are in large, round-topped, terminal clusters. Clusters can be from 4 to 12 inches or more across. Each flower has 5 petals, and both petals and sepals are reflexed. They are fragrant and showy.

Twigs are stout, light tan-colored, and dotted with large, raised lenticels. Because of the lenticels, the twig has a feel of a pretzel. The inner pith forms a substantial portion of the twig diameter and is white. Lateral buds are reddish brown, small, and diverge from the twig. Leaf scars are large, shield-shaped, and contain a single row of bundle scars that lines the edge of the scar. Terminal buds are rarely seen since the twigs either end with a fruiting structure or simply die back.

Small diameter trunks are tannish and rough from raised lenticels. Older specimens develop fissuring along the trunk. Any given specimen may live only 5 years, so the diameter is never very great.

red elderberry, scarlet elder, red-berried elder

Sambucus racemosa **L.**

Sambucus racemosa **L.** **var.** *racemosa*

Sambucus pubens **Michx.**

Sambucus racemosa **L.** **var.** *pubens* **(Michx.) Koehne**

Family: Caprifoliaceae

This elderberry is as common in more northerly regions of the Midwest as common elderberry is in the lower Midwest, where red elderberry is approaching the southern-most extent of its range. It is listed as state endangered in Illinois and Kentucky. This wide-ranging species has several named varieties.

Form and Size: Red elderberry has an upright, open growth form, and it is usually single-stemmed, especially when open-grown. It can be up to 10 feet tall but is usually shorter.

Habitat: It is found in bogs, boggy places, wet woods, forest edges, and roadsides. Red elderberry is rarely in dry sites. It prefers full sun, but it handles partial shade well.

Wildlife Uses: This northern species supplies a small red fruit that is highly preferred by birds, both songbirds and ruffed grouse. Production is usually not as high as in common elderberry, and fruits are quickly completely taken by birds and occasionally mammals such as raccoons. There is some inconsistency in reported use of the species as browse, but it is likely more palatable than common elderberry and taken, sometimes seemingly preferred, by moose and white-tailed deer; bark is eaten in winter by snowshoe hares and porcupines. The species provides little cover; it does not root-sucker, so it is usually present as individuals that have few, stout branches that are not suitable for songbird nests.

Landscaping Value: Hardy from Zones 4-6, this cool-weather shrub is probably restricted in its southern movement by climate. Wild-grown plants are gangly and rather unattractive, except when flowering. Those brought into a groomed landscape are very full, especially when grown in full sun, where the shrub develops robustly. Its flowers have a rather unpleasant odor, and are often tinged with reddish pink before opening. Fruits are bright red, but they soon disappear, thanks to birds. This species could fill the same landscape role as common elderberry. There are several uncommon cultivars, including 'Leucocarpa' that has white fruits, but this species is available through native-plant nurseries.

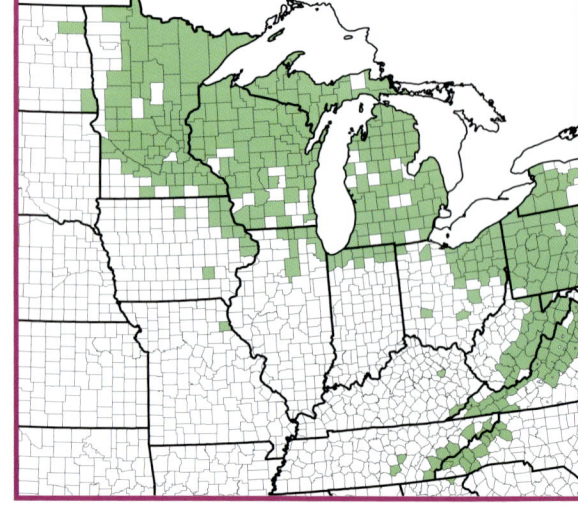

Similar Species Distinctions:
—Common elderberry (*S. nigra* ssp. *canadensis*) is similar, but its leaves lack the dense hairs underneath. Its pith color is white and fruit is purplish.

The opposite, compound leaves have 5 to 7 leaflets and are up to 12 inches long. The leaflets are ovate-lanceolate or oval; their margins are finely toothed. The upper leaf surface is green and mostly smooth, while the lower surface is pale and usually densely hairy. Petioles are often reddish purple.

Fruits ripen much earlier than those of common elderberry, usually by late June. They are in upright clusters that are the same size as the flower clusters. Ripe fruit is scarlet-red, about ¼ inch in diameter, and juicy.

Terminal buds are rarely found, since twigs either end in a fruiting structure or die back during the winter. Buds commonly swell during warm spells in winter, as seen here.

Flowers appear in the first half of May in upright, terminal, compact clusters. The clusters are only about 2 inches tall—a very small size when compared to those of *S. nigra* ssp. *canadensis*. Expanding flowers are commonly rosy-red, but they fade to white upon opening. Flowers

have a somewhat disagreeable odor. Individual flowers have 5 white, reflexed petals, as well as 5 flattened sepals.

Twigs are stout, tannish or brownish, and dotted with raised, orange-colored lenticels; they may be lightly hairy. Pith color is brownish on twigs 1 year old or older. Lateral buds are reddish or purplish, small, and diverge from the twig. Swollen, early spring buds at a node are seen here. Leaf scars are somewhat V-shaped and contain 3 bundle scars.

Bark is grayish and rough from raised lenticels, with some shallow fissuring on larger specimens.

bladdernut, American bladdernut

Staphylea trifolia **L.**

Family: Staphyleaceae

Bladdernut is 1 of our most common native shrubs. It is also 1 of our most beautiful when in flower, with its large clusters of drooping, white flowers that appear before most other shrubs have even broken bud. It is almost always found in the understory of mature woods, where it thrives in the shade during most of the growing season; however, in the springtime, it takes advantage of the sun before the tree canopy has leafed out.

Form and Size: It can reach a height of 13 feet and have a trunk diameter of 4 inches. Bladdernut is a clonal species that can develop sizable thickets through root suckering.

Habitat: Bladdernut is common in moist, rich soils on wooded slopes, in floodplains, on banks of streams, and in poorly drained, upland depressions. It is adapted to heavy shade, but it can grow in full sun—although rarely encountered in such situations.

Wildlife Uses: Bladdernut, an attractive shrub with unique features, is a consummate midwestern plant—its distribution is directly centered on the region. It occurs principally in rich bottomlands where it has important, although targeted, wildlife values. Its characteristic seed-containing bladders cling to shrubs through the winter and appear untouched by wildlife; however, white-footed mice gather large numbers of them and extract the seeds under logs and in other protected niches. White-tailed deer browse substantially on bladdernut. From a cover perspective, it often forms thickets that provide protective cover to deer and smaller wildlife. The limb structure is ideal for songbird nesting and wood thrushes, cardinals, catbirds, and others often choose bladdernut for nest placement.

Landscaping Value: Hardy to Zone 4, this large, common shrub is in some cases too common, as it can become the dominant shrub in the understory of woods where timber management is a priority. It has beautiful chains of creamy flowers in early spring, unusual, persistent pods, and yellow fall color. Its twigs are deep green and especially attractive in the winter months. It prefers rich, moist soil and is a bit sensitive to drought. Bladdernut is rarely available, even through nurseries specializing in natives.

Similar Species Distinctions:
—Poison-ivy (*Toxicodendron radicans*) leaves have 3 leaflets that have only irregular teeth along the margins. Bladdernut has fine, continuous teeth.

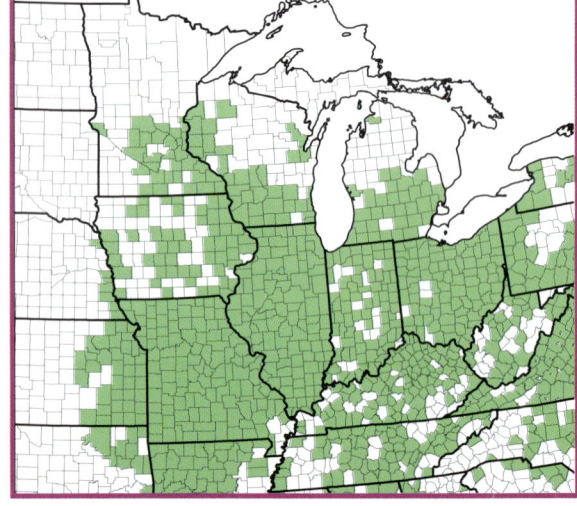

The opposite, compound leaves are trifoliate and up to 10 inches long. Each leaflet is oval, obovate, oblong, or orbicular in shape and has finely toothed margins. Leaves are dark green and smooth above, and slightly paler and smooth beneath. Petioles are long (up to 5 inches in length) and make up half the length of the leaf.

Flowers appear in mid-April with the emerging leaves, and they hang down in large, drooping clusters (racemes) that are up to 4 inches long. Each flower is white, about ⅓ inch long, and perfect. The petals seem to form a tube but are not fused together. The sepals remain closed over the petals, as seen here.

Mature bark is thin and dark brownish gray, with tannish streaking along the trunk.

There is no true terminal bud, because the twig tip dies back during the late fall. What is seen near the tip of the twig is a pair of lateral buds instead. They are ovoid and have 4 or 5 smooth, reddish brown or green scales.

Young twigs are olive green to brownish, smooth, and dotted with brown lenticels. Lateral buds are similar to the "terminal" buds, but smaller. Leaf scars are triangular-shaped and contain 3 bundle scars in the center. Older twigs develop tan-colored fissuring that creates a striking color contrast to the greenish brown twigs.

Alders

Genus: *Alnus*

The Midwest is home to 3 native, large shrub alders that are closely related to the birches. Like most members of the Betulaceae family, alders produce preformed flower buds or catkins, which overwinter on the twigs. These catkins are valuable characteristics for identification. One wintertime alder feature not found on the birches, however, is the overwintering *female* catkins, which are much smaller than the males. Another feature of alders is the persistent, pinecone-like woody fruits that are present most of the year.

Taxonomy of our species can be confusing (especially when referencing older literature), particularly that of *Alnus incana* (L.) Moench ssp. *rugosa* (Du Roi) Clausen, speckled alder, which is commonly determined to be a rough equivalent to the *A. incana* (L.) Moench ssp. *incana,* European alder. Yet another problem is the fact that most alder species will hybridize where ranges overlap. To make matters worse, the introduced *A. glutinosa*, European black alder, has escaped within the Midwest. In its native range it hybridizes with European alder. Considering all this, the possibility of hybridization between the black alder and our native species, particularly speckled alder, is great.

There are an incredible number of homeopathic prescriptions made from alders, including those to cure anemia, diarrhea, toothache, rashes, and swelling.

Alders are beneficial in several ways that warrant their promotion. Their roots, though shallow, stabilize soil along streams and other waterways where they occur. Nitrogen-fixing bacteria grow on the roots, so they also enhance quality of the soil in which they grow.

Because they are often the dominant species in streamside shrub communities, our native shrub alders provide substantial cover, both escape and nesting, to a variety of associated wildlife species. Unfortunately, the principally planted alder in this region is European black alder, a readily available exotic (Weeks, Weeks, and Parker 2010). This species' cover value is somewhat different from that of our natives, because it quickly bolts and grows to tree size. Our 2 most common native species increase in abundance and influence as one moves away from the central Midwest, speckled to the north and smooth to the south, but wildlife value is high everywhere. Nitrogen-fixing, coupled with adequate soil moisture and high organic content, yields high populations of earthworms that serve to attract and support substantial American woodcock populations. Woodcock and alders are linked very much as ruffed grouse and quaking aspen are. Both species are of low preference for herbivores, but they are occasionally browsed by beaver and white-tailed deer; ruffed grouse will feed on male catkins in late winter/early spring, likely because of the high protein content of the pollen.

The ripening fruits of alders are green, but with maturity they turn brown and release tiny, winged seeds from around each woody scale. Seen here are smooth alder fruits.

Both smooth and speckled alders have "stalked" (stipitate) buds—a common feature on most alders. The more northerly *Alnus viridis* ssp. *crispa*, however, does not have stalked buds.

Alders have unusual, triangular-shaped pith unlike the round pith of most woody plants.

Alders all have "preformed" flower buds that begin development during the summer months of the year preceding flowering. The larger males are clustered together at the end of twigs, while the tiny females are clustered along the twigs. The flower buds, or catkins, are present throughout the winter months (and assist in identification) and open in late winter and early spring. Pictured is speckled alder.

speckled alder, tag alder

Alnus incana (L.) Moench ssp. *rugosa*
(DuRoi) Clausen

Family: Betulaceae

Speckled alder is an important shrub along the waterways where it occurs. It provides structure and nutrients for the soil, food and cover for
many wildlife species, and headaches for trout fishermen whose lines
might not avoid the many outstretched limbs. This species' nomenclature is quite confusing, but it usually contains *incana* or *rugosa*. Its
common name in the northern Midwest, tag alder, stems from the constantly attached conelets (the fruit) that are
like tags hanging from the plant. It is listed as state endangered in Illinois.

Form and Size: Speckled alder is a large shrub, sometimes classified as a small tree, though it is nearly always
multi-stemmed at the ground. Its crooked, spreading branches are few, but create a broad form to around 15 feet in
height or more, particularly as one moves northward. Clonal thickets form along waterways when light is abundant
and where flooding helps control other woody species.

Habitat: Habitat is always along streams, lakes, rivers, upland depressions, and open swamps where fresh, moving water occurs. Speckled alder prefers full sun but can tolerate some shade. It has been found to grow in a wide
range of soil pH and can be planted on a more upland site. It tolerates prolonged flooding and drought, and it
quickly invades areas that have been disturbed by fire or logging. When given this opportunity, it forms pure,
dense alder thickets that are impenetrable to humans.

Wildlife Uses: This is the dominant alder in the northern Lake States and is important to wildlife associated with
those steamside shrub communities. In addition to values for the American woodcock, speckled alder is a frequent
nest site for many associated songbirds, such as yellow warblers and alder flycatchers. The seeds that are produced
in the usually abundant fruit are regularly used by birds and small mammals, especially by American goldfinches
and pine siskins. Because of their proximity to streams, they are regularly used by beavers for lodge and dam construction, although they do not appear to be a favored food source.

Landscaping Value: Hardy to Zone 4, speckled
alder should be considered in any wet situation
where fresh, flowing water is available. Its roots
help stabilize stream banks and provide nutrients to the soil. Soil type and pH do not seem
to be an issue; the biggest challenge is locating
the species in a nursery. There are currently no
cultivars available. Alders have very few pests,
but Japanese beetles are fond of this species and
quickly destroy most leaves.

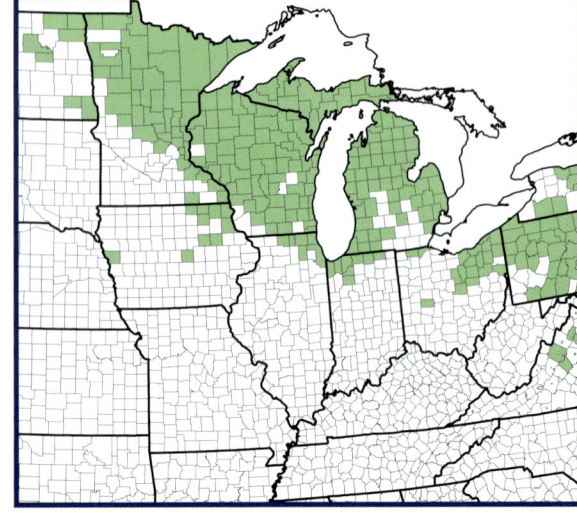

Similar Species Distinctions:
—Other **alders** (*Alnus* spp.) are similar
in appearance. See individual species accounts for details.

Speckled alder leaves are up to 4 inches long and 2 inches wide, oval to obovate, and rounded at the base. The margin is usually doubly toothed, with large teeth at the end of each lateral vein, and smaller teeth in between. The leaf surface is dull, dark green above, pale with varying degrees of hair beneath. Lateral veins are very straight and impressed from above. New leaves have paired stipules that fall after maturity.

Fruit is a cone-like, woody, long-stalked fruiting body with tiny, barely winged seeds inside. The fruit is less than an inch in length and persistent for a year or more. New fruit is green; fruits are brown when ripe.

Alder flowers open in late winter before the leaves. The overwintering catkins are in clusters at the ends of twigs. The large males elongate to release pollen; the tiny, downward facing females have bright red scales.

Buds are reddish to reddish brown with varying amounts of silky white hairs. They have 2 to 3 scales. Twigs are light brown with hairs scattered on the newest growth. Numerous, light-colored lenticels occur on the twig. Leaf scars are raised and shaped like rounded triangles with obvious bundle scars that create the appearance of a face. Alders usually have triangular-shaped pith.

Bark is reddish brown when young with numerous, raised lenticels. Older plants turn gray and lenticels run together creating horizontal, raised "bands."

smooth alder, hazel alder, brookside alder

Alnus serrulata (Aiton) Willd.

Family: Betulaceae

Smooth alder is a common, large shrub that is somewhat of a southern equivalent to speckled alder. Our native shrub alders are overlooked as having any tangible value, but they serve a similar role as willows with regard to soil stabilization along waterways. In addition, they improve soil by fixing nitrogen. This species has had nomenclature issues over the years, and it is often referred to as *A. rugosa*.

Form and Size: A typical size for smooth alder in the Midwest is 8 to 10 feet tall, but in the Southeast, where it is quite common, it can grow much larger—to 30 feet. It has a low, spreading, open crown and tends to grow in small, clonal clumps rather than in the thickets commonly seen with speckled alder. It tends to have a single, straight, slender, unbranched trunk for several feet before branching, unlike the multi-trunked, sprawling speckled alder.

Habitat: Smooth alder is exclusively associated with water. In the central Midwest, it has been found in wet woods, swamps, cold bogs, and streamsides. It prefers full sun but does tolerate partial day's shade. Alder dies out over time in a wooded area as the canopy closes. It tolerates most soil types, but it tends to require acidic to neutral soil pH. The authors have successfully grown this species in an upland site.

Wildlife Uses: Smooth alder has many of the same wildlife values as does speckled alder, largely replacing the latter species as one moves southward in the Midwest. Seeds are regularly used by wildlife, especially American goldfinches. Stands also support large earthworm populations that benefit the dependent woodcock, especially during its migration to more northerly regions where it reaches its maximum breeding density. Because this species occurs most frequently as scattered individuals or clones, rather than in extensive thickets, as speckled alder does, its value as nesting sites for associated shrubland birds, such as yellow warblers, willow flycatchers, and white-eyed vireos, is enhanced.

Landscape Value: Hardy to Zone 5, smooth alder has all the qualities of speckled alder, but it tends to be smaller and less likely to form dense thickets. Its smooth, gray, fluted trunk is an attractive winter feature as are the tiny, overwintering, reddish catkins. Its bright red female flowers are showy in the late winter. Disease and pests are generally not a problem. There are no known cultivars, but it is available through a few nurseries.

Similar Species Distinctions:
—Other **alders** (*Alnus* spp.) all have similar features too lengthy to list here. See individual species accounts for details.

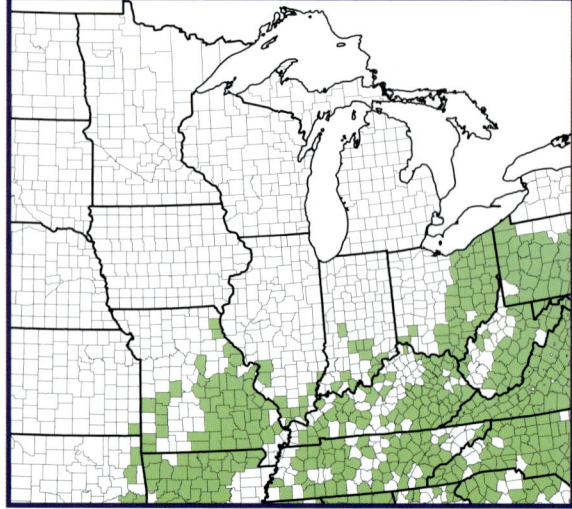

Flowers are separate sex in clusters at the ends of twigs. Males, when open in the late winter, are 3 inches long; the tiny, red females are in an upright position. Winter male catkins are reddish brown.

Smooth alder leaves are up to 4 inches long and 3 inches wide, broadly elliptic to obovate. The leaf base is usually wedge-shaped, and the tip is typically rounded. Leaf margins are usually finely, singly toothed, and generally not wavy. Both surfaces are green, but leaves are commonly "glutinous" beneath—with applied pressure, a leaf will stick to a surface. Leaf hairiness, particularly on the lower surface, can vary, but is greater on new leaves.

Ripe fall fruit is woody, pinecone-like, and less than an inch long. They persist for a year or more once ripe and are very obvious in the winter months. Seeds are tiny, narrowly winged nutlets.

Buds have 2 reddish brown, non-overlapping (valvate) scales that have varying degrees of hairiness. They are stalked (stipitate) like most alders. Twigs are brownish, and eventually gray, with numerous light-colored, scattered lenticels. Leaf scars are raised and shaped like rounded triangles with grouped, face-like bundle scars. Pith is triangular in shape.

Smooth alder's common name is derived from its smooth bark, which is very similar to that of *Carpinus caroliniana*, American hornbeam. Young trees have reddish brown bark that grays with age. The trunk becomes fluted and angled. This is quite different from the rough-textured bark of *Alnus incana* ssp. *rugosa*.

green alder

Alnus viridis (Chaix) ssp. *crispa* (Ait.) Turrill

Family: Betulaceae

This species is also known as mountain alder, and it tends to inhabit drier sites than other native alders. It is, however, able to access high water tables in sandy or rocky sites with its deep root system. There are 3 subspecies of *A. viridis* in North America; the other 2 are subspecies *fruticosa* and subspecies *sinuata*, both of which are found in western United States and Canada.

Form and Size: Green alder is an upright, often multi-stemmed shrub that commonly reaches 10 feet in height, but can grow to twice that height. It is usually not thicket-forming as is typical of speckled alder, but individual plants look very similar in form. Individual stems are probably fairly short-lived, and plants will die out if shaded, as it prefers full sun.

Habitat: In the Midwest, green alder is found in coniferous forests and forest edges and along streams, lakeshores, coastlines, and bogs. It is usually found in acidic, poor, sandy, or rocky soils.

Wildlife Uses: Green alder has many of the same values as do our other 2 natives; it does, however, tend to grow in more upland sites, especially along lakeshores, and it does not tend to form large colonies. In these locations, it provides forage for snowshoe hares and nesting cover for forest shrub-nesters (structure for nesting is good). Ruffed grouse regularly feed on male catkins in late winter. It is, in general, considered to have low browse value for large ungulates, although moose use it heavily in some areas. Winter finches feed on its seeds as they do with other alders.

Landscaping Value: Green alder is very cold hardy and prefers cooler, less humid regions of the Midwest. It is hardy to Zone 1 and not particular about where it is planted. It fixes nitrogen, which improves soil quality, and it reduces erosion with its substantial root system. Seeds are blown in the wind and germinate easily on recently burned sites.

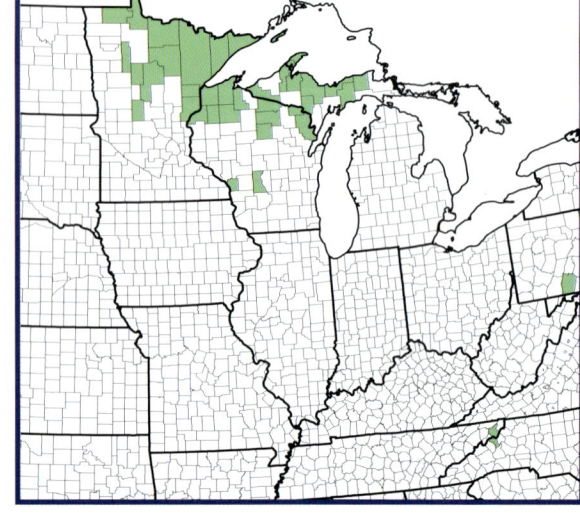

Similar Species Distinctions:
—**Speckled alder** (*A. incana* ssp. *rugosa*) has similar bark and fruit, but leaves are doubly toothed, buds are stalked, and female flowers mostly droop.

Flowers appear with the leaves in May and June. The females are upright on long stalks with numerous greenish or reddish stigma lobes; they are the largest female flowers of native alders at nearly 1 inch in length. Male catkins dangle from the branch tips in clusters.

The alternate leaves are 3 to 4 inches long, elliptic to ovate, unlobed, and finely toothed along the margin. The upper surface is dark green and mostly smooth; the lower surface is paler, hairy along the veins, shiny, and often sticky (glutinous).

Fruit resembles a tiny pinecone, is up to ¾ inch in length, and is upright on branch tips in clusters. They ripen in the fall and gradually open to release tiny, wind-blown, broadly-winged seeds. Cones are persistent, and over time they droop downward.

Green alder is the only native alder without stalked buds. Buds are reddish black, pointed, and covered by 2 scales. Twigs are pale and hairy during the first winter, and they have triangular leaf scars that contain 3 bundle scars.

Bark is thin, gray, and dotted with raised lenticels, giving it a rough texture.

low juneberry, low shadbush, low serviceberry, running serviceberry

Amelanchier humilis Weig. **complex**

Family: Rosaceae

The late Edward G. Voss, curator of vascular plants in the University of Michigan Herbarium and professor of botany for many years, lamented the great taxonomic difficulties with *Amelanchier* species identification in his *Michigan Flora* (1985). There seems to be no consensus among taxonomists about this genus, and regional variation due to hybridization, polyploidy, and asexual reproduction factor into that conundrum. Many authors associate *A. humilis* with *A. sanguinea* and *A. spicata* complexes. The authors of this volume borrow a bit of taxonomic wisdom from Voss by following his classification of *A. humilis* Weig. complex. *A. humilis* is listed as state endangered in Indiana and Pennsylvania. Serviceberries are like harbingers of spring—their early, attractive, white flowers are a vision of spring for winter-weary eyes.

Form and Size: Low serviceberry is a colony-forming shrub that commonly grows to around 3 feet in height. The root system sends up single-stemmed suckers that can spread across a sizable area.

Habitat: It is found growing on sandy ridges near Lake Michigan, in black oak savannahs, and in barren, sandy fields.

Wildlife Uses: Unlike the other serviceberries that tend to be small trees (e.g., *A. arborea* in Weeks, Weeks, and Parker 2010), low juneberry is a shrub that generally is less than 6 feet tall, and the small thickets produced by root-suckering provide some cover for deer and shrub-nesting songbirds. The early spring flowers produce juicy fruits very early (often late May) that are relished by most medium-sized birds, such as catbirds, Baltimore orioles, robins, and other thrushes. Fruits are also regularly taken by mammals, from black bear to squirrels, and occasionally there is browsing by deer and cottontails.

Landscaping Value: Hardy to Zone 4, the serviceberries provide early spring flowers and tasty berries in June. Low juneberry can provide a petite thicket of upright stems given a friable, somewhat acidic soil. It tolerates drought once established and has nice fall color if given mostly full sun. It is easily propagated from seed but is not readily available through nurseries.

Similar Species Distinctions:
—**Serviceberries** (*Amelanchier* spp.) Numerous species occur within the Midwest, and even the best keys often fail to provide an unquestionable conclusion as to species. Authors recommend Voss's *Michigan Flora* (1985) or *Flora of North America*, vol. 9 (in press) for the most recent classification.

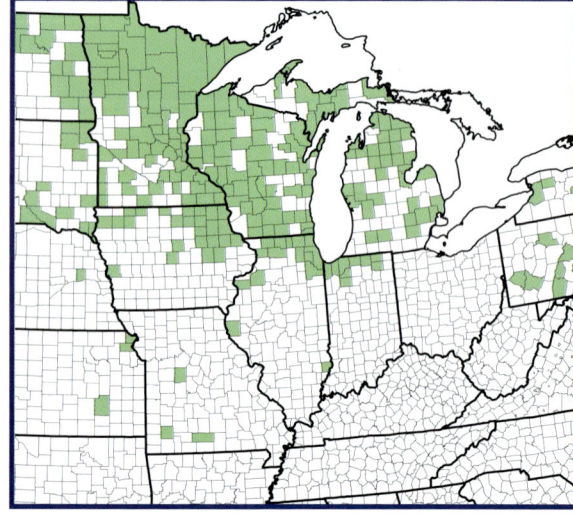

Leaves are alternate, up to 2 inches long, and 1½ inches wide. Shape may be oval, oval-oblong, orbicular, or even ovate or obovate. New leaves are green, folded, and densely hairy beneath, but they become smooth at maturity. Margins are usually finely toothed, especially toward the tip with 12 to 20 teeth per inch when mature. Leaf veins become indistinct near the margin. Leaf surface is dark green and paler beneath.

Serviceberries flower in early spring before most other woody plants—at least any with showy flowers. Low serviceberry flowers appear in late April into May in our region along with the expanding, folded, densely hairy leaves. The 5 broad, white petals are usually between ¼ and ⅓ of an inch in length. The delicate-looking flowers are in sizable clusters. The top of the ovary is quite hairy.

Fruit is about ⅓ inch in diameter and ripens in June. The berry is nearly black (when ripe), fleshy, juicy, and tasty to wildlife and humans alike. There can be a whitish (glaucous) bloom on the fruit when ripe.

Buds are elongate, with numerous, reddish scales partially covered with a silvery coating. White, silky hairs are exposed around the margin of the bud scales. Twigs are reddish, but they quickly develop a silvery-gray coating (epidermis); they are quite slender. Leaf scars are raised and shaped like a narrow quarter-moon.

Though barely large enough in diameter to develop any characteristic bark, it looks like all others within the genus. Bark is smooth and mostly gray, but with vertical, fine, blackish streaks running most of the length of the trunk.

chokeberry

Aronia prunifolia (Marsh.) Rehder
Photinia floribunda (Lindl.) Robertson & Phipps
Photinia melanocarpa (Michx.) Robertson & Phipps
Family: Rosaceae

The difficulties of identifying chokeberry species are too great to report here. The readers are referred to Swink and Wilhelm's *Plants of the Chicago Region* (1994) and Voss's *Michigan Flora* (1985) for thorough discussions. *Aronia prunifolia* is the agreed-upon species by both authorities, and since they are regional, the authors of this volume follow their views. Chokeberry is a beautiful, large shrub with spectacular fall color.

Form and Size: Chokeberry is usually a tall shrub, often reaching a height of 10 feet. It is multi-branched and dense.

Habitat: Chokeberry is commonly found in bogs, in swampy woods, and even on dry wooded slopes. Nearly pure stands of chokeberry have been found that cover an acre or more. It grows best in full sun, but it handles partial shade. Although it prefers acidic soil, it is adaptable. It can grow quickly, and individual stems are fairly short-lived.

Wildlife Uses: Chokeberry is relatively uncommon in our region, and as such, it likely contributes little overall to our wildlife's well-being. Where it does occur, however, it often grows in relatively open situations that produce a dense growth form attractive to shrub-nesting songbirds. Its major value is in its fruits, which are used in the fall and winter by many birds, including grouse, flickers, robins, brown thrashers, and cedar waxwings. However, the fact that many fruits last well into the winter suggests that the preference for the fruits is not high. Chokeberry is preferred browse by white-tailed deer, which are especially fond of new leaves in the early spring.

Landscaping Value: Hardy to Zone 3–4. Chokeberries flower prolifically, especially when grown in full sun. In this setting, they usually produce a large quantity of fruit and develop beautiful fall color. They are most often found in wet soils, but they adapt easily to upland sites. Planting as a hedge or border works well with this large shrub. Some insect and disease problems, similar to those of cherries, are possible. There are several horticultural varieties of *Aronia*, including the dwarf 'Iroquois Beauty.'

Similar Species Distinctions:
—**Serviceberries** (*Amelanchier* spp.) have similar buds, twigs, and leaves. *Amelanchier* species do not have glandular petioles, and their bud scales are lined with soft, white hairs.

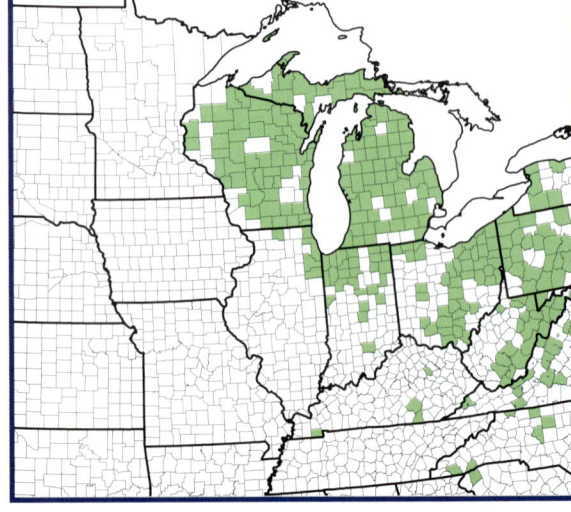

Leaves are alternate, obovate to oval, and commonly 2½ inches long. The margins are very finely toothed, and there is a pair of stipular appendages on the petiole. The leaf surface is dark green and shiny with small black glands running along the midrib; the lower surface is pale. The amount of hairiness on either surface is highly variable, ranging from essentially none to very prominent. The petiole commonly has raised glands of varying size and number.

Chokeberry fruit is a berry-like pome, about ¼ inch in diameter (variable) and blackish (or red) when ripe. It is juicy and often astringent, but these characteristics vary from plant to plant. Most fruits begin ripening in August, but fruit can ripen as late as October.

Flowers appear mainly in the month of May. They are showy, abundant, and in terminal clusters. The 5 white petals create a dramatic backdrop for the bright pinkish red anthers. Flowers are about ½ inch wide.

The elongated buds of chokeberry are covered with 5 rose to burgundy-red, overlapping scales that have varying degrees of hairiness. They are similar in appearance to those of serviceberry. Chokeberry twigs are slender and reddish brown, with scattered, tan-colored lenticels. The surface can be smooth or very hairy, and it is often covered with a silvery coating. Lateral buds are usually appressed against the twig. Leaf scars are thin, V-shaped, slightly raised, and contain 3 bundle scars.

Bark is reddish brown when the plant is young with obvious tan-colored lenticels. With age, it becomes grayish brown and somewhat roughened with enlarged lenticels, similar in appearance to that of small black cherry trees.

American barberry, Allegheny barberry

Berberis canadensis Mill.

Family: Berberidaceae

American barberry is 1 of the Midwest's rarest shrubs, thanks in part to the federally funded Barberry Eradication Campaign that began in 1918. It, along with the European *B. vulgaris*, were found to be alternate hosts for black stem rust, an economically serious disease that is detrimental to wheat and other small grain crops. The program was amazingly efficient and nearly wiped out our native barberry. It is listed as endangered in Indiana, Illinois, and Kentucky, but it is probably now extirpated in the latter 2. The exotic Japanese barberry, *B. thunbergii*, is not an alternate host, and it has been extensively planted in the eastern United States. It is easily distinguished from our native by its single spine at the nodes and entire leaf margins.

Form and Size: Our native barberry is usually less than 3 feet in height, sparsely branched, and spiny. It spreads by rhizomes and forms loose colonies.

Habitat: Barberry was commonly found, often in large colonies, on wooded bluffs along the Tippecanoe and Wabash rivers in Indiana. This species is considered fire-dependent, as it relies on disturbance to maintain its required, open habitat. In other states of our region, it is found along creek banks and even roadsides.

Wildlife Uses: Since the species was targeted for elimination because it served as an alternate host for black stem rust, it is very rare in the region, and the small, scattered populations have no major impact on our native wildlife. Individually, however, spined plants that spread via root-suckering provide very good nest sites for shrub nesters such as indigo buntings and field sparrows. Fruits are taken by songbirds and small mammals and white-tailed deer browse leaves, although relative preference is unknown because of the plant's rarity.

Landscaping Value: Hardy to Zone 3, this is a tough shrub that is adapted to poor soils and can handle drought well. It makes an interesting specimen plant, and given full sun, it becomes full and broad with more upright branches than *B. thunbergii*. Fall color is spectacular. This native, unfortunately, will probably never make its way to a commercial market because of its potential negative impact on agriculture.

Similar Species Distinctions:

—**Japanese barberry** (*B. thunbergii*) usually has single, straight spines, entire leaf margins, and similar flowers and fruit. Young twigs are brownish; older twigs and stems are gray.

—**European barberry** (*B. vulgaris*) has similar leaves with numerous, obvious leaf veins, many flowers and fruits per cluster, and gray twigs and stems.

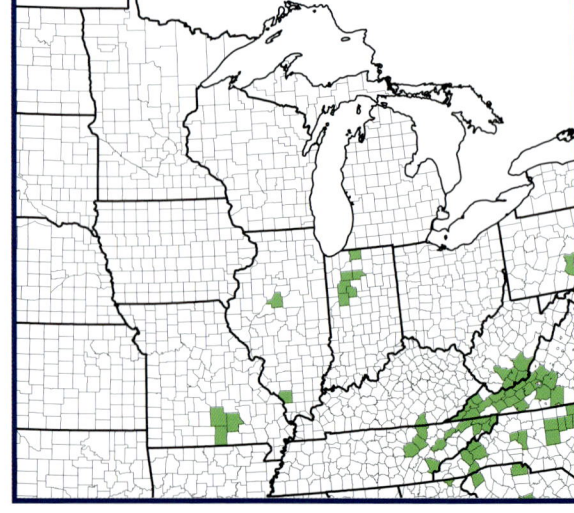

Leaves are up to 2½ inches long and clustered at the alternate nodes at the base of its forked spines. The leaf margins have small, irregular teeth, from 2 to 11 per side. New, late summer leaves on vigorous growth are almost holly-like. The upper leaf surface is pale green, the lower surface pale and glaucous. The petiole varies in length and can be extremely slender.

The dark red, shiny fruit hangs in drooping clusters; each fruit is about ⅜ inch in length. They ripen in late summer and persist through the winter.

Flowers hang down from the nodes in drooping clusters of 5 to 10 flowers. They are yellow and open in late May in the central part of our region. Each of their 6 petals is notched at the tip, another feature that distinguishes this species from the exotic barberries.

The oval-shaped buds have overlapping, reddish brown scales and are surrounded by clustered, peg-like, brownish, flat-topped remnants of leaf petioles. Lateral buds on older twigs appear from a very short shoot. Twigs are purplish red with a 3-branched spine at each node; young twigs rarely may have a single spine. Second year branches remain this color, which is an important distinguishing feature between the very similar *B. vulgaris*, the second year branches of which are brown.

The most unique feature about all barberries is the bright yellow inner bark. Our native has outer bark much like the exotics. It has many thin, vertical, weaving, shallow ridges.

bog birch, dwarf birch, swamp birch

Betula pumila **L.**

Family: Betulaceae

Bog birch is common in the northern Midwest. There are several recognized varieties of bog birch, which are separated by the presence/absence of various combinations of hairs and glands. Bog birch can hybridize with our native tree-sized birches, which produces, as Voss proclaims in his *Michigan Flora* (1985), "beautifully intermediate" specimens. Fall color of bog birch is spectacular if the plants receive full sun.

Form and Size: It is usually a medium-sized shrub, less than 5 feet tall, but more northward in its range, it can reach nearly 10 feet. Its growth form is upright and sparsely branched; it often grows in small clumps.

Habitat: Bog birch occurs in bogs, fens, and occasionally along lake borders, almost always near water. Voss (1985) reports that it is a calciphile—the plant must have lime or alkaline soil. It is moderately slow growing and requires full sun for best growth.

Wildlife Uses: Bog birch is the consummate shrub in wet areas, often forming dense stands. Whether they occur as isolated individuals or in groups, they are good nest sites for songbirds of those habitats, such as yellow warblers and alder flycatchers. Dense stands provide cover for larger animals like deer, moose, and snowshoe hares. Buds and catkins are eaten by grouse in the winter and spring; seeds are taken in the fall by several birds (especially black-capped chickadees) and small rodents. Although this widespread species is browsed heavily by ungulates in the West, white-tailed deer in the East show little preference for it. Streamside shrubs are frequently fed on by beaver, and snowshoe hares take stems as well.

Landscaping Value: Hardy from Zones 2–6, the heat and humidity of the lower Midwest limits its spread southward. Bog birch has several interesting features worth exploring for a landscape plant. Its small leaves are an unusual shape, and they show nice golden yellow-browns in the fall. Its twigs bear tiny, overwintering, clustered catkins that are attractive. Although it prefers full sun, it will handle partial shade. It is adaptable to more upland sites, but unfortunately, it is rarely offered at nurseries. This species is easily grown from seed.

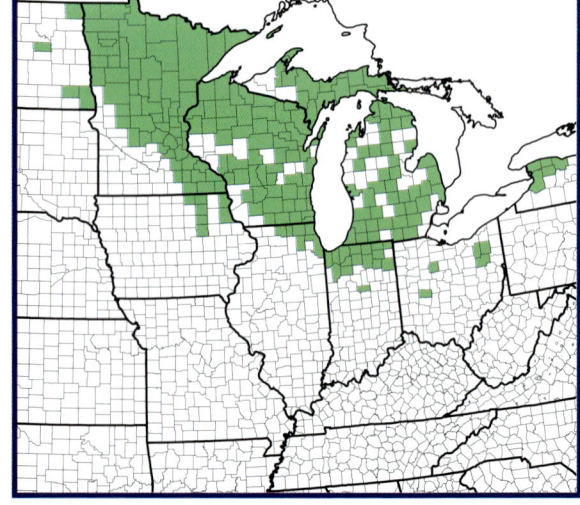

Similar Species Distinctions:
—No other shrub has leaves similar to those of bog birch. Hybrids between it and the tree-sized *B. papyrifera* and *B. alleghaniensis* are common where their ranges overlap.

Leaf shape, size, and surface texture can be highly variable. The usually tiny, alternate leaves can vary from less than 1 inch in length to nearly 2½ inches. Texture ranges from hairy to smooth; they may occur with and without warty glands, and sometimes there is a whitish, glaucous coating on the lower surface. Margins are coarsely toothed; the leaf base is wedge-shaped; and the leaf tip is typically rounded. Mature leaves are usually hairless. Leaf shape is obovate, ovate, or orbicular, and leaves are rather thick. Leaves commonly appear in clusters of 3 from the nodes.

Fruit is a solid, cylindrical cluster of winged seeds that is about 1 inch long. The tiny seeds have wings narrower than the body of the seed. Fruit ripens in the fall and often persists through the winter.

Flowers appear in May with the leaves. Both male (left) and female catkins appear along the twig and are about 1 inch long.

The brownish, preformed, overwintering male flower buds (catkins) are clustered, especially near the tip of twigs (photos 1 and 2). They have numerous, overlapping scales that are tipped with white hairs. Leaf buds (seen in far right twig of photo 3) are few-scaled, smaller, and usually diverging from the twig. The flower buds (seen in photo 3, twig on left; probably female) are often scattered singly along the twig at nodes or paired with leaf buds. Youngest twigs are reddish brown and lightly hairy, with scattered, light-colored lenticels. Twigs of some specimens have roughened, nearly white glands that are very similar in appearance and texture to salt on a pretzel. Older twigs become silvery gray. Leaf scars are raised and contain 3 bundle scars.

Bark is thin and reddish brown, often with a silvery coating. Raised, light-colored lenticels are always present. Older bark becomes dark brown and begins to peel like the bark of other birch species.

New Jersey Teas

Genus: *Ceanothus*

Three species of *Ceanothus* occur in the Mid-west, 1 of which is *C. sanguineus*, a disjunct population found in our region only in the north-ernmost tip of the Upper Peninsula of Michigan in Keweenaw County. Its major distribution is in the Pacific Northwest, where it is used as an ornamental.

All members of the genus have several distinct features that aid in identification. Leaves always have 3 distinctly impressed major veins that begin at the base and run toward the tip. A second feature is the unique fruit remnants seen once the capsules deteriorate. These look much like pie tins that are balancing on a stick.

New Jersey teas are nice additions to any landscape for several reasons; their flowers are obvious, showy, and long-lived, and they produce copious amounts of nectar that attract a nice variety of insects. Leaves of *Ceanothus* have been used as a tea substitute. Its red roots contain many acids and were once used in homeopathic drugs.

The 2 major species in our region have similar wildlife values. They are, in general, rather un-common, occurring on dry prairies, glades, and rights-of-way. These small shrubs supply nesting cover for shrub-nesting songbirds, but their major use by wildlife is as food, both browsing on leaves by white-tailed deer and feeding on seeds by game birds and songbirds. *C. herbaceus* is rare enough in our region to have minimal impact on our wildlife.

A common feature of *Ceanothus* is the fruit remnants that last well into the winter. Once the capsules deterio-rate, what is left is a structure that looks very much like a pie tin that is being bal-anced on top of a stick.

Fruit is a capsule that is separated into 3 segments. Once the fruit ripens, an outer "skin" sheds to reveal the 3-parted capsule that resembles a small, tan-colored pumpkin that has fewer vertical ribs.

Leaves of a *Ceanothus* are very similar, no matter what species is being examined. They always have 3 main veins that run from the leaf base to the tip that are fairly deeply impressed. Pictured here, side-by-side, are *C. americanus* (left) and *C. herbaceus* (right).

New Jersey tea, Jersey-tea

Ceanothus americanus L.

Family: Rhamnaceae

New Jersey tea's name is derived from the fact that its leaves were used as a tea substitute during the Revolutionary War in the Northeast. This beautiful shrub makes a great addition to any dry landscape and is beneficial to many bees, flies, and beetles.

Form and Size: This is an upright, spreading, rounded shrub that is usually less than 3 feet tall. It produces numerous, fine stems and has a dense appearance.

Habitat: This is "the" shrub component of the tall grass prairie where it is commonly (before habitat destruction, much more common) found in dry, sandy, open areas. It is often associated with oak savannahs. Although able to tolerate some shade, it prefers and thrives in full sun. An added bonus is its ability to fix nitrogen.

Wildlife Uses: New Jersey tea is widely distributed in our region, but nowhere very common. It frequently occurs in the open, such as along railroad rights-of-way, and has a dense growth form that lends itself well to use by nesting songbirds, such as field sparrows, indigo buntings, and common yellow-throats. It is occasionally browsed in summer rather heavily by white-tailed deer and has been used as an indicator of deer overabundance. Twigs are taken by deer in winter as well, but not to the degree that twigs are taken by cottontails, which occasionally will cut every twig on a small shrub. It produces relatively large seeds that are used by larger game birds, such as bobwhite and wild turkey, in addition to songbirds.

Landscaping Value: Hardy to Zone 4, this is a lovely shrub when flowering, and it does so over a several week period in July in the heat of summer. Plants without competition become full and spread in a rounded form 3 feet across. It is adaptable to a variety of soils, but it prefers slightly acidic conditions and good drainage. It is drought resistant. New Jersey tea is found through nurseries specializing in natives and in habitat restoration.

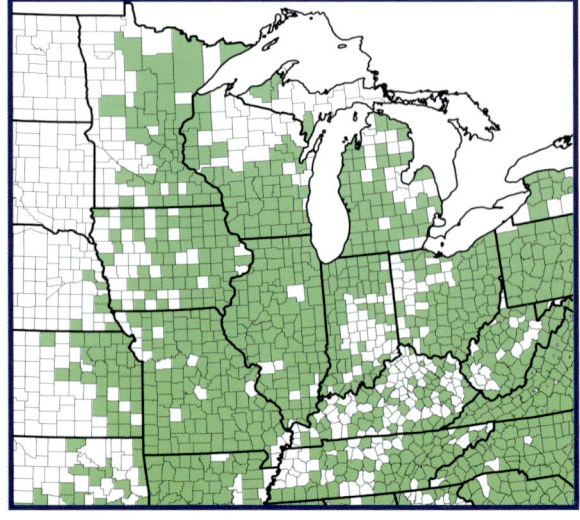

Similar Species Distinctions:
—**Inland New Jersey tea** (*C. herbaceus*) is less common and has more narrow leaves. Otherwise, they are similar.

Leaves are alternate, ovate to elliptic-ovate, and up to 4 inches long. There are 3 main, incised, obvious veins and many deeply impressed secondary veins as well. Margins have numerous, rounded, irregular teeth. The leaf upper surface is dark green and dull; the lower surface is also dark green but often lightly hairy.

The beautiful white flowers appear rather late compared to those of most other native shrubs, usually in June and July. It, thus, has a long flowering period. Flowers are in upright, showy clusters, mainly appearing on stalks from leaf axils. The stalks become progressively shorter toward the top of the plant. Flowers are perfect and produce large quantities of nectar.

Fruit ripens in late summer and is in upright clusters. Each fruit is a 3-parted, domed capsule; each section of the capsule contains a single seed. By winter, the capsules deteriorate, leaving behind the disk-shaped bases that look like pie tins on sticks; these are persistent throughout the winter.

New Jersey tea rarely grows large enough to develop characteristic bark; however, stems of the largest specimens become brown-gray and develop shallow, vertical cracks; otherwise, large stems are greenish brown. Roots of all *Ceanothus* species are red.

Buds are ovoid, greenish brown, and few-scaled. They are the same color as the twig. Dense, whitish hairs can cover the outer scales.

Twigs are slender and greenish brown or reddish, with some degree of hairiness; newer growth has more hair. Leaf scars are raised, broadly heart-shaped, and contain several large bundle scars. "Bear claw-like" bracts may be present at the nodes of the first year's growth.

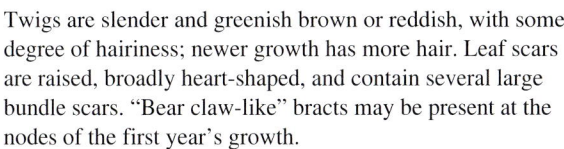

redroot, inland New Jersey tea, prairie-redroot, narrow-leaved New Jersey tea

Ceanothus herbaceus **Raf.**
Ceanothus ovatus **Desf.**
Family: Rhamnaceae

Redroot is an uncommon shrub in the Midwest, and it is currently listed as state endangered in Illinois, Indiana, and Ohio, and state threatened in Kentucky. As 1 of its common names suggests, it does indeed have red, sometimes substantial roots.

Form and Size: This is a small shrub, not more than 2 feet tall, and is very slender with few side branches. Because of these features, it very much resembles an herbaceous plant.

Habitat: Redroot tends to occur in similar habitats across its natural range, occurring in sandy, rocky soil on exposed sites. It is short-lived and requires full sun.

Wildlife Uses: Redroot has essentially the same wildlife values as New Jersey tea, except that its overall impact is substantially less because of its rarity. In Indiana, for example, it occurs only on rocky outcrops along the Blue River, where it likely provides nesting cover for associated species, such as song sparrows.

Landscaping Value: Hardy to Zone 4, this shrub has similar landscaping qualities as New Jersey tea, but it is probably more drought tolerant. It flowers earlier—late May in west-central Indiana—than its counterpart. It prefers well-drained sites in full sun. The authors have grown it for several years in partial shade, and it has become "leggy." Pruning may help with this matter. Fall color is poor, and spring flowers are its best feature. This species is occasionally available from nurseries specializing in natives.

Similar Species Distinctions:
—**New Jersey tea** (*C. americanus*) has broader leaves and flowers a month later.

The fruit is a terminal cluster of 3-celled capsules that are about ¼ inch wide. Each cell is filled with a few dark brown seeds that are pitted, unlike the smooth seeds of *C. americanus*.

Leaves are elliptic-lanceolate to oblong with 3 prominent, deeply impressed, main veins. Secondary veins are also impressed. Leaf margins are sharply, glandular-toothed; the leaf tip is usually slightly pointed. Leaves are dark green above, somewhat paler and lightly hairy beneath, especially along the veins. Leaf length is a maximum of 2½ inches.

The delicate white flowers appear terminally on short stalks in early June in the central part of our region. Some clusters will be on short stalks from the upper leaf axils. Each 5-petaled, perfect flower is about ⅛ inch wide.

Buds are basically like *C. americanus*, with 4 to 5 reddish scales that are tipped with tufts of whitish hairs. Most buds sit on protruding nodes that give the appearance of their being on stalks. Twigs are reddish green and very slender, with a sparse cover of fine hairs. Leaf scars are somewhat half-moon shaped with a row of 3 to 4 bundle scars. Paired, red, leafy bracts sit at the top of many first-year leaf scars.

dwarf hackberry

Celtis tenuifolia **Nutt.**
Celtis pumila **Pursh.**
Family: Ulmaceae

Although usually considered a shrub, this species of hackberry is rarely multi-stemmed. It truly looks like a dwarf version of a large hackberry tree. You will know you have found a dwarf hackberry if you find a 5-foot-tall specimen that has flowers or fruit. Several varieties of this species have been recognized, including *C. pumila* var. *Deamii*. The included photo was taken in southern Missouri and illustrates the gnarly nature of trees that grow in dry, poor soils.

Form and Size: Dwarf hackberry is a small tree or large shrub that is irregularly branched with slender twigs that make up a narrow crown. Maximum height is 18 to 20 feet. It often has a gnarled, almost grotesque appearance, probably partially attributable to very slow growth in poor soil.

Habitat: There are 2 rather distinct habitats where it is found. One is in sandy, wooded dunes along Lake Michigan; the second is in the unglaciated areas of the Midwest, where it occurs on dry, wooded ridges. These woods usually have fairly open canopies that allow in enough light for the species to survive.

Wildlife Uses: This large shrub has many of the same wildlife values as its congeneric trees, *C. occidentalis* and *C. laevis* (Weeks, Weeks, and Parker 2010). The fruits are eaten by many birds, including cedar waxwings, robins, and, unfortunately, European starlings, but often are not used until well into the winter. Mammals such as chipmunks, squirrels, and raccoons also relish fallen fruits. Although it does not often have the witches'-brooms common in *C. occidentalis*, the often gnarled growth form presents equally good nest placement opportunities for birds.

Landscaping Value: Hardy to Zone 4, this miniature, tree-like shrub is a great, durable plant for small places. It is not particularly showy any time of the year, but it is a "steady" shrub that is extremely drought tolerant. Its disfigured limbs are an interest year-round thanks to small leaves that allow the passerby to see them. If grown in good quality soil, or under less harsh conditions, the gnarled limbs may not develop. There are no horticultural varieties, but it is available occasionally through nurseries specializing in natives.

> **Similar Species Distinctions:**
> —**Sugarberry** (*C. laevis*) is tree-sized and has similar small leaves; however, it is a bottomland species. Their habitats are exactly the opposite.
> —**Hackberry** (*C. occidentalis*) is also tree-sized and has larger leaves that sport nipple galls. It is also considered a bottomland species, although it can be found in more upland sites.

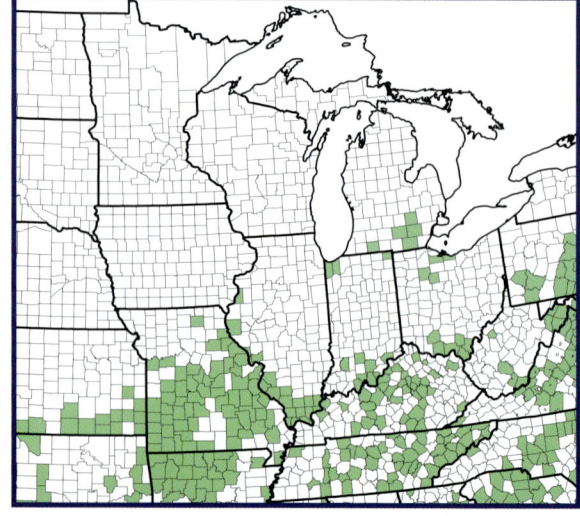

Flowers appear in mid- to late April, are greenish, and develop with the leaves. They are a maximum of ½ inch across and have no petals. Flowers are either single-sex or they have both male and female parts; individual trees have either kind or both. They are short-lived and easy to overlook.

Leaves are alternate, very elm-like, broadly egg-shaped, and thin. The size of the leaves is quite variable, from ¾ to 4 inches in length. The margins are barely toothed and rarely are there any teeth near the base. Both sides are dark green, but the upper surface is usually rough textured, and the lower surface has varying degrees of hairiness.

Fruit is a rounded, reddish orange or purplish, berry-like drupe that ripens in the fall Fruit often persists throughout the winter. Fruit is about ¼ inch in diameter and contains a single hard pit.

Buds have few brownish scales and tend to be more "plump" than those of the other *Celtis* species. The twig tips often die back overwinter, so usually only lateral buds are found.

Twigs are very slender and grayish, with scattered, raised lenticels. Leaf scars are raised and somewhat triangular-shaped. Lateral buds are appressed against the twig.

The bark is smooth and gray with a few raised "warts" that grow thicker and larger as the plant ages.

sweetfern, fern bush

Comptonia peregrina (L.) Coulter
Myrica aspleniifolia L.
Myrica peregrina (L.) Kuntze
Family: Myricaceae

Sweetfern is wonderfully aromatic. All the plant's parts are covered with yellow resin dots that produce this fragrant compound. Currently, sweetfern is on Indiana's "Watch List," is state endangered in Illinois, Kentucky, and Tennessee, and state threatened in Ohio. Unfortunately, white-tailed deer relish its leaves and twigs and are helping to keep sweetfern populations low in the lower Midwest.

Form and Size: It is usually less than 3 feet in height with unbranched, arching stems. Growth is best in full sun, and it is a nitrogen fixer. It is an alternate host of sweetfern blister rust, *Cronartium comptoniae*, which can infect all hard pines. In the Midwest, those pines affected are jack, Virginia, and red.

Habitat: Sweetfern is a colonial, suckering species that is nearly always found in dry, open, sandy areas along roadsides or forest edges. Occasionally, it is found in moist habitats, growing with pin oak. Individual stems are fairly short-lived, but suckers are produced annually.

Wildlife Uses: Sweetfern sometimes forms rather dense stands on sandy soils or in blueberry barrens. In such situations, it provides valuable nesting cover for ground-nesting northern birds, such as hermit thrushes and white-throated sparrows. Occasionally shrub-nesting species, such as field sparrow or clay-colored sparrow, will select these very small shrubs as nest sites. There is only limited use of this species by browsers, although moose and white-tailed deer browse it in winter, sometimes heavily. Cottontail rabbits also browse stems, and ruffed grouse use buds and catkins to a degree. Several species of mammals and birds, including the flicker, have been reported using the seeds.

Landscaping Value: Hardy to Zone 2, this low-growing shrub makes an ideal, hands-off landscaping plant for a border or along walkways. It is low maintenance, easy to grow, and adaptable to many soil types. It has year-round interest with its unusual leaves, its cinnamon-brown, winter-curled leaves and male catkins, and its wonderful scent. Suckers from established plants are easily pulled. Sweetfern is available through some nurseries specializing in native shrubs.

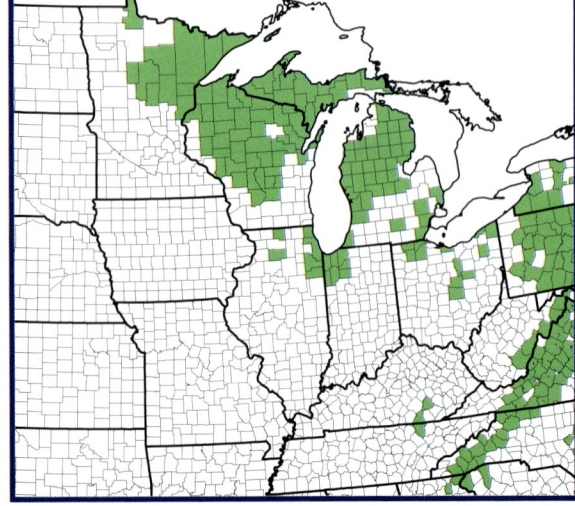

Similar Species Distinctions:
—No other native shrub or even herbaceous plant resembles sweetfern. In any season, leaves or twigs will give off its characteristic scent that immediately identifies it.

Sweetfern leaves resemble fern leaves to some extent. They are alternate, 4 inches long, and linear-shaped, with odd, segmented lobing their entire length. The petiole is very short and hairy. Leaves are dark green above, lighter green beneath, with yellow resin glands and hairs scattered on both sides; glands are very small and even with a hand lens, they are easier to see on the upper surface.

Fruit is a brown nut in a burr-like cluster that matures in early summer. The burr is actually 8 linear bracts and contains 4 nuts. Once ripe, the seeds fall, and the burr deteriorates quickly.

Although sweetfern never achieves a large diameter, its stems are fairly distinctive. The color quickly becomes a deep cinnamon-brown, and the lenticels expand and become more obvious. With age, all the hairs and resin glands fall, and small, gray lines develop along the stem.

The terminal buds are mostly over-wintering flower buds (left). They are larger than the lateral, vegetative buds (right). Male and female buds are mixed along the twig tip, but the larger male catkins are $\frac{1}{3}$ inch long, while the females are somewhat shorter, sometimes only $\frac{1}{8}$ inch long. Both are covered with elongated scales, hairs, and yellow resin glands. Twigs are reddish brown with scattered hairs and resin glands. There are also numerous, scattered, tan-colored lenticels. Lateral buds are ovoid, reddish brown, and covered with 5 to 6 scales that are tipped with silvery hairs. Leaf scars are raised, half-moon shaped, and contain 3 bundle scars in the center.

Flowers appear before the leaves in late April and early May. Males (left) are an inch in length and droop as they stretch and open. Females (right) are nearly that long at maturity and develop deep red, obvious stigma lobes. Individual plants can be separate sexes or have both male and female flowers.

alternate-leaf dogwood, pagoda dogwood

Cornus alternifolia **Linnaeus f.**

Family: Cornaceae

This beautiful, often small-tree-sized dogwood has a unique layered branching that gives it a tiered look. A common name, often found in nurseries, derives from this branching pattern—pagoda dogwood. Fall color has multiple hues, usually ending in a deep burgundy.

Form and Size: This is a large dogwood, often reaching heights of 18 to 20 feet, but normally it is smaller. It suckers and creates sizable clumps over time, with the oldest stems dying back over the years. Its unique, layered branches with the sympodial growth form create an impressive shrub in both natural and horticultural landscapes.

Habitat: Alternate-leaf dogwood is usually a woodland species growing in mesic soils, often at the base of slopes. It can be found on drier sites and even at the edges of poorly-drained, seepy areas. It handles partial shade to full sun and is commonly found growing in the understory of pines.

Wildlife Uses: This species has 2 characteristics that impact its utility to wildlife—its large size and alternate branching. It is a common moist-forest understory species in the northern part of the region, but it occasionally grows to 20 or more feet, especially in forest edge situations—here it is used for nest sites similarly to how flowering dogwood is used, by species such as wood thrush and cedar waxwing. Its blue-black fruits are large enough that small songbirds do not frequently take them, but larger birds, such as thrushes and ruffed grouse, regularly use them. Black bears are said to be especially fond of the fruit. Leaves are often taken by white-tailed deer and twigs by cottontail rabbits.

Landscaping Value: Hardy to Zone 3, this shrub has year-round appeal. Everything about it is pretty, and there is something unique for every season, from the deep reddish purple, up-reaching twigs, to its layered branching, abundant white flower clusters, and gorgeous fall color. It handles just about any soil type and almost any moisture regime. It is a perfect specimen "tree," and it can fill that role in a small yard or patio area. It easily reaches 12 feet tall in 8 to 10 years. A cultivar, 'Argentea,' has variegated leaves. This is 1 of our easiest native dogwoods to find in nurseries throughout the Midwest.

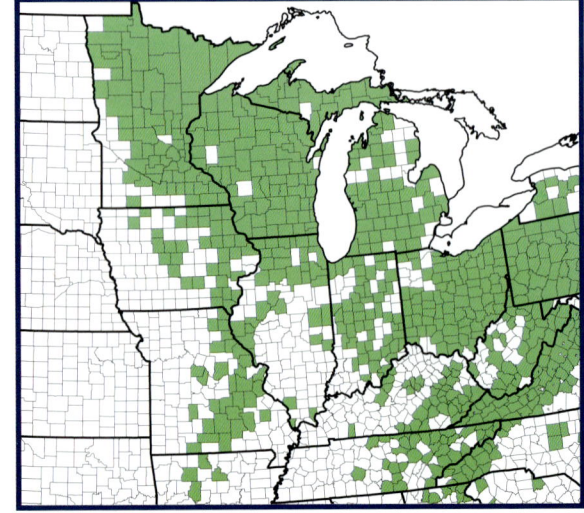

Similar Species Distinctions:
—This is our only native dogwood with alternate branching. Its leaves, with their dogwood-style arcuate venation, have petioles twice the length of any of our other dogwoods.

This dogwood has mostly oval leaves that are about 4 inches long. This is our only native dogwood with alternate leaves. Petioles are quite long, nearly 2 inches or longer, and sometimes as long as the leaf itself. Leaves are grass green, shiny above, and pale beneath. Leaf margins are entire but somewhat wavy. There are usually 6 pairs of veins that grow toward the leaf tip (arcuate venation). Leaves are sometimes clustered at the branch tips.

The fleshy, single-seeded drupe matures in late summer and is bluish black when ripe. Its red pedicels sit in striking contrast to the nearly black fruit. Birds do not allow the fruit to last long. The white, flat-topped flowers appear in May in upright clusters at the end of the branches. The perfect flowers are somewhat ill-scented.

Up-reaching, purplish branches are attractive and an obvious feature in the winter. This growth form is known as sympodial branching.

Buds are ovoid, 3-scaled, and the same dark purplish color as the twig. Twigs are an unusual dark wine-purple, smooth, and shiny. They have scattered, light-colored lenticels. Lateral buds are the same color as the twig, 3-scaled, and somewhat divergent. Leaf scars are V- or U-shaped, tan-colored, and slightly raised. Pith is white throughout.

Bark is smooth throughout a plant's lifetime. When young, the bark is wine-purple with vertical, elongated, tan-colored lenticels. Over time, the bark becomes gray with fine vertical cracks.

Hazels

Genus: *Corylus*

Hazelnuts are the Midwest's only common native shrubs that produce a nut, similar to those produced by oaks, beeches, and hickory trees. A European shrubby species, *C. avellana*, is the hazel (or filbert) of commerce, and it can be found in most supermarkets. Both our native shrubs have been domestically crossed with this European filbert, resulting in larger nuts overall. Fruits of midwestern native species are quite small, and as is usually the case, bigger fruit is considered more desirable.

Two species occur in the Midwest, and 1 of them, beaked hazel, has 1 of the largest natural ranges of any shrub in the United States and Canada. Both species, however, are very com-

mon—so common, in fact, that they are overlooked by all but the wildlife that relish the sweet nuts.

The hazelnuts are multi-stemmed shrubs that can at times produce important cover for wildlife, especially beaked hazel, which at times occurs in relatively dense thickets. Nesting use by songbirds seems less than one would anticipate, but the multi-stem bases often hold the nests of brown thrashers or towhees. Perhaps the most obvious benefit of this genus to wildlife is its production of nuts, as mentioned above. These nuts are heavily used by wildlife, principally mammals such as white-footed mice and chipmunks, but also larger birds, including blue jays, red-bellied woodpeckers, and wild turkey.

Hazels gradually develop next year's male flower buds (catkins) during the summer months. Once the leaves fall, these catkins become very obvious. American hazel, shown here, has winter catkins more than 2 inches long.

Fruit is a meaty nut inside a thin shell that is enclosed in a husk (actually an involucral bract). Our 2 native hazels have very differently shaped husks. Pictured here is American hazel.

Leaves of the hazels are quite similar with doubly toothed margins and hairs, at least on the underside.

The abnormally large bud (on the right) of this American hazel is caused by feeding damage of an *Eriophyid* bud mite. This is a common bud infestation of hazels.

beaked hazel

Corylus cornuta **Marsh.**

Family: Betulaceae

Beaked hazel is *the* hazel of the Northwoods, and it is probably found in every forest and woodlot in the Upper Lake
States—anywhere the tall grass prairie was not the dominant
cover type. It has many characteristics similar to American
hazel, *C. americana*, including an edible nut. Both species are
common in the upper Midwest, but distinguishing between the 2 is usually fairly straightforward. Beaked
hazel lacks the hairs and stalked glands that American hazel has on its leaves and twigs. It is currently
listed as state endangered in Illinois and presumed extirpated in Ohio.

Form and Size: Beaked hazel is a clonal species that expands from a central plant via underground rhizomes. Growth form is basically the same as American hazel, but overall, it is a bit shorter. It can reach
13 feet in height, but it is usually much shorter.

Habitat: Beaked hazel is a generalist of disturbed areas, but it prefers more moist sites than American hazel. It is found in aspen stands and pine stands; it also mixes with northern hardwoods and conifers of all
kinds. It prefers full to partial sun, which is when the best fruit crop occurs.

Wildlife Uses: The wildlife value of beaked hazel is similar to *C. americana*, serving a somewhat different clientele as it becomes more common in the northern latitudes of our region. Beaked hazel has a
greater tendency to produce large clones in areas where it has been long-established, and these thickets
provide considerable protective cover, even for large animals like moose and white-tailed deer.

Landscaping Value: Hardy to Zone 3, this species is more cold-hardy than our other native. Its growth
form is well adapted to a naturalized setting. It handles a fair amount of shade but thrives in full sun. Soil
type is not problematic, but it is more sensitive to drought than *C. americana*. Beaked hazel has been
crossed with the European filbert, *C. avellana*, the fruit of which is known as filazel. Nurseries that sell
beaked hazel are usually in the West where
it is also common.

Similar Species Distinctions:
—**Paper birch** (*Betula papyrifera*)
leaves are similar but not usually hairy
except for the petiole. Its buds are pointed. It is tree-sized.
—**Eastern hophornbeam** (*Ostrya virginiana*) is also tree-sized, and its hairy
leaves are more egg-shaped with elongated tips. Buds are pointed.
—**American hazel** (*C. americana*) has
hairy leaves, petioles, and twigs. Fruit
husk is not beaked.

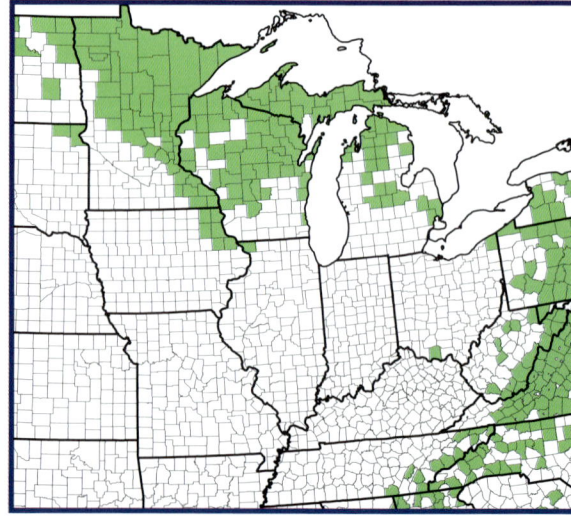

Leaves are up to 4 ½ inches long, dark green, and smooth. Its leaf margins are doubly-toothed and similar in most respects to *C. americana*, except they lack the amount of hairs and glandular hairs, especially on the petiole. The underside of the leaf is paler and may have hairs, especially along the veins.

Fruit is a nut about ½ inch in diameter that is enclosed in a thin, elongated (beaked) husk that is covered with stiff hair. The husk can be several inches long, and caution should be taken when handling it—the stiff hairs can cause a painful irritation. Fruit ripens in the fall.

Flowers appear in the early spring before leaves. Only the reddish stigmas are visible from the top of this female bud. The overwintering tan-colored male catkins stretch to release pollen; they are half the length of those of American hazel—about 1 inch long before stretching.

Twigs are slender, brownish, and smooth, with scattered, pale lenticels. Buds are ovoid to rounded and covered with brownish scales that are mostly smooth, except for the margins that have a row of hairs.

Beaked hazel does not develop much bark character because of its small diameter, but it is dark brown and dotted with pale lenticels that expand slightly with age.

hazelnut, American hazel, filbert

Corylus americana Walter

Family: Betulaceae

This ubiquitous shrub is common throughout most of the Midwest, particularly in roadside fencerows. Its fruit is smaller than the commercial filberts (*C. avellana*) but worth the collection effort if you can beat the squirrels and mice to them. Hazelnut produces fruit sporadically because of its early flowering; it is similar in that respect to oak trees. Bumper crops may be produced only once every 4 to 5 years.

Form and Size: Hazelnut is a sizable, spreading shrub to about 15 feet tall. Typically, it reaches a height of perhaps 10 to 12 feet. Thickets are created by the numerous root suckers that create a dense growth form.

Habitat: This is an unparticular native shrub that is adapted to a wide variety of habitats from wet to dry. It grows quickly and can provide screening when grown as a hedge. It prefers full sun but will tolerate some shade. Individual clumps can last for decades through the annual production of new suckers.

Wildlife Uses: Hazelnut occurs in many early successional habitats, especially along fencerows and roadsides, where its multiple-stem growth form provides valuable cover, especially for nesting songbirds like brown thrashers and cardinals. The nut that it produces is an important food source for mast-feeders in its early successional habitats; they are quickly eaten by white-footed mice, chipmunks, fox squirrels, and ruffed grouse and dispersed by blue jays. Ruffed grouse also eat the winter catkins, and white-tailed deer browse the twigs in winter, although it is not considered a highly preferred species.

Landscaping Value: Hardy to Zone 4, this is an easy-to-grow shrub that could be utilized in a naturalized setting. Size can be controlled if desired by mowing over new growth. It is very adaptable to most soil types, except for those compacted by construction and poorly drained areas. Fall color of hazelnut is various subtle shades of oranges and yellows, and it is probably its best feature aesthetically. Free food is there for the taking, if you can beat the wildlife to it. Several hybrids are available that are crossed with the commercial *C. avellana*; many native-plant nurseries carry hazelnut.

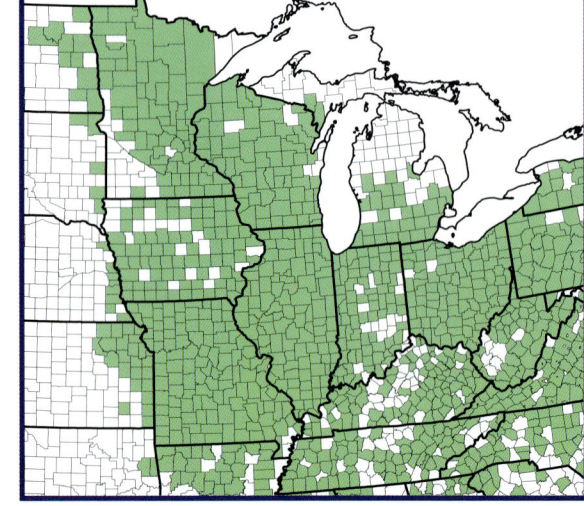

Similar Species Distinctions:
—Several northern members of the birch family can be mistaken for hazelnut; see beaked hazel for comparisons.

Buds are of 2 kinds. The obvious, overwintering male flower buds (catkins) (pictured on page 134) are about 1½ inches long and covered with many reddish scales. They can appear terminally or scattered along the twigs. The vegetative buds (left), and those that produce the female flowers, are plump and covered with numerous reddish scales that have abundant grayish white hairs, particularly along the rims. Twigs are light brown and often densely covered with stiff, dark, glandular hairs, especially on the newest growth. Leaf scars are shaped like rounded triangles and have 3 groups of bundle scars.

Leaves are alternate and may be up to 6 inches long, although more often around 3 or 4. They have an irregular shape but generally are ovate to orbicular. They are long-tipped with irregular lobing on the toothy margin. Leaves are dark green and slightly hairy above, somewhat paler and hairy-glandular below. The petiole is fairly short and covered with stalked glands.

Hazelnut can be the first native shrub to flower in the spring; depending on the weather, they occasionally will flower as early as late February. Our alders and leatherwood have very early flowering habits as well. Both sexes are on the same plant; the male catkins (far left) stretch to release their yellow pollen, while the female buds (near left) open enough to allow numerous, bright red stigma lobes to protrude.

The fruit is a nut that is enclosed by 2 leafy bracts that retract at maturity, exposing the ripe fruit. Fruits are usually in clusters of 3 or 4. Each nut is about ¾ inch long and has what appears to be a "bald head," which is actually the point of attachment within the bract.

Bark is very thin and gray with scattered light-colored lenticels visible on younger stems.

hawthorn

Crataegus **species**

Family: Rosaceae

Hawthorns, generally speaking, are easy to recognize as haw-thorns. However, determining a species of hawthorn, within the genus *Crataegus*, can truly "drive botanists to distraction," as the late Edward Voss stated in *Michigan Flora* (1985). His lengthy and detailed discussion on the genus, as described by him and other experts over the last century, is worth reading, although his taxonomy is somewhat outdated, having been based largely on work done by E. J. Palmer more than 50 years ago.

Identification is as difficult as understanding the scientific explanation for why identification is a prob-lem. It has been described by T. A. Dickinson (2011) this way: "Taxonomic complexities in genera like *Crataegus* involve species concepts, population structure, and reproductive biology. The complexities are especially bound up with the occurrence of gametophytic apomixis and polyploidy." The most current taxonomic work done on the hawthorns is available in *Flora of North America*, Vol. 9 (in press).

For this publication, we chose to simply introduce the genus *Crataegus*, and for those readers truly interested in a challenge, refer you to the *Flora of North America* account. We have to concur with the renowned botanist C. S. Sargent, who, after returning to the United States from South America in 1906, wrote, "you can form no idea how delightful it is to travel in a country where there is no *Crataegus*." For accurate identification, one must follow a specimen (or population of specimens) through the entire grow-ing season, collecting or observing flowers, fruits, flowering branches, and non-flowering branches.

Form and Size: Hawthorns are mostly small trees, but there are several species that grow to 12 to 15 feet in height and produce small colonies by root suckering, traits more typical of shrubs. They are densely branched and usually have spines of some size scattered along the branches.

Habitat: Most hawthorns are disturbance-oriented and are found in areas such as overgrazed or aban-doned pastures, brushy thickets, and forest edges. They prefer full sun, where they have the most abun-dant flower and fruit production.

Wildlife Uses: Because of its typical abundance of limbs and substantial spines, *Crataegus* is an excellent wildlife cover, both escape cover and nesting cover for songbirds. It is frequently used by woodland-edge nesters, such as yellow-billed cuckoos and cardinals. Fruits are generally rather large and dry, not par-ticularly favored by wildlife; however, they often persist into the winter and are used then principally by mammals and larger birds, such as squirrels and wild turkey. Small-fruited species are used in late winter by songbirds, especially robins, cedar waxwings, and car-dinals.

> **Similar Species Distinctions:**
> —**Crabapples** (*Malus* spp.) are small trees that produce fruit about the size of plums, much larger than those of hawthorns. Their leaves usually are not doubly toothed, and they produce thorns (modified branches, usually with buds) rather than spines (long, hard structures, without bark or buds).

Landscaping Value: Hawthorns are utilized in many ur-ban areas, particularly the small-tree-sized species. Shrub-by species could be used in more naturalized settings, but they would probably have to be held in check by mowing. They have beautiful springtime flowers and red fall fruits that often last well into the winter.

Leaves are simple, usually doubly toothed, and often lobed. They are typically ovate in outline.

Fruits ripen in the fall and are usually red. They vary in size from species to species, and they can be smooth or hairy. They look like tiny, red apples.

Flowering occurs in the spring after leaf out. They are usually white and 5-petaled, with numerous (10 to 20) stamens; they are about 1 inch across.

Buds are round, plump, and covered with 6 red scales that are smooth or hairy. First year twigs are reddish brown with an occasional nodal spine. Leaf scars are small, narrow, and contain 3 bundle scars.

Mature bark is grayish or brownish and scaly.

leatherwood

Dirca palustris **L.**
Family: Thymelaeaceae

Leatherwood is 1 or our most unusual native shrubs. As its name suggests, its twigs are as supple as soft leather and are easily twisted into baskets. It makes a nice ornamental with its compact form and pretty yellow fall leaf color. White-tailed deer find it very difficult to browse because of the suppleness of the twigs.

Form and Size: It attains a maximum height of 6 feet and has a thick-branched, uniform appearance. It becomes more common as one moves north into the Upper Lake States, where it stands out in woodland understories with its early springtime flowers.

Habitat: Leatherwood is considered an indicator of a rich, moist site, and it is rarely found growing else-where. It is a woodland species that is not common in the lower Midwest. Partial to full shade is preferred.

Wildlife Uses: The plant's general toxic qualities have been credited by some for relative low wildlife use; this is not likely. The fruits are taken to a degree by birds, but the seed kernel and not the pulp seems to be the focus on use by small mammals, including squirrels. Observers have reported low browse use by white-tailed deer, even in heavily browsed woods; we generally concur, but we have found several cases in which an individual shrub is "ravaged" by severe browsing; almost every twig, including those of large diameter, are chewed off, and stringy, tough bark hangs in shreds. The species seems little used as a nest site by songbirds.

Landscaping Value: Hardy to Zone 3–4, leatherwood is a compact, upright shrub, well suited for a for-mal setting where neatness counts. It retains a single trunk throughout its life and never has branch break-

age or messy fruit to clean up. Fall color is a pretty yellow, and its smooth, gray bark gives wintertime appeal. It is adaptable to heavy soil but susceptible to drought. Full sun is not tolerated by leatherwood. The ear-ly spring flowers and fruits are not usually abundant. Leatherwood is rarely available through native-plant nurseries.

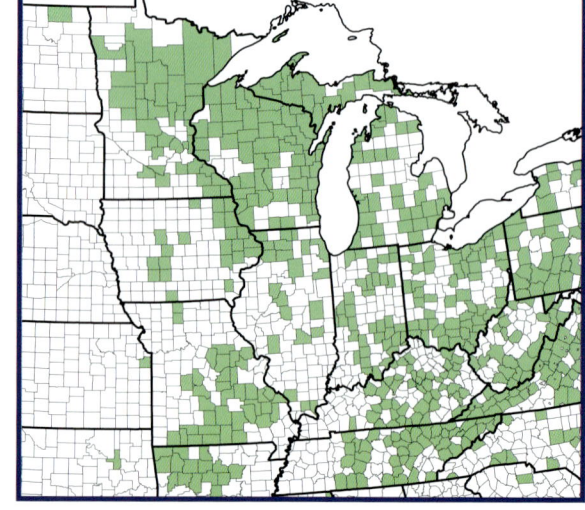

Similar Species Distinctions:
—Leatherwood is so distinct that noth-ing resembles it.

Leaves are very plain-looking, alternate, sometimes oval but mostly obovate, and up to 4 inches long. They have entire margins, and are pale green above, somewhat paler beneath. The very short petiole hides the buds during the growing season; thus, it is hollow at its base where it fits over the bud.

The ½ inch long fruit is a green, single-seeded drupe that ripens in May. Each seed is dark brown. Its fruit is rarely seen, since it ripens so much earlier than those of most woody plants.

The perfect flowers appear as early as mid-March, long before leaf development. The hairy bud scales appear to be leaves grouped at the node behind the cluster of 3 flowers. The light yellow, tubular flowers are just under ½ inch in length and have 8 exerted stamens.

 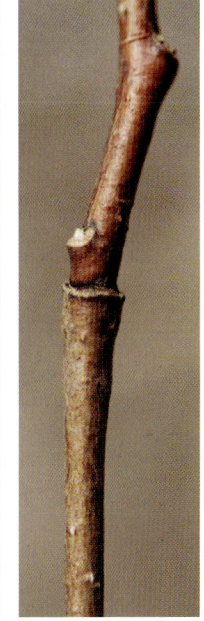 

There are no true terminal buds. All buds are small, somewhat sunken into the twig, dark brown, and pointed at the tip. The bud scales are covered with wavy, brownish white hairs. Twigs are "socket-jointed" from node to node (right). They are smooth and light brown with scattered, light-colored lenticels. Twigs are very leathery and will not break off, but rather will tear. Buds are surrounded by a circular leaf scar that has a single row of bundles scars within it. Older twigs become darker brown and develop some fine fissuring.

The bark of older specimens is usually gray, but it can remain brown like the twig color. As it ages, it becomes roughened with raised, scattered lenticels near its base.

black huckleberry, highbush huckleberry

Gaylussacia baccata **(Wangenh.) K. Koch.**

Family: Ericaceae

Huckleberry is commonly mistaken for our native blueberries. There are several major distinctions, the most obvious of which is the yellow, glistening glands that cover most of the huckleberry plant. The leaf backs are particularly yellow. Springtime brings an interesting color combination with its lime-green-colored new leaves and pink flowers that are dotted with the yellow glands.

Form and Size: Huckleberry is usually less than 3 feet tall, but when growing in moist sites, it can be nearly 5 feet. It is rather spindly and few-branched. It grows well in partial to full shade, and spreads via underground rhizomes, which helps to create small colonies.

Habitat: Huckleberry occurs in a variety of habitats, from dry ridges to bogs. It is common in sandy, black oak woods where it is almost always associated with several species of blueberry.

Widlife Uses: This species is frequently found growing with blueberries, and it has similar wildlife values as small species like *Vaccinium pallidum* and *V. angustifolium*. Its usual low growth form provides nesting cover for many ground-nesting birds, such as wild turkey and eastern towhee, and for low-shrub nesters, such as field sparrows as well. The fruits are eaten by most gallinaceous game birds and several songbirds. Plants are normally browsed very lightly by deer and cottontails, although the latter often heavily impact individual plants by winter browsing.

Landscaping Value: Hardy to Zone 4, huckleberry is so commonly associated of blueberries that most people are unaware it is something different, but its yellow glandular leaves and fruits make it stand out. Grown en masse, the lime-green new leaves along with the bright pink flower clusters are truly beautiful. The higher the quality soil in which huckleberry grows, the larger it will become, but acidity to some degree is necessary. Fall color is spectacular, and winter twigs and buds are deep red trending toward black. No known cultivars are available, and it is only occasionally found at native-plant nurseries.

> **Similar Species Distinctions:**
> **—Blueberries** (*Vaccinium* spp.) can look very similar in most characteristics. If leaves are present, check for yellow, glandular dots on the underside of leaves. If it is not too late in the growing season, huckleberry will have them; blueberries will not. Huckleberry fruits have fewer, larger seeds than those of blueberries.

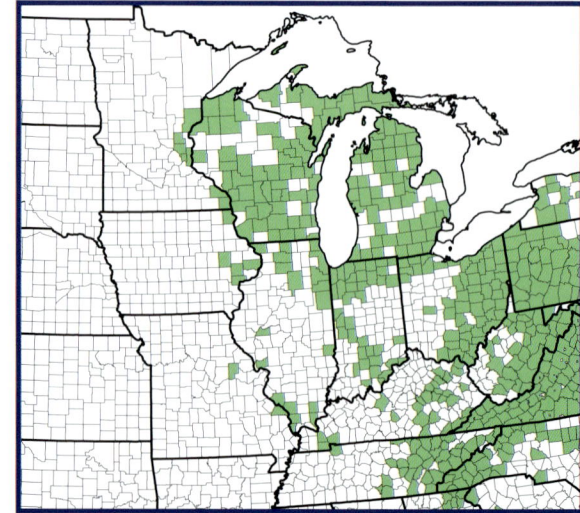

Leaves are alternate, up to 1¾ inch in length; they are typically oval to oblong, but they can vary in shape. The tip is abruptly pointed, and the margin is entire. The upper surface is light green and somewhat hairy with impressed lateral veins; it has yellow glands, but they are sometimes difficult to see, especially later in the growing season. The lower surface is somewhat lighter and hairy with yellow resin dots.

The dark blue-black berry ripens from mid-July into September. It is about ⅓ inch in diameter and has no white, glaucous coating as do some native blueberry fruits. It has 10 hard nutlets inside compared to the fine seeds of blueberries. Fruit is not as palatable as are blueberries because of these larger seeds, but they are still sweet and tasty.

The small, perfect, urn-shaped flowers are only ¼ inch long on short stalks. They are rosy-pink and dotted with yellow resin glands. Flowers appear in mid-April through mid-June.

Bark remains thin, smooth, and dark brown, and it does not develop the scaly pattern of our largest blueberries.

Terminal buds (left) are of 2 kinds. Larger buds are ovoid and covered with several pinkish orange or greenish pink scales; these are flower buds. The smaller, more slender vegetative buds have loose-fitting scales near the tip. Twigs are slender, pinkish red or greenish pink, and covered with scattered hairs. Lateral buds are similar to the terminal buds and diverge from the twig. Leaf scars are raised and somewhat triangular in shape with a single, central bundle scar.

witch hazel, common witchhazel, witch-hazel

Hamamelis virginiana **L.**

Family: Hamamelidaceae

Witch hazel is often found in tree identification guides because of its large size. Though commonly 15 feet tall, it is nearly always multi-stemmed at the base with small diameter trunks—suggesting simply a very tall shrub. An astringent is made from steeping the bark in water.

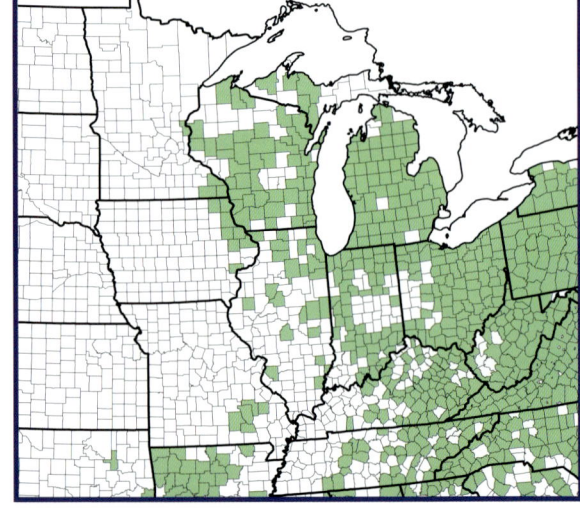

Form and Size: This is a large shrub, sometimes as tall as 20 feet, which has a spreading form when shade-grown. It always has numerous stems of various ages from the same base. Where there is 1 witch hazel, there are usually more, and they often form loose thickets.

Habitat: Witch hazel is common and tends to inhabit wooded slopes. It is found on sandy slopes of black oak woods and in wooded ravines. The authors have never seen it growing in a location other than shaded woods. It is very shade tolerant, but it can handle nearly full sun. It tends to be sensitive to high winds.

Wildlife Uses: Witch hazel occurs in a great variety of wooded locations, and its wildlife value is quite variable as a result. White-tailed deer browse it to a degree, substantially more in the northern part of our region than the southern. Its buds and flowers are eaten by ruffed grouse and gray squirrels; seeds are taken by these species in addition to other game birds and songbirds. Bark is eaten by rabbits and beaver on occasion. When witch hazel forms thickets, as it often does in bottomlands, it provides cover for even large animals, such as black bear and deer; in addition, limbs overhanging streams are occasionally used as nest sites by eastern kingbirds and acadian flycatchers.

Landscaping Value: Hardy to Zone 3, this easy-to-grow large shrub is not particular about soil or sun, but it is a little sensitive to drought. It makes an interesting specimen plant in a shaded area but needs room to spread. Witch hazel can grow in nearly full sun, where it develops a dense, upright form. Fall color is a beautiful, deep yellow. Flowering begins in mid-October, but if flowers open while the leaves are turning, they get lost in the foliage. Both witch hazel and *H. vernalis* are available through nurseries specializing in natives.

Similar Species Distinctions:
—**Ozark witch hazel** (*H. vernalis*) is found in the Ozarks of Missouri, Arkansas, and Oklahoma. It has velvety-hairy twigs, hairy lower leaf surfaces (sometimes), and flowers in early spring.

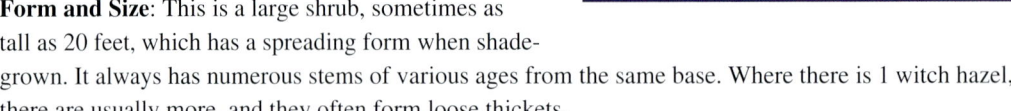
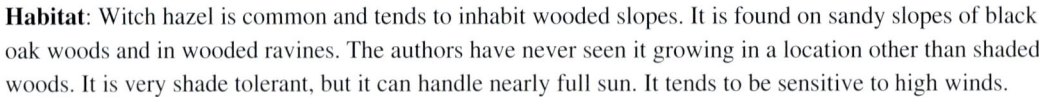

Witch hazel leaves are alternate, oddly shaped, and look much like a lopsided fan. They are oval to obovate in outline with irregular, wavy, or toothed margins. They are up to 6 inches long with a blunt tip and lopsided base. The leaf surface is green above, paler below, and mostly hairless. Petioles are short.

The buds are tannish brown and covered with mostly very short hairs that range in color from beige to rust. The buds are termed "naked," because they lack true scales. There are nearly always 2 buds per node; the largest 1, the vegetative bud, is elongated, up to ⅓ inch long, and appears to be stalked. The other, the flower bud, is very small, only ⅛ inch, and is positioned at the base of the stalk. The 2 buds give the appearance of a boot when turned upside down and made to "walk" on a flat surface. The more vigorous the twig growth, the more elongate the vegetative bud. Twigs are slender, tannish-colored, and covered with hairs of various lengths and colors ranging from tan to rust, especially on the newest growth. Scattered, light-colored lenticels are not obvious because of the twig color, but they become more apparent over time. The older twigs turn gray, and shaded, slow-growing twigs grow in a zig-zag fashion from node to node, giving a stairstep appearance. Leaf scars are shaped like a rounded triangle with 3 bundle scars.

Fruit is a woody, hairy capsule that ripens during the fall a year after flowering. Each capsule is about ½ inch long, and when ripe, splits down the middle and forcefully ejects its shiny black seeds many feet from the plant.

The perfect, 1 inch wide, yellow flowers appear in late October after leaf-fall, which gives this witch hazel the distinction of being our latest flowering shrub. Each flower has 4 long, slender, crinkled petals.

Mature bark is gray or sometimes brownish and mostly smooth. The oldest shrubs develop a roughened, scaly texture near the base. Tan-colored lenticels become obvious along the gray trunk with age.

woolly hudsonia, false heather, beach-heath, beach-heather

Hudsonia tomentosa **Nutt.**

Hudsonia tomentosa **var.** *intermedia* **M. Peck**

Family: Cistaceae

Hudsonia grows in pure sand on dunes mainly along the Great Lakes. It is a less than spectacular shrub to behold, *except* when in flower. Then in June it lights up the sand with its bright yellow profusion of flowers. In the more southerly regions of its range, it has been subject to habitat loss through housing and industrial construction. It is listed as threatened in Indiana, endangered in Illinois and Iowa, and presumed extirpated in Ohio.

Form and Size: Hudsonia is of short stature, rarely a foot tall. It is most often wider than it is tall with numerous, densely packed branches. Growth is slow—no more than 2 inches a year.

Habitat: It grows in pure, well-drained, sandy soil along the Great Lakes. It is adapted to shifting sand dunes along the Lakes, but it is sometimes found in more stable sand deposits of old glacial lakes and on sandy terraces along the Mississippi River. This is a species adapted to disturbance, and it does not compete well with other vegetation.

Wildlife Uses: This diminutive plant, which occurs in the most barren of habitats, has minimal wildlife value. While it is likely that some browsing and seed use occurs, it has not been observed. It does, however, represent the only vegetation in most of its habitats and thus serves as escape and nesting cover for deer mice, song sparrows, vesper sparrows, and other small, dune inhabitants.

Landscaping Value: Hardy to Zone 2, this plant is a specialist of acidic, sandy, well-drained soil. It is breathtaking in late spring when in full flower, but otherwise it is a gray-green, low growing shrub that has little seasonal interest. In fall and winter, it appears rather dead. Pruning helps it develop a more dense growth form, and full sun is required for best appearance. According to William Cullina in his book *Native Trees, Shrubs, and Vines* (2002), semi-hardwood tip cuttings can be taken in the summer and rooted in a 1:1 ratio of peat and sand if mist or humid growing conditions are provided. They root easily if treated with 1000 to 2000 ppm IBA. Hudsonia is not available in midwestern nurseries.

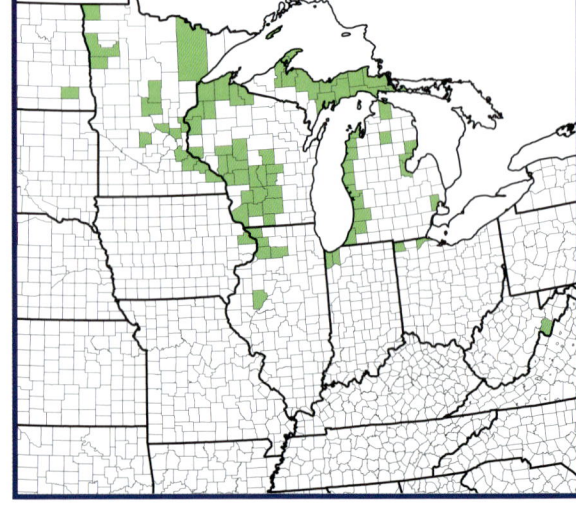

Similar Species Distinctions:
—There is no plant that resembles hudsonia at any time of the year. Another species, *H. ericoides*, which is essentially hairless, is found along the East Coast.

Leaves are small (only ⅛ inch long), overlapping, and scale-like. They are numerous and cover much of the tiny twigs. They are dark green during the growing season with many whitish gray hairs covering them. These dense hairs give the plant a gray-green appearance that helps it blend in well with its sandy habitat. Scattered plants are often difficult to find even in the summer months. Leaves are persistent throughout the winter but turn brown in the fall. New growth is mainly from or near the tips of the twigs, leaving the inner portion of the plant brown.

The perfect, bright yellow flowers are less than ¼ inch across but are produced in large quantities on established plants. Each flower has 5 petals and numerous yellow stamens. They can occur over an extended flowering period, but mainly flower in June. Records from the Chicago region list flowers occurring from May 23 to August 22.

Fruit is a small, 3-angled capsule that is nearly stalkless. It matures shortly after flowering and contains 1 or 2 seeds. Fruit is not persistent, and it is gone before the summer is through.

The plump, winter buds blend in with the numerous, overwintering leaves because of their brown, hairy scales. They are mostly terminal or along the newest twig growth.

The youngest twigs are extremely slender, brown, and covered with leaves. Older growth is reddish brown and shiny with a few dead leaves or scales along them. Older growth has no lateral buds.

Deciduous Hollies

Genus: *Ilex*

If asked for a description of a holly, most people would automatically think of a plant that has evergreen, sharp, spiny leaves. However, in the Midwest, all native shrub hollies are deciduous. We have 4. One species, mountain-holly, has been "lumped" by some taxonomists into the genus *Ilex*, altering its long-standing designation as *Nemopanthus*. A species not covered here, *I. montana*, occurs mainly in the Appalachian Mountains, and it sneaks into eastern Kentucky and Tennessee. It closely resembles *I. verticillata,* but it is usually a large shrub or small tree and has fruit twice as large.

Hollies are beautiful additions to a landscape, especially in the late fall and early winter when their abundant, bright red fruit seems to signal the coming of Christmas. In other seasons of the year they are less spectacular. When planting hollies for fruit production, one needs both male and female plants in proximity so cross-pollination can occur. Most hollies require a slightly acidic soil, and leaves will become chlorotic if this need is not met.

Our deciduous hollies are strongly associated with water and/or areas of moist soils. As such, they supply food and cover to a broad spectrum of wildlife species, occasionally including waterfowl. Growth form is generally dense and limby, yielding a substantial amount of cover, although they only rarely occur in colonies that would enhance that value. Fruit production is high, but fruits do not seem highly preferred. They last into the winter when their value and use is increased as more palatable species are usurped, and the fruits are rendered more palatable through repeated freezing and thawing.

The dry fruit, which is rarely taken by wildlife until at least early winter, is the main attraction from a landscaping standpoint. Hollies, like this possumhaw, always produce red fruit in small clusters along the twigs.

Older twigs, such as those on this possumhaw, are gray and no-
ticeably bare in the winter because of their deciduous nature. They
develop spur-shoot-like growth, seen here, that produces closely
spaced leaves.

Mature bark of all the hollies is very similar in appearance. It is essentially light gray
with numerous, raised, rough lenticels. Pictured is mountain-holly.

possumhaw, possum-haw, swamp holly

Ilex decidua **Walter**

Family: Aquifoliaceae

Possumhaw is a southern deciduous holly. It occurs along the southern border of Illinois and Indiana in the swampy regions of the lower Wabash Valley and Ohio River, where it reaches the northernmost extent of its range. Simpson Nursery in Vincennes, Indiana, sells beautiful, locally collected forms of possumhaw. Quite often possumhaw retains its fruits into the spring; they can be found persisting after leaf out and while the plant is in flower.

Form and Size: It can become 10 feet tall in the Midwest, but it is larger in the South. It is much branched with sometimes strongly angled lateral twigs. Root suckering can produce small clumps of possumhaw, but single-stemmed specimens are commonly found.

Habitat: Possumhaw is restricted to swamps, sloughs, low woods, and borders of ponds. It grows best in moist, acidic soil, but it can be found, especially farther south, on drier, rocky sites. It tolerates shade but becomes gangly. When growing in full sun, it becomes much fuller and produces larger quantities of fruit.

Wildlife Uses: Although it occurs in other habitats, possumhaw is most frequent and abundant along edges of streams or ponds. In such areas they supply nesting cover for water-associated birds, such as red-winged blackbirds—their form is conducive to such uses. Major value for wildlife, however, has to be in food production. It is browsed by white-tailed deer, but generally only in the spring; new sprouts that are stimulated by any type of disturbance are especially attractive. The generally large quantities of fruit produced by females are not highly preferred and remain into the winter when they are heavily used by mammals (including opossums) and many songbirds and game birds. Once use begins in the late winter/early spring, the fruits, softened and fermented by freezing and thawing, are quickly devoured.

Landscaping Value: Hardy from Zones 6–9, this holly becomes tree-sized in the southern portion of its range. In the Midwest it is a suckering, multi-stemmed, large shrub with beautiful, gray bark. It normally grows in wet sites but adapts well to uplands. Wintertime is when possumhaw shows off with its clustered red fruit against the gray branches. There are several cultivars available, including 'Warren's Red,' which has deep green, shiny leaves and is a zone hardier.

Similar Species Distinctions:
—**Winterberry** (*I. verticillata*) is common in the upper Midwest and has finely toothed, slightly hairy leaves that have longer petioles.

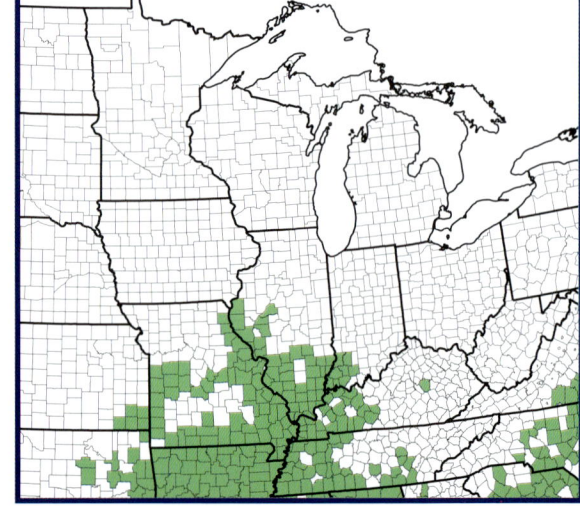

Fruit is a red, berry-like drupe that is only ¼ inch across. It matures in early fall and can persist throughout the winter. Each fruit has 3 or 4 ribbed seeds.

Leaves are alternate, lanceolate to obovate, and up to 1½ inches long. The leaf tip is blunt, and the base gradually tapers to a short, purplish, grooved, hairy petiole. Leaves are green and shiny above, paler and hairy beneath, especially along the midrib. Leaf margins are shallowly scalloped. Leaves tend to be clustered on the older spur shoots.

Flowers are perfect or single sex and appear in early May. There are 4 to 6 (usually 4) white petals on the tiny, ¼ inch diameter flowers that appear in clusters. The tiny sepals are entire. Flowers appear before or with the leaves. Males pictured on left; females on right.

The youngest twigs are greenish, but they soon become silvery gray or brownish gray and smooth. Older, slow-growing twigs develop spur shoots (pictured on page 151), where leaves grow close together. There are scattered lenticels along the twigs. Leaf scars are somewhat U-shaped and contain a single, central bundle scar. Small black stipules (use hand lens) flank the base of the petiole or the leaf scar.

Buds are reddish brown and few scaled. They can be very blunt-tipped or have a rather elongated, sharp point, especially on vigorously growing twigs.

Bark is gray and smooth, except for the numerous, raised lenticels and warty protuberances that are scattered along the trunk.

mountain-holly, common mountain-holly, catberry

Ilex mucronata (L.) Powell, Savolainen & Andrews
Nemopanthus mucronatus/mucronata (L.) Trelease
Family: Aquifoliaceae

This species was recently placed in the genus *Ilex*, but it does not really resemble other midwestern hollies. Its unique leaf feature is a very slender, purple petiole that, along with the leaf blade, resembles a canoe paddle. Spring color is an overall pale appearance because of the very light green new leaves and the numerous yellowish flowers.

Form and Size: Mountain-holly can be a large shrub, up to 16 feet in height. It commonly forms thickets from root suckers.

Habitat: It is found in swampy places, on lakeshores, and commonly in tamarack bogs. It does tolerate partial shade, but it grows best in full sun. Mountain-holly requires an acidic soil. It can be found in, and will tolerate, more upland sites.

Wildlife Uses: Mountain-holly generally grows in moist areas, such as bogs, lakeshores, and streamsides where it provides nesting cover for species like red-winged blackbirds and song sparrow that occupy such areas. Deer browse the species to a degree, but the major food value is its fruits, which are taken by a wide variety of birds, including ruffed grouse, pileated woodpeckers, and several thrushes, and by mammals, from black bears to white-footed mice.

Landscaping Value: Hardy from Zones 3–6, this species does not grow well in the heat and humidity of the lower Midwest, and it is limited by habitat as well. It prefers moist, acidic sites, and it is difficult to establish elsewhere. It has been described as having "stubby" twigs. The bright fruits are beautiful, but they do not last long before they are eaten by wildlife. Flowers appear with the leaves and are not particularly showy. This is just an unusual native shrub that would make a nice addition to any moist setting for naturalizing. It is rarely found commercially, even in nurseries specializing in natives. Authors purchased it from Klyn Nursery in Perry, Ohio.

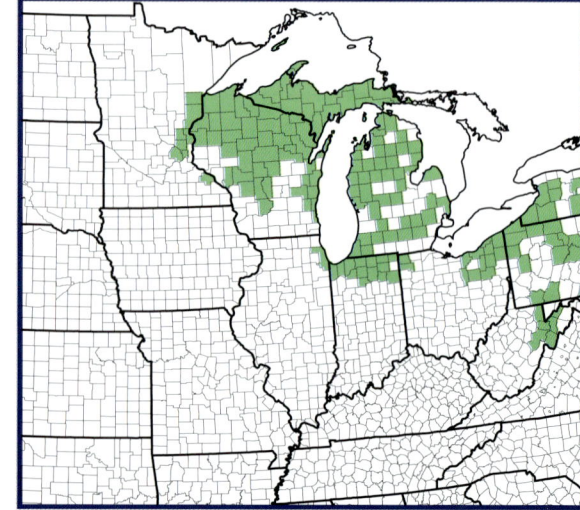

Similar Species Distinctions:
—There is nothing like it from a leaf standpoint. **Appalachian mountain-holly** (*I. collinus*) is a rare shrub of West Virginia to North Carolina that has somewhat larger, veiny leaves with glandular, toothy margins.

Leaves are alternate, elliptic-oblong or oblong-oval, and up to 2 inches long. They are green and smooth above, pale and smooth beneath. The margin is usually entire, but there can be a few small teeth. The leaf tip is blunt, and the base is tapered. Petioles are up to 1 inch long and purplish with no hairs.

Fruit is a bright red (occasionally yellowish) drupe, on long peduncles of an inch or more. It is ¼ inch in diameter and ripens by late July. Fruit can persist throughout the winter if not taken by wildlife.

The tiny, ¼ inch wide flowers appear in May with the leaves and are much less showy than our other hollies. They appear on long pedicels, up to 1 inch or more, and have tiny, yellowish-white petals. Flowers are in axillary clusters. Individual specimens usually have single-sex flowers, but flowers can be perfect. A female is pictured to the far right; male flowers are shown on near right.

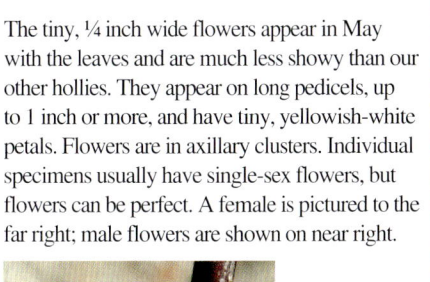

New growth is dark burgundy-red, shiny, with many scattered, light-colored lenticels. Lateral buds diverge from the twig and are the same color as the twig. Older twigs develop a silvery-gray coating. Leaf scars are V-shaped and contain a single bundle scar.

Buds are similar to those of *Ilex verticillata*. They are ovoid, with 2 to 3 reddish scales, and somewhat teardrop–shaped. They are more pointed than in winterberry.

Bark is thin, gray, and smooth, except where the numerous lenticels have become raised and rough, appearing almost Braille- or barnacle-like.

winterberry, black alder

Ilex verticillata (**L.**) **A. Gray**

Family: Aquifoliaceae

Winterberry is 1 of the Midwest's better known native shrubs, and it is available from many nurseries. The female plants are best recognized in the winter when leaves are gone (this is a deciduous holly) by the bright display of clustered red berries. It is particular about where it grows, however, as it likes acidic soil. Fall color is an unusual yellow, which creates quite a visual impact when combined with the red fruit. Winterberry is listed as endangered in Iowa.

Form and Size: Winterberry develops large clumps in the wild, with individual specimens often reaching a height of 15 feet. It is a dense, multi-branched species. Those growing in full sun develop a full, rounded form.

Habitat: It is commonly found bordering lakes, in swamps and marshes, and in bogs and wet woods. It tolerates partial shade, but it grows more densely and produces more fruit when given full sun. It can become chlorotic if not provided an acidic soil.

Wildlife Uses: The nesting cover supplied by winterberry is substantial and used by many shrub-nesting songbirds, especially species associated with edge and moist habitats. Protective cover is even supplied in winter, especially if shrubs are growing in a group or grouped with other wetland species. Fruit use is similar to that of possumhaw, a delay in major utilization until mid- or late winter, when use by birds and mammals seems to explode. Because it is often associated with water, use by wood ducks and other waterfowl is common.

Landscaping Value: Hardy to Zone 3, this species prefers moist, acidic sites, but it is adaptable to drier sites with more neutral pH. Shrub borders and mass plantings with a mix of both sexes work well, especially to ensure that females produce fruit. Fall color can vary from a sickly greenish yellow to burgundy. In winter it is particularly delightful when females are covered with bright red berries in contrast to the gray branches. There are many horticultural varieties of this species.

Similar Species Distinctions:
—**Possumhaw** (*I. decidua*) has leaves with scalloped margins, short, purple petioles, and mostly hairy leaves. Its range is more southerly in the Midwest.
—**Bigleaf holly** (*I. montana*) is a species of the Appalachian Mountains with leaves nearly twice the size of winterberry.

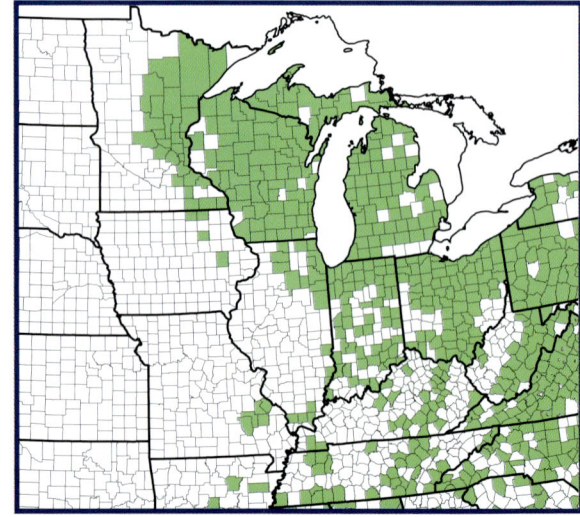

Leaves are alternate and oval, lanceolate, oblanceolate to broadly obovate. They can be up to 4 inches in length but are usually shorter. They are dark green and mostly hairless above, with numerous, deeply impressed veins over the upper surface. The lower surface is paler and often has hairs running along the veins. Margins have sharp teeth, and both leaf tip and base are somewhat abruptly pointed. There is a defined, lengthy, green, hairy petiole.

Fruit is a berry-like drupe that is bright red when ripe in the fall. They persist throughout the winter unless taken by birds. Fruits are ¼ inch in diameter and clustered.

The tiny, white flowers appear mainly in June in axillary clusters. They have 4 to 8 petals and are only ¼ inch across. Male (left) and female (right) flowers are on separate plants.

New twigs are deep red-purple, but they quickly develop a silvery-gray coating. There are numerous light-colored lenticels scattered along the twigs. Leaf scars are basically U-shaped and contain a single, central bundle scar. Lateral buds are like tiny nubbins that project away from the twig and are often superposed (stacked). Small black stipules (use a hand lens) occur on either side of the leaf scar (or petiole attachment).

Buds are covered with several reddish brown or tannish-brown scales; some of the outer ones occasionally appear loose. They are more ovoid-shaped than those of possumhaw. On older twigs, the buds can become silvery-gray.

Bark is gray and smooth between the lumpy lenticels that enlarge with age. Older specimen's thin outer bark splits and creates almost a diamond-like pattern.

Virginia willow, sweet-spire, Virginia sweetspire

Itea virginica **L.**

Family: Grossulariaceae

Virginia willow is not a true willow, and it is a southern shrub of the swamps. It has made its way into the horticultural trade because of its ability to thrive in upland sites and for its gorgeous flowers. It is at the northernmost limit of its range in the Midwest, and it is listed as state endangered in Indiana and extirpated in Pennsylvania. This species can be found under the family Saxifragaceae in older literature.

Form and Size: In the Midwest, Virginia willow probably never exceeds 4 to 5 feet in height, but it can grow much larger in the subtropical South. Plants are usually rather spindly when found in the wild, but they do sucker from roots, so small clumps can be found.

Habitat: It is a shrub of shady swamps, where it grows in acidic, poorly drained soils. In the wild it is usually found growing in nearly full shade, but it does well in nearly full sun, where it becomes much more branched. It will tolerate upland sites but requires acidic soils.

Wildlife Uses: This bottomland species is widely scattered and seems to be common nowhere in our region. It typically grows under a forest overstory, and its spindly form gives little cover; when it grows in the open, it is occasionally used as a nest site by songbirds. White-tailed deer browse it heavily and have likely eliminated it from some sites in our region. Its seeds are used sparingly by wintering sparrows.

Landscaping Value: Hardy to Zone 5, this is a very adaptable species and seems to thrive in upland sites. It becomes a bit chlorotic if soil is too alkaline. Full sun to partial shade is acceptable—it will flower just fine. But the more sun it is given, the better show Virginia willow puts on, especially in the fall. It has a flowering period of several weeks and is quite attractive, even at a young age. Winter twigs are an unusual red-green combination. There are several horticultural varieties available, including the more petite 'Little Henry.'

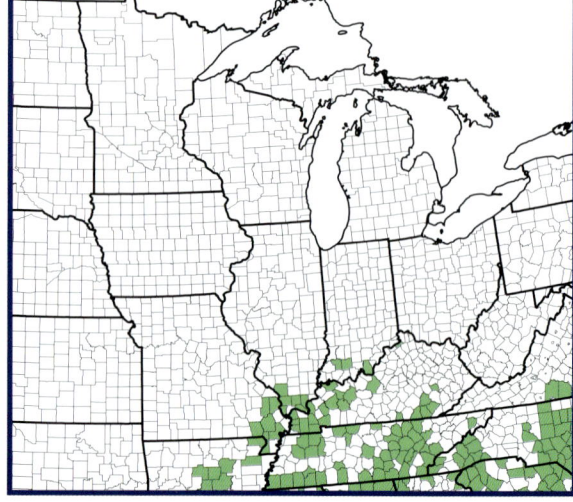

Similar Species Distinctions:
—There is nothing quite like Virginia willow. From its leaves to its twigs, it is unique.

Leaves are alternate, up to 3 inches long, and elliptic to oblong-lanceolate with finely toothed margins. They are deep grass green and smooth above, somewhat paler and smooth beneath. There are usually only 4 pairs of lateral veins per leaf. The leaf tip is long-pointed, and the base is gradually tapered to a short petiole.

Fruits are dry, 2-celled, pointed capsules that populate drooping clusters that are about 5 inches in length. Fruits ripen in the fall, but they persist throughout the winter. Capsules are covered with very short hairs.

The mostly white (occasionally pinkish), perfect flowers appear mainly from mid- to late May. They hang in terminal, drooping clusters (racemes) that are up to 5 inches long. They gradually open from base to tip over a 2-week period.

Buds are small, ovoid, few-scaled, and the same color as the twig, which can be greenish to reddish. They are more or less lightly covered with short hairs.

Twigs are slender and reddish, greenish, purplish, or a combination of these. They can be smooth or somewhat hairy, especially on the youngest growth. There are scattered, tan-colored lenticels along the twigs that often develop "cracks" from top to bottom that give the appearance of an elongated lenticel. Leaf scars are similar to a rounded triangle that contain 3 bundle scars that look face-like. Lateral buds diverge from the twig and are commonly superposed (stacked) above the leaf scar. Twig pith is chambered and white.

Bark is thin, brownish or grayish, and generally smooth, especially this far north where individual specimens do not acquire much diameter. Farther south, where it grows much larger, bark can become somewhat scaly. Since plants are usually growing in areas of frequent flooding, bark is often covered in mud.

spicebush, spice-bush

Lindera benzoin (L.) Blume
Benzoin aestivale (L.) Nees
Family: Lauraceae

Spicebush is 1 of our most common woodland shrubs, and it is an indicator of rich, moist soils. Its fire engine-red fruit was used as a substitute for allspice during the Revolutionary War, when it was known as allspice bush. Over the years, the name was contracted. All parts of the shrub have a spicy, "lemon pledge" scent. A recently introduced ambrosia beetle that entered the United States through the Savannah, Georgia, port in 2002 attacks and is killing members of the Lauraceae family by spreading laurel wilt. This includes those in the genus *Lindera*. Whether the beetle will continue to spread and be able to survive midwestern winters has yet to be seen.

Form and Size: Spicebush is a large shrub that can reach 15 feet in height. It develops a wide-spreading, open crown from several large basal stems.

Habitat: It is common in moist-to-wet woods, even poorly-drained areas, and along streams. Rarely is it seen growing in the open; it is a shade-loving species. It does handle partial sun and may form dense thickets, especially after disturbance. It prefers moist sites and is sensitive to drought.

Wildlife Uses: This is a woodland species, occurring most commonly in bottomland forests, where it is found as a scattered shrub that has little cover value for wildlife. Larger shrubs are occasionally used as nest sites by mid-canopy species, such as wood thrush. The fruits are quickly taken by ruffed grouse, wild turkey, and many songbirds, including the thrushes, which seem especially fond of them. Deer and rabbits occasionally browse leaves and twigs, but it is not considered an important browse species.

Landscaping Value: Hardy to Zone 4, spicebush is a large, common midwestern species that has several worthy attributes. It flowers well before the woods "green up," making the yellow flower clusters easy to see. Spicebush has pretty, yellow fall color, and its leaves are preferred food of spice-bush swallowtail butterfly larva. The brightly-colored fruit is especially attractive but does not last long. Winter twigs are usually greenish and nicely scented when scraped. Branches may be pruned in late winter to promote new, full growth. No cultivars are available, but spicebush is available through some native-plant nurseries.

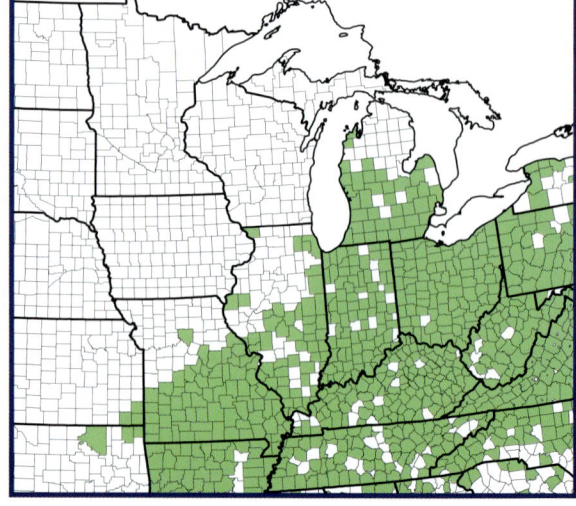

Similar Species Distinctions:
—No other natives are similar to spice-bush. Flowers are similar to those of **sassafras** (*Sassafras albidum*), a tree.

The bright yellow flowers emerge in clusters in March, well before leaves appear. They are only ¼ inch across and have 3 petals and 3 sepals, which are indistinguishable from each other. Flowers are unisex and usually occur on separate plants.

The largest leaves, which tend to be toward the tips of twigs, are up to 6 inches long, but they are usually a more modest 2 to 3 inches. They are oblong-ovate to oval; however, the lower leaves are much reduced and oval to nearly orbicular. All leaves are pointed at both ends and have entire, slightly wavy margins. They are dull green and smooth on the upper surface and pale green beneath. There can be some degree of hairiness beneath, especially along the main veins.

The bright, shiny, fire engine-red fruit ripens in mid-fall and is about ⅓ inch long. They are clustered in groups at the nodes. Spicebush fruit is highly preferred by birds, so it rarely lasts into the winter.

Buds are of 2 kinds. The vegetative buds are ovoid and few-scaled. They are pinkish green to blackish green and are usually mixed with a few to several preformed flower buds. The flower buds are rounded to ovoid, few-scaled, and appear to sit on top of short pegs. They are the same color as the vegetative buds. The more light the plant gets during the growing season, the more over-wintering flower buds it sets for the following spring.

Twigs are slender, smooth, and greenish tan or greenish brown; they quickly become dotted with light-colored lenticels. Leaf scars are U-shaped and have 3 bundle scars. Lateral flower and vegetative buds sit clustered at the top of each scar.

Mature bark becomes dark brown with a hint of green in the right light. There are many raised, light-colored lenticels occur in vertical columns along the stem.

Allegheny pachysandra, Allegheny spurge

Pachysandra procumbens **Michx.**
Family: Buxaceae

Allegheny pachysandra is a small sub-shrub that grows in extensive colonies along wooded streams. Its leaves remain somewhat green throughout the winter, but they are pretty much flattened and worn out from the elements by spring, at least in the northern part of its range. Farther south, it is evergreen.

Form and Size: Pachysandra is a small, colonial ground cover that never grows taller than 8 to 10 inches.

Habitat: Pachysandra prefers moist, rich, wooded sites, and it is found at the base of, and on the lower parts of, wooded slopes and along wooded streams. It does not do well in full sun, especially hot, summer sun, and it appears to dislike substantial direct light of any kind except what it receives in early spring before canopy closure. It spreads to form an open ground cover—a novice would probably mistake it for an herbaceous plant.

Wildlife Uses: This species is so rare throughout our region that its overall impact on wildlife is minimal. However, where it does occur, it usually covers the ground with partially evergreen foliage less than a foot tall that supplies protective cover for small mammals and nesting cover for ground-nesters such as the Kentucky warbler. Deer evidently do not browse this species, likely because it has secondary compounds (as does its congener, Japanese pachysandra, *P. terminalis*) that deter herbivory.

Landscaping Value: Hardy to Zone 5, this small sub-shrub is just barely woody. It has large, upright spikes of flowers in the early spring before any leaves can block them from view. Full sun turns the leaves yellowish green by mid-summer. It grows well in upland sites and makes a nice border plant along a walkway. This is a highly desired, native ground cover that is a great replacement for the exotic and commonly planted Japanese pachysandra.

Similar Species Distinctions:
—This species would most likely be mistaken for herbaceous species, but which ones, we are not sure.
—**Japanese pachysandra** has truly evergreen leaves, terminal inflorescences, and white, fleshy berries.

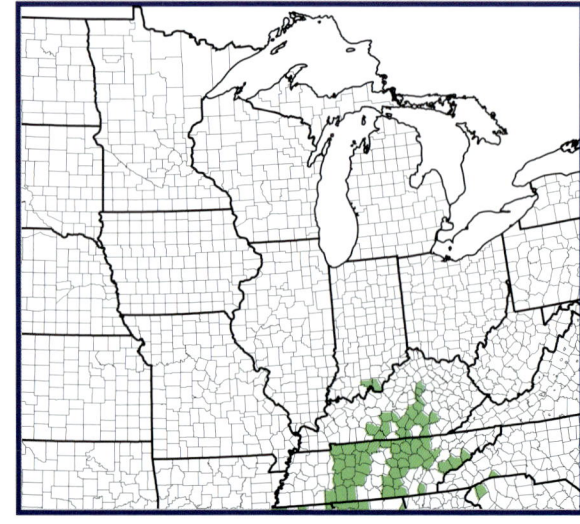

Stems are slender, green, and covered with very short hairs. There are small lateral buds that diverge from the stem. Buds have reddish, overlapping scales. Flower buds seen here begin developing during the summer months and continue to enlarge into the fall.

This semi-evergreen, alternate-branched sub-shrub develops new leaves from the base of the plant in the spring after flowering occurs. Leaves are mostly 3-lobed, but can be 2- or 5-lobed, and are clustered near the top of the stem. An average size is 2 inches wide by 2½ inches long, with a 1½-inch-long petiole. They are green and very lightly hairy above, somewhat paler and lightly hairy beneath, mainly along the veins. The petiole is lightly hairy as well. Seen here (right) are the new spring leaves.

Fruit ripens in mid- to late summer and is seldom produced in the wild; the 3-pointed, fleshy capsule splits along each segment to release shiny, black seeds. Capsules are ½ inch in diameter. The authors have seen ants chewing into the capsules and carrying away seeds.

P. terminalis, Japanese pachysandra

If produced, clusters of flower buds lay near the base of plants over winter.

Flowers appear in April in 2- to 4-inch-long, brownish spikes that originate from the center of the plant. Male and female flowers are separate on the same spike; the females (a single one seen in photo on left) are positioned only nearest the base. The males have 3 to 4 white, fleshy petals that are ⅓ inch long.

common ninebark

Physocarpus opulifolius **(L.) Maxim.**

Family: Rosaceae

Ninebark is usually not a common shrub, but it occurs locally in scattered sites, often along stream banks. It is a beautiful shrub in flower, and it has interesting, red-colored, ripening fruits in the summertime. This native grows very large if protected from competition in a managed landscape, but there are several horticultural varieties that have much smaller stature.

Form and Size: Ninebark can become quite tall, occasionally reaching 9 feet in height. It has long, arching branches that produce a multitude of fine branchlets, giving the shrub a dense summer appearance. Plants sucker vigorously at the base, and mature plants become very full. Winter appearance is more ragged, but the peeling bark and persistent fruit add appeal.

Habitat: Ninebark is fairly intolerant of shade, and it is found along marsh edges, in low, open woods, and on the slopes and banks of streams. It tends to tolerate a wide variety of soil types and moisture regimes. When grown ornamentally, it can vigorously spread in the surrounding area through seed germination.

Wildlife Uses: One of the most distinctive features of this shrub is the large clusters of seeds contained within papery bladders (the origin of the generic name); while moderate use of seeds by birds has been reported, the authors have never observed use of these seeds by songbirds. Perhaps dealing with the bladders makes foraging inefficient. The plant is rarely browsed by cottontails or white-tailed deer. However, from a cover perspective, ninebark is excellent, often growing densely enough to supply cover even for deer and wild turkeys. The shrub's limb structure makes it outstanding nest cover for many shrub-nesting songbirds, and it is highly preferred.

Landscaping Value: Hardy from Zones 2–7, this is a large shrub with long, curving branches. It is very adaptable and hardy in almost any location, and it tolerates full sun to partial shade. Its spring flowers are the highlight of this species, but developing fruits during the summer months are an attractive red color. Fall color is yellowish. Prune the shrub to ground level before any spring growth appears for a less gangly form. There are several horticultural varieties, including 'Diablo,' which has reddish purple leaves.

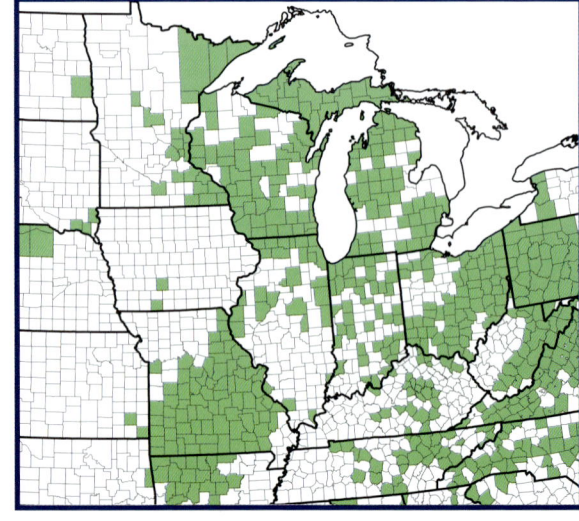

Similar Species Distinctions:
—This species should not be confused with another native species. Flowers are similar to those of *Aronia*, but no other characteristics are the same.

Leaves average about 2 inches in length and usually have 3 lobes (sometimes none or 5), which gives the plant a maple-like appearance. They are dull, dark green above, and paler beneath, with varying degrees of hairiness. The margins are irregularly toothed. Leaves of non-flowering branches tend to be larger than those of flowering branches.

Flowering begins in May and continues into June in our more northerly regions. The perfect flowers are white and about ½ inch wide. Each flower has 5 hairy petals and sepals, and it is supported by a hairy pedicel. Stamens are numerous and often have purple-colored anthers, which strongly contrast with the white filament. Flower clusters can be 2 inches wide.

Fruit (both left photos) begins to ripen in late July and continues into the fall. Each fruit is made up of 3 to 5 pointed follicles that split along the sides when ripe. Each follicle is 2- to 3-seeded. Fruits are persistent and commonly found on plants until mid-winter, when they begin to shatter.

Terminal buds are ovoid and covered with loose-fitting, golden-brown scales that are often tipped with short, white hairs.

Mature plants develop thin, fibrous bark that exfoliates in long strips, revealing the reddish brown, smooth inner bark. Young plants have smooth, reddish brown, thin bark.

Twigs are golden brown and usually smooth. There are numerous vertical ridges, many of which seem to originate on either side of the leaf scars. Lateral buds have few, reddish brown scales that are often tipped with white hairs. The buds are appressed against the twigs. Leaf scars are distinctly raised and commonly heart-shaped. There are 3 prominent bundle scars per leaf scar.

Cherries

Genus: *Prunus*

There are many *Prunus* species in the Midwest, both native and introduced. Several are hard to categorize as to whether they are trees or shrubs, such as *P. virginiana* and *P. pensylvanica,* or the wild plums, such as *P. americana*. Sometimes it is a matter of where in its range one finds the plant, or in what habitat it is growing. For more complete coverage of *Prunus* within a certain area, check local guides or experts.

Cherries are attractive ornamentals for several reasons. Spring brings a profusion of white flowers, followed by red to nearly black summer fruit. Winter bark can be interesting as well, with obvious, numerous rows of lenticels along the stems.

One problem with the cherries in general is the abundance of diseases to which most are susceptible, including black knot, leaf blight, leaf spot, leaf scorch, cankers, powdery mildew, and more (Sinclair, Lyon, and Johnson 1987). These not only impact their aesthetic value as ornamentals, but potentially their life expectancy.

It is hard to imagine a genus more widely used for both food and cover for wildlife than *Prunus*. This is true whether we are considering the more diminutive shrub forms covered here or the larger species that are more frequently placed in the "tree" category (Weeks, Weeks, and Parker 2010). All produce a fleshy fruit used by both mammals and birds, and the fruits are often small enough to be used by even smaller songbirds. Both *P. pumila* and *P. virginiana* are pioneering species, more common in the northern portions of our region, and they are spread largely by the birds that feed on the fruits. Shrub species provide some cover, but they are not exceptional in doing so.

Cherries are a sight to see when in flower, as there are a plethora of white petals in the spring. This is choke cherry in full bloom along a roadside in the Upper Peninsula of Michigan in late May.

Members of the genus *Prunus* always have lenticels scattered along the twigs and stems. As their diameter grows, the lenticels stretch and become elongated horizontally. This is a key identification tool to use with this genus, as well as with *Betula*, the birches. Seen here is choke cherry.

Fruits of most cherries ripen in mid- to late summer, and they are usually produced in great abundance. Choke cherry fruit is pictured.

sand cherry, sandcherry

Prunus pumila **L.**
Prunus susquehanae **Willd.**
Family: Rosaceae

This interesting cherry is a frequent shrubby component of the flora of the dunes situated closest to the Great Lakes. On the open, barren dunes facing the lakes, it is almost a harbinger of spring, as its flowers appear before most things green up in that habitat. Its sizable fruit is among the largest produced by any of our native cherries. This is a variable species that is thoroughly discussed by Voss in *Michigan Flora* (1985). Sand cherry is listed as rare in Pennsylvania, threatened in Tennessee, and presumed extirpated in Ohio.

Form and Size: Sand cherry is an erect shrub that may reach a height of 9 feet but is usually much smaller. On older plants, large limbs will straggle to the ground, which gives the plant an open, few-branched appearance.

Habitat: It is most often found in pure sand within ¼ mile of the Great Lakes, but it is also found in wet to dry prairies and savannahs, on bedrock outcrops, and in open pine forests. It requires full sun and tolerates extreme conditions in the hot summer sands of the dunes.

Wildlife Uses: Sand cherry, as the name implies, grows in sandy areas where it is often a petite plant, supplying cover for birds using small-shrub nest sites and overhead cover for ground nesters. It occasionally grows densely and tall enough to provide cover for larger species such as cottontails, snowshoe hares, and white-tailed deer. It produces a fruit that is surprisingly large for the usually diminutive character of the shrub; it is juicy and taken by mammals, songbirds, and game birds.

Landscaping Value: Hardy from Zones 3–6, this species' range is probably restricted by environmental factors, particularly humidity. It is a beautiful sight to see when in flower, described as having "clouds of flowers." This is a tough plant that commonly grows in the harsh, hot sand dunes in full sun. It does require well-drained soil that is probably somewhat acidic. Fall color is a deep burgundy. There are no known horticultural varieties available, and it is occasionally sold by nurseries specializing in natives.

> **Similar Species Distinctions:**
> —**Willows** (*Salix* spp.), at least some species, have similar leaves that may be mistaken for sand cherry. However, all other characteristics do not match those of willow, including twigs, flowers, and fruit.

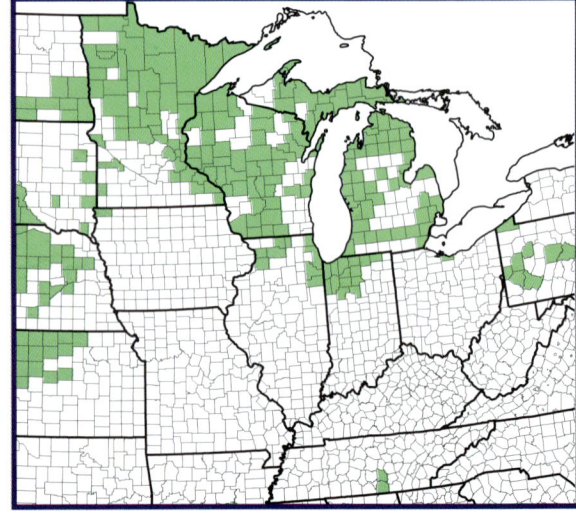

Leaf shape is variable, but leaves are usually about 1¾ inch long and ½ inch wide. They are oblanceolate or spatulate to elliptic, with an obvious, long-tapered base. They are smooth and dark green above, and pale beneath. The margin is finely toothed and gland-tipped. The petiole is up to ½ inch long, often glandular, and can have linear stipules attached. Stipules are also finely toothed and gland-tipped. Like all *Prunus*, sand cherry has alternate branching.

The perfect flowers appear in late April through much of June. Branches closest to the hot sand have flowers first. They appear in short, axillary clusters and have 5 mostly white petals. The throat is often reddish, and there are many protruding stamens in each flower.

Terminal buds are ovoid-shaped and covered with numerous deep reddish scales. On older twigs, a partial silvery coating can develop on some scales. Twigs are bright reddish color, especially when young and growing vigorously. Older twigs develop a silvery coating. Scattered, tan-colored lenticels are especially visible on the vigorous red twigs. Leaf scars are somewhat oval-shaped and contain several bundle scars. Lateral buds are similar to the terminals except that they are smaller. Multiple buds are often found at the nodes, as flower buds develop in the fall and are present in clusters during the winter.

Mature bark is similar to that of most other cherries. It is dark gray to black with small, flaky ridges or plates. Horizontal lenticels are scattered along the trunk; they elongate as the trunk diameter expands.

Fruit begins to ripen by late July and turns black when fully ripe. It is large for a native cherry, to about ¾ inch diameter, and has a sizable, single pit of about ½ inch. Fruit taste is variable, from sweet to sour, but the fruits make good jam.

choke cherry, common chokecherry

Prunus virginiana **L.**

Family: Rosaceae

Choke cherry, named for its tremendously astringent-tasting (think green banana) fruit, is a common midwestern cherry. Its best distinguishing features are its very broad leaves, compared to other cherry species, and its almost 2-toned buds. Its spring flower show and fall color are both spectacular.

Form and Size: Choke cherry is a medium to large shrub that rarely reaches a height of 18 feet. It occasionally reaches tree size and has been documented to reach 60 feet in Michigan. It freely suckers from the root collar and rhizomes, and it forms dense colonies where found.

Habitat: Common habitats where it occurs include woods and forest edges, where it can receive partial sun, fencerows, streamsides, and open, established dunes near Lake Michigan. It is commonly found growing in partial shade but tolerates full sun.

Wildlife Uses: The value of choke cherry as a fruit producer is high, rivaling that of its tree-sized congeners, *P. serotina* and *P. pensylvanica*. The lists of songbirds that use choke cherry are long, including all the usual suspects; they are used by mammals as well, from white-footed mice to black bears. Leaves and twigs are browsed by deer and moose, which, as is usually the case, are unaffected by the cyanide toxins that occasionally impact livestock. Ruffed grouse feed on buds, and snowshoe hares chew on bark to a degree; palatability seems to vary from place to place. Choke cherry is a good cover plant, especially when it grows in thickets; it is regularly used as a nest site by songbirds.

Landscaping Value: Hardy from Zones 2–6, this cherry has a somewhat limited range in the lower Midwest but is very durable. It is found in a variety of soil types and moisture regimes. Choke cherry in full bloom is a wonderful sight; its fruits are an unusual shade of red and usually abundant; and its fall color is the best. Even winter buds and twigs are uniquely colored. No known cultivars exist, and choke cherry is occasionally offered from native-plant nurseries.

Similar Species Distinctions:

—**Canada plum** (*P. nigra*), from a leaf standpoint, can be similar; however, Canada plum has numerous, deep veins and larger teeth on the margin.

—**Black cherry** (*P. serotina*), from a flower standpoint, can be similar, but black cherry is tree-sized. Its leaves are more linear.

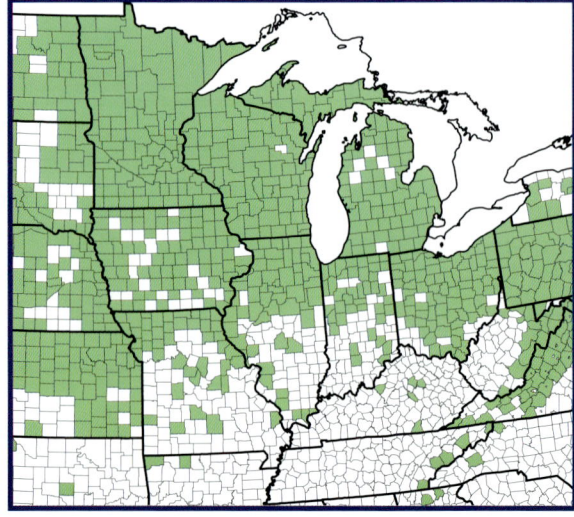

Fruit ripens in August, is about ¼ inch in diameter, and hangs in long clusters that look similar to those of black cherry. The fruit color is usually a deep red, but it can become dark purple when fully ripe. It is very astringent, but it is edible if enough sugar is applied.

Leaves are alternate, about 2 inches long, usually obovate (broadest above the middle), and have an abruptly sharp tip. The margin is sharply and finely toothed, often doubly so. The upper surface is dark green and smooth, and the lower surface is paler. There can be hairs along the midrib or in the lateral vein axils beneath. Petioles usually have a pair or pairs of prominent glands near the leaf blade.

Flowers appear in early April in the southern parts of its range, but they can be found into May in more northerly latitudes. The perfect flowers have 5 white petals and occur in dangling dense clusters (racemes) that are up to 6 inches long.

Twigs are thick for a cherry; they are reddish when very new, but they are nearly always covered with a silvery coating (epidermis). There are numerous scattered, tan-colored lenticels. Lateral buds are similar to the terminal but smaller and closely appressed to the twig. Leaf scars are oval to V-shaped and contain 3 bundle scars. When scraped, twigs emit a rather unique odor that is similar to that of black cherry but has less of an almond scent.

Terminal buds are ovoid but with a fairly elongated tip. They are covered with numerous scales that are light brownish with wide, creamy-white margins. No other native cherry has this scale color combination. Scales are hairless.

Mature bark is usually dark brown, thin, and finely scaled or fissured. Scattered lenticels are present but are not elongated as in other cherries.

dwarf chinkapin oak, dwarf chestnut oak

Quercus prinoides **Willd.**

Family: Fagaceae

The status of this unusual shrubby oak throughout the Midwest is currently unknown. It is so similar in appearance to the tree-sized chinkapin oak, *Q. muehlenbergii*, that the best way to distinguish between the 2 is to check for sexual maturity at a small size, say 3 or 4 feet in height—this species will have flowers or fruit at that size. Dwarf chinkapin oak tends to grow in dry, sandy, open sites compared to the woodland locales of chinkapin oak.

Form and Size: This is a clonal shrub that can reach 12 feet or more in height but is generally smaller. Without flowers or fruit, the best way to distinguish between this species and the tree-sized chinkapin oak is to be aware of the habitat in which the specimen is growing. It might have a single stem, or it can be a multi-stemmed plant.

Habitat: Dwarf chinkapin oak is usually found in very dry, sandy, or gravelly habitats. It can be found in open woods, but it is very shade intolerant.

Wildlife Uses: This shrub oak is used by wildlife in much the same way as are its larger congeners; its small acorns are taken by a myriad of mammals, from white-footed mice to white-tailed deer, and an equal array of bird species, including ruffed grouse and woodpeckers. Its twigs are browsed in winter by the cottontail, which is the principal herbivore in open woods and prairies where it normally occurs. It frequently forms thickets, which often complicate prairie/savannah management, but at the same time, yields cover for terrestrial wildlife and nest sites for birds like eastern kingbirds and orchard orioles.

Landscaping Value: Hardy to Zone 2, this oak is very adaptable to almost any site with the exception of poorly drained. The authors have a specimen growing in a slightly alkaline, rocky clay-loam in full sun; it has 2 main trunks that are nearly 15 feet tall, and it is fruiting at 5 years. No pruning has been done, and it looks suspiciously tree-like! So it can grow quickly—and resemble a tree. This oak is found occasionally in nurseries specializing in natives.

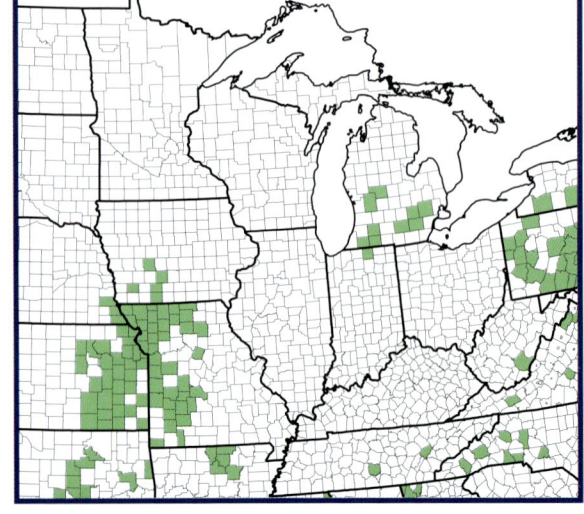

> **Similar Species Distinctions:**
> —**Chinkapin oak** (*Q. muehlenbergii*) is tree-sized, and its leaves have more teeth and veins than dwarf chinkapin.

Fruit is a woody acorn that ripens in late summer and early fall of the first growing season. They are small—only ½ inch across—with a warty cap that covers half the acorn. The acorn is short-stalked and sweet, like those of chinkapin oak.

Leaves are alternate, variable in shape, and 4 to 5 inches long. They are thick, dark green, and glossy above, paler and lightly hairy beneath. The margins have a variable number of teeth, but 4 to 8 are commonly found. In addition to having teeth, the margins are quite wavy. The leaf is usually broader in the upper half and is similar in appearance to those of several tree-sized oaks, including chinkapin and swamp white, *Q. bicolor*.

Flowers appear in early May from the leaf axils of the new growth. Males (far left) hang downward in clusters; the inconspicuous females (near left) are in tiny spikes—both on the same plant.

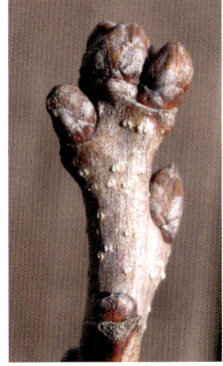

Bark is ashy to dark gray and, with enough age, can develop some fissuring and flaking near the base.

Terminal buds are clustered as in all species of oaks, and they are rounded with numerous, pale-colored, overlapping scales. The scales are partially "painted" with a silvery-white coating.

Twigs are a tannish color, smooth, and often partially covered with a silvery-white coating. There are scattered, raised lenticels along the twigs. Lateral buds are the same as the terminal, but they are not clustered and commonly diverge from the twig. Leaf scars are somewhat V-shaped with scattered bundle scars.

Buckthorns

Genus: *Rhamnus*

Most woody plant families contain species with very similar characteristics. As an example, the native maples, all members of the maple family, have opposite branching, buds with overlapping scales, and the same type of flowers. Buckthorns, however, do not seem to follow any rules. Branching can be opposite, alternate, or sub-opposite; buds can have numerous overlapping scales or have no scales at all; flowers can be perfect or unisex. Of the 3 buckthorns native to the Midwest and the 6 (at least) introduced species, one finds all the variability mentioned. The 2 most aggressive exotic buckthorns in the Midwest are discussed in the "Introduced Species" section of this book.

Rhamnus caroliniana, Carolina buckthorn, is 1 of those borderline tree-shrub species that tends to be single-stemmed. If classified as a tree, then it is a very small tree. The authors of this volume chose to include it in *Native Trees of the Midwest* (Weeks, Weeks, and Parker 2010).

Generally speaking, buckthorns are not particularly dramatic when used as landscape plants. They are rather the plain Janes of the native shrubs. Additionally, with the invasion of the exotic soybean aphid (*Aphis glycines*) into the Midwest, which overwinters in the egg stage on *Rhamnus* species, agronomists and farmers are pressing to eradicate any buckthorns growing near agricultural fields.

Our native shrubs in this genus have relatively low wildlife value, similar to the situation with Carolina buckthorn. Fruits are only moderately attractive to game and songbirds; cover value is quite variable, depending on species. Unfortunately, this genus contains 2 (in our region) exotics that have devastating effects on natural habitats and thereby negatively impact the well-being of many of our native wildlife species dependent on those habitats (see the "Introduced Species" section).

Leaves are somewhat similar on most buckthorns in that they have fine serrations along the margin. Lance-leaf (right) looks very much like a cherry, but alder-leaf (left) is very distinct with its deeply impressed veins.

Most buckthorn fruit is black when ripe. It is apparently not overly appealing to birds, and often it ends up dropping to the ground underneath the parent plant. Shown is alder-leaf buckthorn.

Flowers of most buckthorns are small, greenish white, and difficult to see since they appear with the leaves. Seen here is lance-leaf buckthorn in flower.

alderleaf buckthorn, alder buckthorn, American alder buckthorn

Rhamnus alnifolia L' Her.

Family: Rhamnaceae

This native buckthorn is usually restricted to bogs, calcareous fens, or low, wet woods in northern portions of the Midwest. It is very attractive when grown in a "domestic" setting, as its compact growth form and deeply imprinted leaf veins present a formal appearance.

Form and Size: Alderleaf buckthorn is a petite shrub that reaches a height of 3 feet. It is rather spindly in the wild, but it becomes a rounded, compact plant without competition.

Habitat: It is common on the borders of tamarack bogs and along wet woods and marsh edges.

Wildlife Uses: Wet areas in the northern part of our region often have this native buckthorn as a principal component. These small, attractive shrubs are browsed lightly by deer and moose, and the fruits are sparingly taken by songbirds; many often remain on the plants into the winter. However, its structure is ideal for songbird nesting, and it is often chosen by wetland-associated species like song and swamp sparrows, red-winged blackbirds, and common yellow-throats.

Landscaping Value: Hardy to Zone 2, this is a wet-site specialist that grows nicely on well-drained, upland sites. Its form is its best feature when grown ornamentally, and it makes a nice specimen plant along building foundations. Leaves are dark green with prominent veins and are quite attractive. Fall color is a dull yellow, and flowers are unspectacular. It has no insect problems and grows well in partial shade. Several exotic buckthorns, *R. cathartica* and *Frangula alnus*, have invaded the wetlands where this native occurs, and both threaten to replace it in the wild. Alderleaf buckthorn is occasionally available through northern nurseries that sell natives.

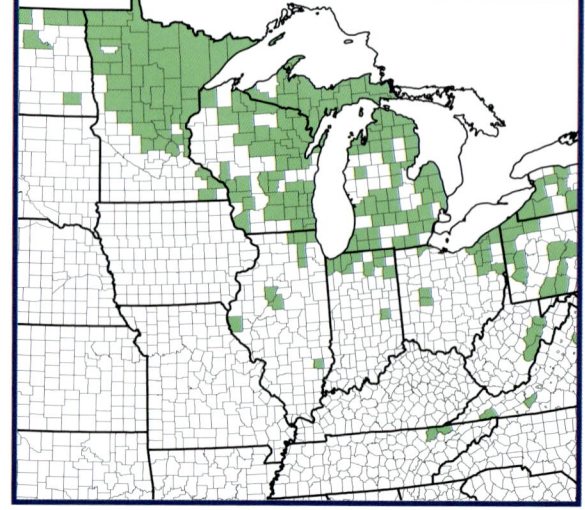

Similar Species Distinctions:
—The common name suggests that the leaf of this species looks like that of an alder. That may be in reference to **green alder** *(Alnus viridis)* which has singly-toothed leaf margins. All other alder species have doubly-toothed leaf margins.
—**Choke cherry** *(Prunus virginiana)* winter buds are similar in shape and color, but twigs have numerous pale lenticels.

Many members of the Rhamnaceae family have sub-opposite leaf arrangement. Alderleaf, however, nearly always has alternate branching. Its leaves average about 3 inches in length, but they can be much larger. They are oval-shaped with an abruptly pointed (usually) tip. Leaves are dark green and smooth above, with obviously impressed main and secondary veins. Beneath, the leaves are lighter green and have small, pale hairs along the veins. Leaf margins are finely toothed.

Stem diameter is so small that the bark remains thin and brownish, reddish, or gray.

Flowers are small, only ¼ inch across, and inconspicuous. They appear in mid-May in the leaf axils of the new growth in clusters of 2 or 3. There are no petals, but just 4 (or 5) greenish sepals that help them blend in with the nearly fully developed leaves. Flowers are separate sex on different plants (females right, males left).

Buds are ovoid, pointed at the tip, and covered with 5 or 6 multi-colored scales that are variable shades of brown with a silvery "coating." Scale margins often have a row of fine, whitish hairs. Twigs are slender and brown when they first appear. Over the first growing season, they develop a silvery-gray coating that remains on the twig. Twigs are smooth. Leaf scars are U-shaped and contain 3 bundle scars. Lateral buds diverge from the twig somewhat.

Fruit is a fleshy drupe that ripens in August and is about ½ inch in diameter. It is black at maturity and contains 3 seeds (usually).

lance-leaved buckthorn

Rhamnus lanceolata **Pursh.**

Family: Rhamnaceae

There is nothing particularly dramatic about this shrub, whether in fruit or flower. It just blends in, and it actually has several features that resemble cherries more than buckthorns. Within its range, this species seems to be declining in numbers, and in some areas where it was once common, it is now gone. It is currently listed as endangered in Pennsylvania.

Form and Size: Lance-leaved buckthorn is a single-trunked shrub that can become quite large in the right habitat. It can grow to 12 feet in skunk cabbage seeps, but it is much smaller on rocky cliffs. It has a fairly sizable crown made up of numerous small limbs.

Habitat: Lance-leaved buckthorn is found in poorly-drained wet sites along rivers (such as seeps), on wet roadsides, and in forest openings. It is also found on rocky bluffs and cliffs. Full sun to partial shade is required, as it declines and dies when light conditions are poor.

Wildlife Uses: From a wildlife food perspective, this species is similar to the other buckthorns—low browse preference by white-tailed deer and marginal use of its often prolifically produced fruit by birds and small mammals. Its limb structure, however, makes it excellent cover; many limbs close to the ground shield small mammals and birds from the elements and predation. Higher in the shrub, dense branches and small branchlets yield numerous potential nest sites for songbirds.

Landscaping Value: Hardy to Zone 4, this buckthorn is an uncommon shrub that can be rather picky about where it grows. It prefers moist, calcareous soils where it grows to small tree size. Grown on dry sites, it remains shrubby and 4 to 5 feet tall. It has no major seasonal features to speak of and is not available in nurseries.

Similar Species Distinctions:

—**Cherries** (*Prunus* spp.). The bark and leaves of many cherries are very similar to those of lance-leaved buckthorn, but many features differentiate them. See the 2 *Prunus* species in this book.

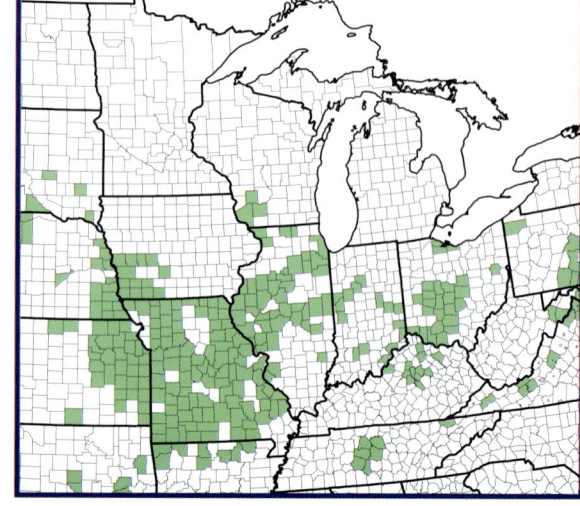

Leaves are opposite to sub-opposite, up to 4 inches long, ovate-oblong or ovate-lanceolate, and usually widest above the middle. The upper surface is dark green and smooth; the lower surface is pale green. Leaf tips are usually elongated. The lateral veins run upward toward the leaf tip rather than to the leaf margin, a common trait seen in dogwood leaves. The margins are edged with fine, incurved teeth. Fine, short hairs are common along the petiole and leaf margin, especially near the leaf base.

The fruit is a drupe that is about ¼ inch in diameter. Fruits turn from green to red and finally black when fully ripe in August, and each contains 2 grooved nutlets.

The tiny, inconspicuous flower clusters appear in early May in the leaf axils on new growth, and individual flowers are only ⅛ inch across. They have 4 greenish petals and individual plants are single-sex. A pleasant fragrance is noticeable when the shrub is in flower. Females are seen above; males to the left.

There are no true terminal buds. Each bud has 5 to 6 dark brown scales that are tipped with a narrow line of white hairs. Buds are small, more plump, and less pointed than those of most buckthorns. The new year's growth is greenish and eventually dark brown and hairy. Older twigs, however, are dark reddish brown and covered almost entirely with a silvery-gray coating, similar to the twigs of many cherries. Branching is an unusual sub-opposite. Leaf scars are raised and U-shaped.

Mature bark (left) is moderately thick for a shrub, dark brownish with patches of silvery-gray coating. It breaks up into small, somewhat curling plates and develops shallow fissuring near the base. Lenticels that are only somewhat darker than the bark are scattered along the trunk and large branches.

Gooseberries and Currants

Genus: *Ribes*

Gooseberries and currants are common in the Midwest, particularly in areas with cooler, less humid environments. In Indiana, for instance, there are 4 known natives; in Michigan, there are twice as many natives, plus a handful of cultivated species originating from Europe and the western United States, which have escaped to varying degrees.

Ever since the exotic rust pathogen *Cronartium ribicola* was introduced into the eastern United States around 1900, *Ribes* species have been somewhat persecuted. The rust kills eastern white pine (and several western pines), but it needs an alternate host to complete its life cycle, which is where *Ribes* comes into play. "Rust busters" hired by federal and state agencies tried to eradicate *Ribes* for decades, from the East Coast to the West Coast, with little results to show for their efforts. Today the fungus is controlled through forestry practices, such as limb pruning of pines for increased air movement, which reduces the number and viability of fungal spores. However, there are still states, including Michigan, that ban the commonly cultivated *Ribes nigrum*, the black currant of commerce.

Native gooseberries and currants are usually found in open woods and forest edges, where the lack of sunlight usually prevents fruit development. However, plants growing in the open become full, flower profusely, and set much fruit. If you enjoy the summertime fruit, then they are worth planting. Otherwise, they are less than spectacular as ornamentals.

Ribes are notoriously difficult to identify simply by leaf, as they all look pretty much the same. Flowers are essential, and fruit helps to some degree with identification.

Several *Ribes* species have quite similar growth forms and value to wildlife. Plants in the understory of woods are somewhat spindly and supply less nesting cover than individuals that are growing in the open. The spines that characterize this group probably have little impact on their wildlife cover value. The fruits are small enough to be attractive to songbirds, but they are also taken by game birds and small and large mammals.

The fruit of *Ribes* is always an added bonus to help identify a species. Some are smooth as seen on this Missouri gooseberry, while others, like *R. cynosbati*, are quite prickly. All species have the longitudinal lines that give the fruit a basketball-like appearance.

Leaves on all *Ribes* have a maple-leaf-like shape with 5 (sometimes 3) lobes and somewhat rounded teeth. A few have features that differ enough to make a species determination, such as *Ribes americanum* with its yellow-dotted glands on both leaf surfaces. Pictured is *R. cynosbati*.

Ribes glandulosum Grauer, skunk currant, is a common species of the Upper Lake States in moist habitats. Its ovaries, flower stalks, and fruits are covered with beautiful, red, gland-tipped hairs. It exudes a skunk-like odor that is easily smelled.

Spines and bristles are common on the gooseberries, but they are usually absent on the currants (the 2 were once in separate genera, partly for this difference). Pasture gooseberry, shown here, has bristles along the base of the plant, but many more on newer growth.

Ribes triste Pall., swamp red currant, is another common species of the Upper Lake States, also found in moist habitats. It is similar in appearance to skunk currant, but it has tiny, dangling, pink flowers and no gland-tipped hairs.

American black currant, wild black currant

Ribes americanum/americana **Mill.**

Family: Grossulariaceae

When in flower, black currant can rival exotic horticultural currants in beauty. The substantial flower clusters are produced prolifically for a several week period. In the past, currants and gooseberries were designated into separate genera (*Ribes* and *Grossularia*, respectively), based mainly on presence or absence of glandular dots and spines. Currently, they are all within the genus *Ribes*. All have now been placed in the Grossulariaceae family; formerly they were classified under Saxifragaceae.

Form and Size: Black currant can grow to around 4 feet in height, and develop an upright, spreading form. Given full sun, it develops a broad, dense crown. In heavy shade, it is few-stemmed and rather spindly.

Habitat: It is found in many habitats, but it is most often in moist places, such as on borders of swamps, lakes, and streams, in wet, open woods, and in ancient marshes and bogs. It prefers partial sun and tends to show signs of stress with too much exposure to sun and heat.

Wildlife Use: Wild black currant occasionally grows to 4 feet, especially when growing in the open where it also has sufficient limb and leaf structure to supply nesting sites for songbirds. It is more common in the northern reaches of our region and grows most frequently in wet areas where it is occasionally browsed (preference not high) by deer and moose. Its fruits differ from most *Ribes* in our region in being in drooping clusters, which have fruits that progressively ripen, so are available for extended periods for the songbirds and game birds (e.g., gray catbird, robin, ruffed grouse) that regularly use them. Fruits are also eaten by mammals, such as chipmunk, squirrels, and raccoons.

Landscaping Value: Hardy to Zone 3, black currant has several nice attributes worthy of note. Its glistening yellow glands dot the leaves (especially when new), buds, and twig tips, and are especially attractive on the pinkish twigs. The bark is spine-free and dark brown with tan-colored, linear stripes. Given half-day sun, black currant is robust, flowers profusely, and produces numerous chains of small, black, edible fruits. Fall color is yellowish; it is easy to grow, and it prefers moist, well-drained soil. Several nurseries specializing in natives sell black currant in the Midwest.

Similar Species Distinctions:
—Gooseberries in general look the same from a leaf standpoint. It is nearly impossible to identify species accurately using them alone, and flowers are a must; fruits are helpful as well, but they are often difficult to find.

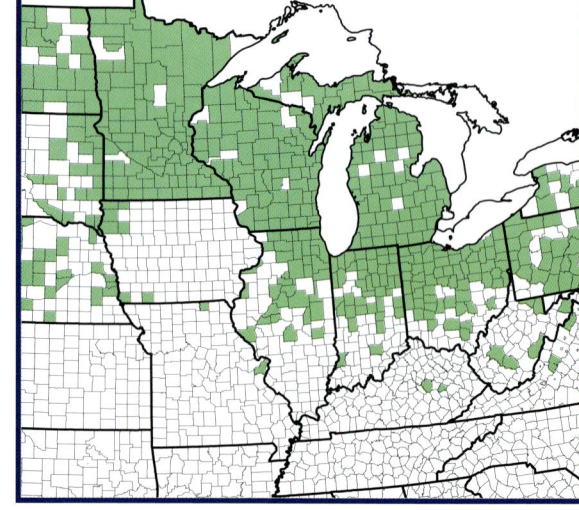

Leaves are alternate, up to 2¾ inches long, and usually have 3 (sometimes 5) lobes. They are orbicular in outline and can have a petiole as long as the leaf itself. Leaf margins are single- or double-toothed. The upper surface is green with scattered resin dots. The lower surface is slightly paler, hairy along the veins, and dotted with golden resin. Swink and Wilhelm, in *Plants of the Chicago Region* (1994), proclaim: "The student is encouraged to view, particularly on the lower leaf surface, the gorgeous, golden, globular, glistening, glittering glands."

The fruit ripens in mid-summer and does not persist. The berry is about ¼ inch in diameter, the smallest of our native *Ribes*. Fruit is black when ripe; it is sweet and hangs in drooping, chainlike clusters.

This is our showiest native *Ribes* when in flower. The perfect flowers hang in clusters that are 2 to 3 inches long and composed of up to a dozen flowers. Individual flowers are about ⅓ inch long and have 5 yellowish-green petals. Flowers appear in late April and May.

Mature bark is dark brown with vertical, off-white "stripes" that are raised and ridge-like. Some scaly fissuring occurs over time. Younger stems retain much of the pale color that characterizes them; it is slowly replaced with the brown of older bark.

The terminal bud is ovoid and covered with 5 to 6 somewhat loose-fitting, whitish pink scales. The scales are hairy and covered with golden glands. Twigs are pale-colored, somewhat hairy, and covered with rounded, golden resin glands. The lateral buds are similar to the terminal but smaller. Leaf scars are raised and triangular-shaped with 3 bundle scars. Twigs develop distinct ridges originating from the leaf scars that eventually become prominently pale in comparison to the dark stem.

pasture gooseberry

Ribes cynosbati **L.**

Grossularia cynosbati **(L.) Mill.**

Family: Grossulariaceae

This is the most common gooseberry in the central Midwest. It is also our most prickly. Gooseberries, in general, look very much alike from a leaf standpoint; to identify them accurately, it is best to use flowers and fruit. Pasture gooseberry fruit has the most prickly fruit of our natives, and immature fruits especially do not look overly appealing.

Form and Size: This is a sizable gooseberry, often reaching 5 feet in height and spreading to 4 feet or more. It is usually a solitary species, but it can form small, clonal groups. It is very shade tolerant, and does not handle full sun well, although it occasionally occupies open areas.

Habitat: Pasture gooseberry is found in many habitats, but it prefers rich, moist sites. It can be found infrequently on rocky, wooded slopes and in rock crevices. It is common in the northern regions of the Midwest.

Wildlife Uses: Pasture gooseberry occurs not only in open areas, as the name implies, but in woodland understories as well; in all areas it is a low, spreading shrub that provides nesting sites for songbirds, such as indigo buntings. As is true for all in this genus, it rarely grows in groups, so in spite of its prickly structure, it rarely supplies much protective cover. The fruits have soft spines as well, which do not deter their being taken by songbirds, game birds (such as ruffed grouse and bobwhite), and small rodents. Leaves are sparingly browsed by white-tailed deer.

Landscaping Value: Hardy to Zone 3, pasture gooseberry is 1 of the largest in the Midwest. It is also 1 of the spiniest. Ripening fruits gradually lose their prickles and are edible, but its numerous stem prickles are a bit intimidating; it is probably not a good idea to plant pasture gooseberry in an area where children play. This species is rarely available through nurseries specializing in natives.

Similar Species Distinctions:
—Gooseberries all look very similar, but the numerous spines on this species help rule out all but a few in the Midwest.
—**Missouri gooseberry** (*R. missouriense*) is very similar but has lengthy flowers and smooth fruits.
—**Bristly black currant** (*R. lacustre*) has gland-tipped hairs on the flowers and fruits. Fruits are the size of *R. americanum*—quite small. This species occurs in northern Wisconsin, Michigan, and Minnesota.

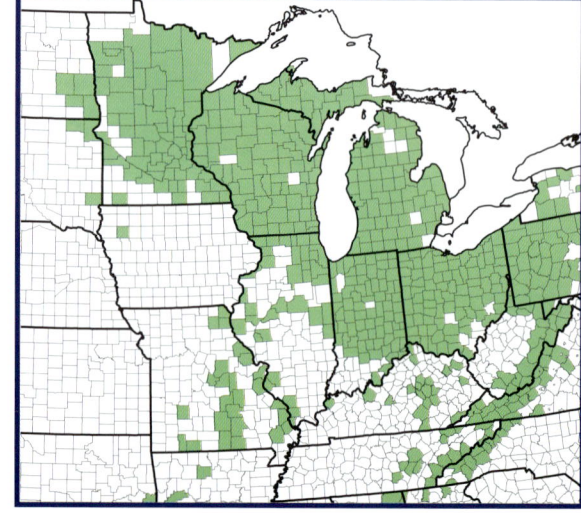

The perfect flowers appear in late April into May in small clusters. Each flower is about ¼ inch long, and has 5 tiny, white petals. The ovary is covered with bristles, and the calyx lobes are reflexed.

Leaves are alternate, about 2 inches long and orbicular in outline. Like most gooseberries, leaves tend to be broader than long. They are 3- to 5-lobed with toothed margins; they are green and lightly hairy above, and paler with numerous white hairs beneath. Leaves can develop singly along the stem or in small clusters from short shoots. Leaf petioles are hairy and up to 2 inches long.

The fruit, a berry, ripens in early to mid-summer, and is deep red-purple when fully ripe. It is about ½ inch in diameter and edible. It has varying amounts of soft bristles.

Twigs are pale with thin outer bark that quickly splits linearly along the length of the twig and branches. Inner bark is dark brown and dotted with tan-colored lenticels. Twigs have scattered hairs, especially on the newest growth. Bristles are usually present, but in varying degrees; vigorously growing twigs may be covered with them, while others may have almost none. Lateral buds are similar to the terminal, and they diverge from the twig. There are 1 to 3, reddish spines at each node that are about ⅓ inch long. Leaf scars are thin and V-shaped; they contain 3 bundles scars.

Terminal buds are ovoid and are covered with numerous, loose-fitting, overlapping, tan-colored scales. Scales have scattered hairs as well.

Mature bark (right) is thin, dark brown, and dotted with tan or rusty-brown-colored lenticels. Only slight fissures develop on these relatively small diameter stems. Spines and bristles fall over time, leaving the stems essentially smooth. Younger stem bases are pale-colored and densely bristly.

swamp gooseberry, northern gooseberry

Ribes hirtellum/hirtella **Michx.**

Grossularia hirtella **(Michx.) Spach.**

Family: Grossulariaceae

Correctly identifying gooseberries is best done using flowers, comparing the various lengths of some parts relative to other parts (e.g., petals, sepals, stamens, etc.). In this species, the sepals are roughly the same length as the white petals, while the stamens are about twice as long. Swamp gooseberry is becoming rare in the southern portions of the Midwest, mainly because of habitat loss.

Form and Size: This is a rather small gooseberry—a 3-foot-tall plant is considered large. It has a very upright, non-spreading growth form with few branches.

Habitat: It is found in tamarack bogs, fens, wet roadsides, swampy woods, and lakeshores. It is shade tolerant, but specimens growing in full sun thrive and become full-looking from the vigorous leaf growth.

Wildlife Uses: Swamp gooseberry becomes increasingly common as 1 moves northward in our region; it occurs especially in wetland areas where its upright limbs provide very good nest sites for associated songbirds, such as red-winged blackbirds and alder flycatchers. As with other gooseberries, it rarely grows in groups that would provide good protective cover. Its fruits are used regularly by birds and mammals; foliage is moderately browsed by white-tailed deer, and snowshoe hares take twigs and bark in the winter.

Landscaping Value: Hardy to Zone 2, this is a cool weather species that thrives in the northern portions of the Midwest. It is a nearly prickle-free species with pretty, somewhat more deeply divided leaves than other native *Ribes*. Although it is usually associated with wet sites, it is growing fine in a well-drained upland site for the authors. Bees are crazy for the flowers of swamp gooseberry, and they remain in the vicinity throughout the several-week flowering period. Fall color is unimpressive. The authors have not seen this gooseberry offered through native-plant nurseries.

Similar Species Distinctions:
—As with all gooseberries, look for flowers and fruits to make positive I.D. This species' lack of spines allows one to rule out those that have them.
—**Maples** (*Acer* spp.) have similar leaves, but leaves are usually much larger on mainly tree-sized plants. Maple leaves usually have pointed lobes; gooseberry leaves have rounded lobes.

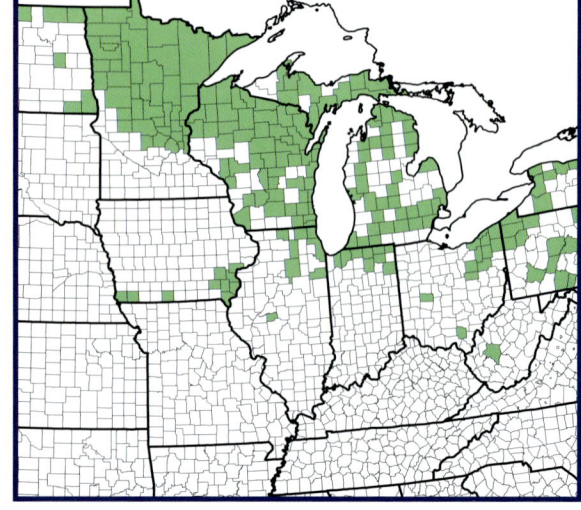

Leaves are somewhat orbicular in outline, but they are the most consistently 3-lobed of our native gooseberries. Because of the 3 lobes, the shape tends to be longer than wide. The leaves are the smallest of our gooseberries at up to 1½ inches long. The upper surface is green and nearly smooth when mature. It is somewhat paler and hairy beneath, especially along the veins. The margins are toothed, and the petiole is about 1 inch long and hairy. Like all gooseberries, leaves are alternate.

Flowers appear in late May in small clusters. They are perfect and have 5 short, white petals with long, protruding stamens. The ovary can have hairs, but that is not the case here.

The fruit, a berry, ripens in mid-summer and is usually purplish when ripe. Like many gooseberries, there are often longitudinal stripes on the fruit—think green or purplish basketball here. Fruit is smooth, about ⅓ inch in diameter, and edible.

Terminal buds are an elongated ovoid with overlapping, papery-thin, pale-colored scales. Young twigs are slender and pale tannish gray. They develop ridges that begin at the raised leaf scars that run the length of the twigs and branches, similar to those of *Ribes americanum*. With time, the thin outer bark layer splits open to reveal the darker inner bark that is dotted with lenticels. Lateral buds are similar to the terminal, except smaller; they tend to be more appressed against the twig than in our other gooseberries. Pictured here is an older twig that has a short shoot growing perpendicular to the twig.

This is a small gooseberry, and therefore, it does not achieve much diameter growth. The thin, tannish outer bark eventually falls off, revealing the dark brown, thin inner bark that is dotted with pale lenticels. This species can have spines and bristles, but they are never abundant. Some still remain on this mature plant.

Missouri gooseberry, wild gooseberry

Ribes missouriense **Nutt.**

Grossularia missouriensis **(Nutt.) Cov. & Britt.**

Family: Grossulariaceae

This gooseberry is usually associated with moist, open habitats in the prairie region of the Midwest. Its flowers are very beautiful, elongated, and graceful in appearance. In general, the gooseberries' good names have been besmirched because of the fact that they are an alternate hosts to an introduced rust fungus—*Cronartium ribicola*, eastern pine blister rust. Extensive eradication programs have reduced gooseberry populations in regions where eastern white pine is native and grown for timber.

Form and Size: It can grow to 5 feet in height and has an upright, somewhat spreading form. Like most gooseberries, it is very shade tolerant, but this species can to handle more sun than most.

Habitat: Missouri gooseberry is found on wooded bluffs near streams, on steep slopes of wooded ravines, in wooded floodplains, and in open old-fields, brushy thickets, and roadsides.

Wildlife Uses: Missouri gooseberry, as the name indicates, is most common in the western reaches of our region, where it occurs in both disturbed sites and woodland understories. In both locations, it has a multi-stem, broad growth form that gives it value as protective cover for birds and mammals, as well as nesting sites for songbirds. Its fruits are taken by the same cadre of birds and mammals that use the other species of gooseberries, especially *R. cynosbati*, with which it often shares habitat.

Landscaping Value: Hardy to Zone 3, this is 1 of the most adaptable gooseberries in the Midwest. It prefers rich, mesic sites, but it handles very dry, poor quality areas as well. Its numerous spines are daunting, but they have a kind of Jekyll and Hyde appeal when considered with the elongated, graceful flowers. Flowers are usually abundant, whether grown in full sun or partial shade, since they appear before overhead trees leaf out. Fruit is spineless and edible. This gooseberry is occasionally offered through native-plant nurseries.

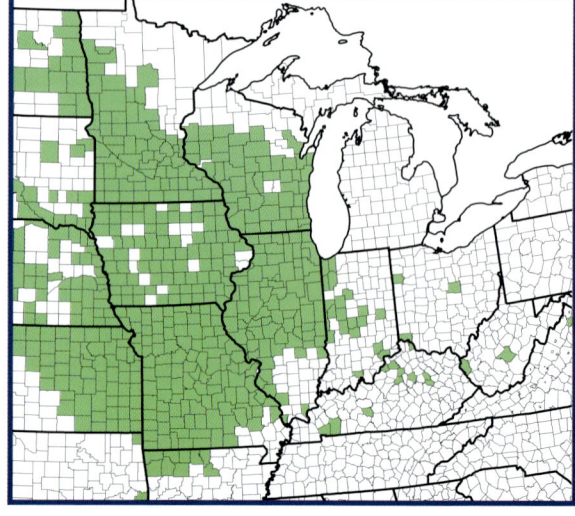

Similar Species Distinctions:
—The spiny gooseberries mentioned under *R. cynosbati* are the major species with which Missouri gooseberry would be confused.

Leaves are alternate, up to 2¾ inches long, and usually 3-lobed. Like most gooseberries, the leaf outline is orbicular in shape. The upper surface is green and mostly smooth; it is somewhat paler and hairy beneath. Leaf margins have irregular, large teeth. Petioles are up to 1½ inch long and hairy.

The perfect flowers appear in late April into May in clusters of 3 or more. Each flower has 5 elongated white petals that reflex back, which helps to highlight the long stamens. Flowers are about 1 inch in length. The peduncles are hairy and slightly glandular.

The fruit, a berry, ripens in mid-summer and is deep purple when ripe. Ripe fruit is smooth and up to ½ inch in diameter. Like most gooseberries, this species' fruit is also edible.

Buds are more elongated than in our other native *Ribes* and are covered with thin, loose, golden-brown scales. The scales can be nearly white near the tips, and they usually have some degree of hairiness.

Twigs are slender, pale, and covered with varying amounts of reddish brown bristles. New, vigorously growing shoots tend to be densely covered with bristles. Lateral buds are like the terminal, and they diverge from the twig. At each node are reddish brown spines, numbering from 1 to 3, that are about ½ inch long. Leaf scars are narrow, V-shaped, and contain 3 bundle scars.

Mature bark (left) is dark brown, thin, and scaly. With age, the bristles and spines fall. Younger stems are pale, with outer bark that splits (and eventually falls off) to reveal the dark inner bark. New stems are usually densely covered with reddish brown bristles.

purple-flowering raspberry, flowering raspberry

Rubus odoratus L.

Family: Rosaceae

This raspberry is fairly common in the northeastern United States, but farther south it has become quite rare. It is currently listed as endangered in Illinois and threatened in Indiana. It is a very interesting plant, and it is 1 native that has drawn attention from horticulturalists. The 1 drawback with this shrub is that it attracts Japanese beetles.

Form and Size: It is fast growing on moist sites and can become 4 feet tall within 5 years. It is a spreading, suckering species that is usually as wide as it is tall. It thrives in nearly full shade, but it tolerates morning or filtered sun.

Habitat: Purple-flowering raspberry prefers rich, moist soil in shaded woods or clearings of conifer woods, but it is also found on rocky, gravelly, wooded slopes. It is very adaptable to well-drained sites.

Wildlife Uses: This raspberry has shorter, more upright stems than *R. occidentalis*; plants are very similar to *R. idaeus* in cover value for wildlife. The fruits are produced throughout the summer and are quite dry and tasteless to the human tongue. Although other authors say that they are eagerly taken by songbirds, game birds, and mammals, we find that they are not nearly as coveted as are *R. occidentalis* and *R. idaeus* fruits.

Landscaping Value: Hardy to Zone 5, this raspberry is similar in appearance to the northern species *R. parviflorus*, thimbleberry. Its large flowers continue to be produced over an extended period, and flower buds are densely covered with reddish, glandular hairs. It is readily available from nurseries in the Midwest that specialize in natives, and it makes a great addition to a landscape.

Similar Species Distinctions:
—Thimbleberry (*R. parviflorus*) is found in the Upper Peninsula of Michigan, northern Michigan, northern Wisconsin, and northeastern Minnesota. Their ranges do not overlap at all. Leaves are similar, but thimbleberry has glandular hairs only along the lower leaf veins.

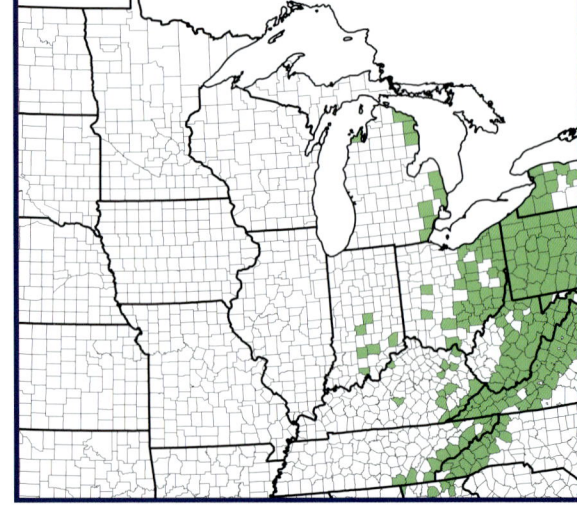

Flowering begins in early to mid-June and continues through July or later. Its flowers are rose-purple, with 5 petals, and numerous, tan-colored stamens. Flowers are large for a *Rubus*—up to 1½ inches in diameter. The sepals are covered with glandular hairs, which are obvious on flower buds. The long pedicels are densely covered with glandular hairs as well. Pictured on left are flower buds.

Its alternate leaves have 3 to 5 elongated lobes and look much like a maple leaf. They are as large as 7 inches wide. The green leaves have an extensive, fine venation pattern and are lightly hairy on the upper surface. Beneath, they are paler and lightly hairy, with glandular hairs running along the veins. Petioles are densely covered with glandular hairs and may be as long as the leaf blade itself.

R. parviflorus, thimbleberry

Fruits ripen in mid- to late summer. They are pinkish red when ripe and about ¾ inch in diameter. The dry, aggregate drupe is not as deep as in other *Rubus* species, and it is shaped like a shallow bowl. It is not palatable to humans.

Twigs are slender, brittle, light brown, and densely covered with glandular hairs. Lateral buds diverge from the twig and are plump, with a few scales that are the same color as the twig. Older twigs begin to exfoliate their outer "bark," although it remains attached to the stem. Beneath the exfoliating bark, the twig is smooth. Stems do not have terminal buds, and they often die back at the tip during winter.

Purple-flowering raspberry does not have biennial canes like all our other native *Rubus* species. Over time, its oldest stems become an inch or so in diameter and exfoliate outer layers of bark. It continues this process throughout its lifetime. Bark is thin and reddish brown.

Willows

Genus: *Salix*

One of the most daunting groups of shrubs to identify—at least without the help of someone familiar with them—is the willows. To most people, it is similar to a life-long urbanite visiting a dairy farm where many head of cattle have been named by the farmer. The urbanite asks how recognizing each 1 by name is possible, since all the cows look the same! Like Holstein cows, willows have distinguishing features that set them apart, but the more features available, the better. Female flowers or fruit are often a requisite for accurate identification. One difficult thing for the authors to accept is that willows cannot reliably be identified by winter twigs and buds. They are notorious for their variable characteristics, and hybridization is common, which only complicates matters.

The 14 native shrub willows included in this book represent the majority found throughout the Midwest. A few are not included, most of which are found on the northern fringes of the Lake States:

S. pellita (Anderss.) Anderss.ex Schneid.
   satiny willow, found in U.P. Michigan and
   n. Minnesota.
S. planifolia Pursh.
   planeleaf willow, Isle Royale, Michigan,
   and n. Minnesota.
S. pyrifolia Anderss.
   balsam willow, n. half of Minnesota and
   Wisconsin, U.P. Michigan.
S. pseudomonticola Ball
   false mountain willow, a few counties in
   Minnesota.
S. maccalliana Rowlee
   McCalla's willow, a few counties in
   n.w. Minnesota.

Midwest willows, with the exception of *S. discolor*, pussy willow, are not usually considered acceptable additions to a domesticated landscape, but there are several that are particular favorites of the authors and worthy of consideration; they are discussed on the species pages.

This section on *Salix* would not have been possible, and certainly not as accurate, without the generous help of Dr. George Argus, Salicologist and Curator Emeritus of the Canadian Museum of Nature in Ottawa, who is pictured on the opposite page with *Salix cordata* at the Dunes National Lakeshore in northwestern Indiana in 2006. He reviewed all the willow images and provided all the keys. For much more in-depth information on willows, see Argus in *Flora of North America*, Vol. 7 (2010).

Willows have long been recognized in the American West as being important browse, in both summer and winter, for large ungulates such as moose, elk, and mule deer. Because it is of frequent occurrence in riparian areas in many arid western landscapes, it also serves as important cover for multiple species, including songbirds. In our midwestern region, some of the same generic generalities likely apply, but the importance of *Salix* as a browse species is more limited. Willows are browsed by white-tailed deer, moose, snowshoe hares, and beavers, but they are, in general, not predominant dietary components. Because of their association with wetland sites (necessary for germination), they do supply important nesting cover for birds of those habitats, such as yellow warblers, alder and willow flycatchers, and red-winged blackbirds. Unfortunately, difficulty with species identification has greatly limited wildlife biologists' ability to distinguish species–specific differences in wildlife use. Thus, since we are cursed with some of the same inadequacies, linked with sporadic (and often rare) occurrences of individual *Salix* species, we make species-specific observations of only a few of our included willows, and we refer the readers to this general evaluation of wildlife values.

The leaves of many willows look so similar that it leads to frustration for those basing decisions strictly on that 1 characteristic. Willows have several different leaf shapes based mainly on what time of year they are observed. Collecting them throughout the growing season is important. Shown are the leaves of 2 similar species, *S. serissima* (top) and *S. lucida* (bottom).

Buds of most willows have a rather unique feature: they have a single bud scale that sits over the bud like a cap. The leaf scars are shaped like a shallow V and contain 3 bundle scars. These are buds of autumn willow, *S. serissima*.

Dr. George Argus, Salicologist and Curator Emeritus of the Canadian Museum of Nature in Ottawa.

Balsam willow, *S. pyrifolia*, occurs in the Lake States and has leaves that are not "willow-like," but rather shaped like a tongue with numerous, deeply impressed veins. It is named for the odor given off by dried specimens, not live ones.

Female flowers of willows are important for correct identification. Their characteristics are as stable and reliable as any. The prairie willow female, seen here, has densely hairy ovaries.

Bebb's willow, long-beaked willow, beaked willow

Salix bebbiana **Sarg.**

Family: Salicaceae

Bebb's willow is a fairly common, large willow with 1 of the largest ranges for a willow in North America. It has several characteristics that are most helpful in identification. First, it is quite distinctive when in fruit. It is 1 of only a few native, shrubby willows that produce fruits before the leaves are much developed. Its fruits are large, and each ovary is on a long stalk that is subtended by a light brown bract. Secondly, its leaves have numerous, fine, impressed veins (reticulate venation) that is fairly unique for our native willows. That feature is easily seen here.

Form and Size: Bebb's is 1 of our largest shrub willows, and it can grow to 15 feet or more in the Midwest. It has 1 of the widest distributions of any North American willow, and it has been measured to 30 feet in height elsewhere.

Habitat: It can be found in low, wet ground, boggy areas, fens, wet thickets, and many other wet habitats throughout its large range. It prefers full sun.

Wildlife Uses: This large willow has many of the same characteristics and wildlife values as pussy willow. Reader is referred to the *Salix* genus page 192.

Landscaping Value: Hardy to Zone 1, this species is also classified as a small tree. It has interesting, large flowers and fruit with beaked ovaries that are obvious with no leaves present. The leaves have reticulate venation that is a bit unusual. It is also a major willow taken for "diamond" willow walking sticks. Trunks or stems are scarred from an infection by the fungus *Valsa sordida*, which causes color contrast between the heartwood and softwood. This large shrub would be a nice addition for a pond or lake margin if any kind of structure is desired. Bebb's is occasionally available through native-plant nurseries.

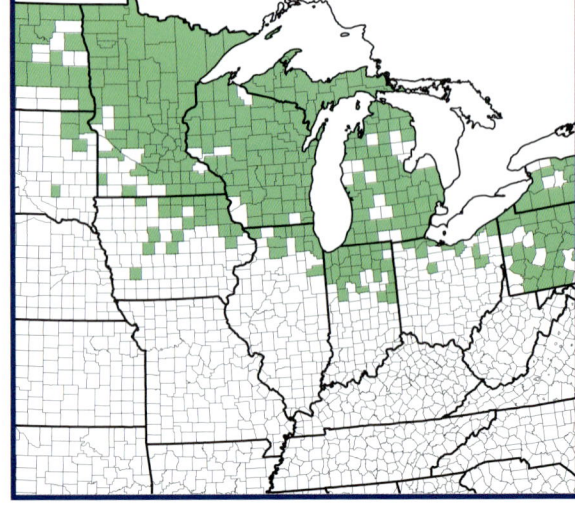

Similar Species Distinctions:

—**Pussy willow** (*S. discolor*) is a large willow that has similar-shaped leaves, but they lack the reticulate venation. Catkins appear about the same time in the spring and are large like those of Bebb's; however, they have dark floral bracts.

Leaves are alternate, narrowly oblong, and up to 4 inches long. Leaf width is around 1¼ inches. Stipules can be present, but they generally are not. Leaves are dull or shiny and impressed with much venation. The upper surface is green and hairy or not, while the lower surface is pale and commonly covered with white hairs. Leaf margins are entire to slightly toothed.

Flowers appear in early to mid-May. Females emerge just before the leaves; males emerge with the leaves. Flowers are about 1 inch in length in upright clusters (aments), with males (near right) and females (far right) on separate plants. The hairy ovaries are elongated and have a long stalk (stipe), which is subtended by a tawny brown bract. Female aments are loose, rather than densely packed.

Fruits mature from late May into June. The fruit is comprised of many capsules in an upright cluster that is nearly 2 inches long. Each capsule is elongated and tapers at the tip like a beak. Capsules split open and release many tiny seeds that are attached to white, fluffy "down" (pappi).

Twigs are brown and slender, and they may have hairs, especially on the youngest growth. Buds are all the same size and have varying amounts of hair. There is a single bud scale that can be various colors. Leaf scars are somewhat V-shaped and contain 3 bundle scars.

Willow bark, particularly on shrub willows, is not necessarily unique by species. Bebb's willow achieves a large enough diameter that it produces an interesting, interlacing bark pattern that is fairly distinctive.

sage-leaf willow, sage willow, hoary willow

Salix candida Flügge.

Family: Salicaceae

Sage-leaf willow is probably the most distinctive native shrub willow in the Midwest. Its leaves are usually densely covered with short, white, woolly hairs that are much like the leaves of the herb after which this plant is named. It is much more common in the northern portions of the Midwest, and it is currently listed as threatened in Indiana and Ohio, and endangered in Pennsylvania.

Form and Size: It is a petite shrub, usually less than 3 feet tall, with a rather open branching. It becomes larger in more northerly climates, and it prefers full sun for best growth.

Habitat: Sage-leaf willow is found in bogs, sometimes fens, sedge meadows, shorelines, and stream borders—it is limited by its requirement for calcareous habitat. It tends to be common where large, shrubby species do not dominate.

Wildlife Uses: The reader is referred to the *Salix* genus page 192.

Landscaping Value: Hardy to Zone 1, this is truly a beautiful shrub, especially when in flower or fruit, and the authors are testing plants for their ability to adapt to a more domesticated landscape. So far, pampering helps. One specimen that was planted in dark "prairie soil," mulched, and ignored, is not much over 1 foot tall. Another is a specimen plant in front of the house along a sidewalk in clay-loam amended with much sand. It is waist-high and needs pruning to control growth that occasionally overhangs the walk. Japanese beetles have been the biggest problem to date after 3 growing seasons. This willow is unfortunately unavailable from even those nurseries specializing in natives.

Similar Species Distinctions:
—**Russian olive** (*Elaeagnus angustifolia*) and **autumn olive** (*E. umbellata*) both superficially, from a distance, look similar just because their leaves are gray-green. Both are introduced species that have linear leaves and silvery scales covering the leaves and twigs. Russian olive is a small tree; autumn olive is a large shrub. Both produce perfect flowers, instead of unisex catkins, and fleshy fruits.

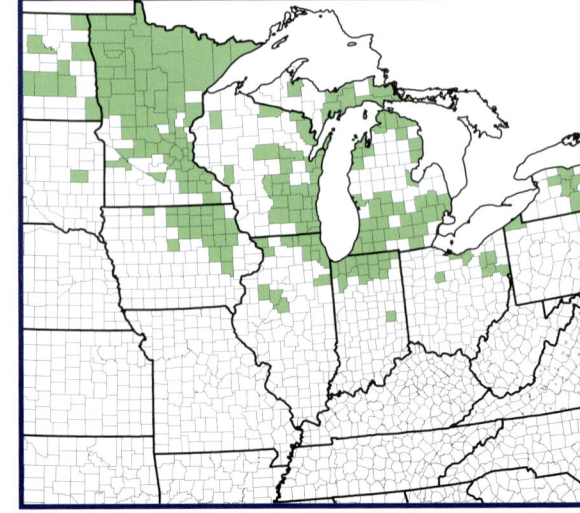

The alternate leaves are linear-oblong, about 2½ inches long by ½ inch wide. The upper surface is dark green, finely and deeply veined, and shiny or densely hairy. The lower surface is usually extremely white with dense hairs. Margins are commonly curved under and entire. Stipules may be present. This species hybridizes with several other natives, so variability of the leaf shape and texture can be expected in some cases.

Fruits ripen in early June and are usually woolly white all over. They are densely packed with capsules that split open and release many tiny seeds that are attached to white fluff (pappi).

Flowering occurs in mid- to late May with the emerging leaves. Female aments (right) are densely flowered and hairy. The stigma lobes are often deep red and stand out against the dense, white hairs. Both male and female aments look like balls of fuzz before they open. Male (left) and female flowers are usually on separate plants.

Twigs are slender and vary in color and texture. They are generally brownish and can be smooth or densely white-hairy. Lateral buds are all the same size and vary in texture as well. They are covered with a single bud scale. Leaf scars are somewhat V-shaped and contain 3 bundle scars.

This willow does not achieve a diameter large enough to develop any unique bark. The stems are usually brownish.

Carolina willow, Ward's willow, coastal plain willow

Salix caroliniana **Michx.**
Salix longipes **Shuttlew.**
Family: Salicaceae

This is a small tree of the South, and it has a restricted range in the Midwest. Here, it grows on gravel bars in rivers and large streams where flooding causes hardships on individuals. It is fairly easy to recognize when in leaf by its elongated leaves with finely toothed margins, its nearly white leaf backs, and the numerous, sizable stipules. In the South, it hybridizes with *S. nigra*, black willow, which it resembles in many characteristics. It is listed as endangered in Pennsylvania, threatened in Ohio, and was recently removed from endangered status in Indiana.

Form and Size: Plants are often only a few feet tall, bent over, and covered with some amount of flood debris in the Midwest. The farther south one moves in its range, the larger it is. It prefers full sun for best growth, but it seems to handle some shade along the waterways.

Habitat: Carolina willow is found on rocky, gravelly streambeds where it seems to be adapted to flooding events. Farther south, it grows along waterways, road ditches, and roadways.

Wildlife Uses: The reader is referred to the *Salix* genus page 192.

Landscaping Value: Hardy to Zone 5, this is an attractive willow when in leaf because of the contrasting green and white leaf surfaces that flash when blown in the wind. It grows well in upland sites, where it quickly becomes large. The authors started several from seed, and after 4 growing seasons on an upland site, they are 5 feet tall and nearly as wide. Flowers, fruit, and fall color are drab. It is doubtful that it can be purchased, at least in the Midwest, from any nursery.

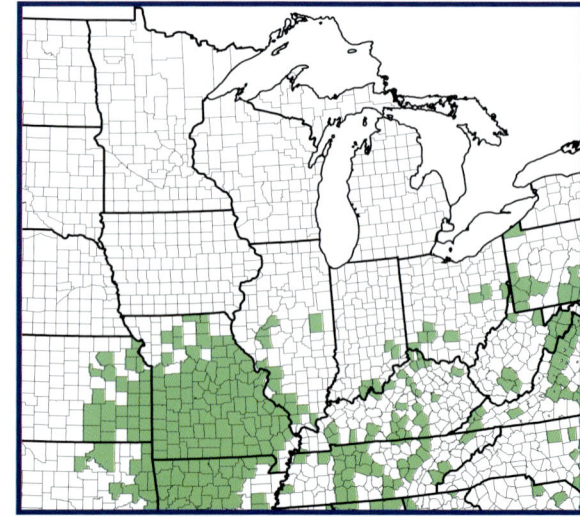

Similar Species Distinctions:
—**Black willow** (*S. nigra*) is tree-sized, but it has similar leaves without the glaucous back. Its leaves are usually narrowly linear.

The alternate leaves are narrowly lanceolate, 4 to 5 inches long, and ½ inch wide. The upper surface is highly glossy, green, and smooth, while the lower surface is usually coated with a white, glaucous bloom. The margins are finely toothed, and both leaf tip and base are pointed. The stipules create a nearly complete circle around the twig at each node. Stipules have small teeth along the margins.

Fruits are capsules that ripen late May and early June. They are in loosely-fruited aments that split open and release the tiny seeds that are attached to cottony fluff (pappi). Ovaries are smooth.

Flowers appear in mid-May along with the emerging leaves, and they are very elongated for a willow ament; they are extremely slender and up to 9 inches long. Aments are somewhat loose-flowered and essentially hairless when open, except for the floral bracts at the base of the ovaries (which are hairy). Flooding debris is seen on both male (left) and female (right) aments in these pictures.

Bark is light brown to gray, with flat-topped, narrow ridges along the trunk.

Twigs are slender and brownish or yellowish, and they vary from smooth to hairy. Lateral buds have a single bud scale, are the same color as the twig, and are all 1 size.

heart-leaf willow, sand dune willow, dune willow

Salix cordata **Michx.**
Salix adenophylla **Hook.**
Salix syrticola **Fern.**
Family: Salicaceae

This interesting willow, with its unique leaves, is usually very distinctive. Its thick, heart-shaped leaves, as well as its extremely hairy twigs, make this willow easy to identify. Heart-leaf willow is listed as threatened in Indiana, and endangered in Illinois and Wisconsin. Pictured is heart-leaf willow on the north shore of Lake Michigan near Manistique, Michigan.

Form and Size: It is rather few-branched and commonly only 4 feet tall. It can, however, become twice that tall. It requires full sun for best growth.

Habitat: Heart-leaf willow is restricted to sand dunes and beaches along the Great Lakes.

Wildlife Uses: Reader is referred to the *Salix* genus page 192.

Landscaping Value: Hardy from Zones 2–6, this willow is limited by habitat, much of which has been altered. It is an interesting willow with such unique leaves that it is never mistaken for another. Winter twigs and buds are usually densely hairy and quite attractive. It has grown well in a deep loam for the authors, and its thick leaves are not overly preferred by Japanese beetles or other insects. This is 1 native willow that can easily be purchased through nurseries that specialize in natives.

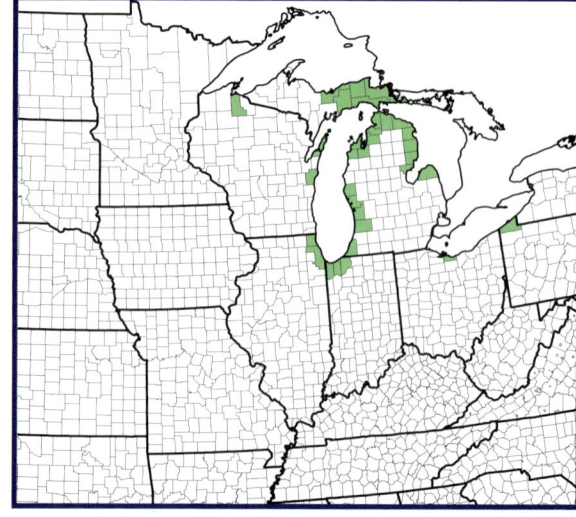

Similar Species Distinctions:
—No other willow has leaves like heart-leaf, and its habitat eliminates many woody possibilities.

The alternate leaves are thick, leathery, and roughly heart-shaped. They are about 3½ inches long and have a short, pointed tip and rounded base. They are dark green on both sides, with silky, white hairs all over. Margins have fine, gland-tipped teeth. Stipules are obvious and have fine, gland-tipped teeth as well.

Flowers appear in mid-April at the same time as leaves are emerging. Aments are about 1½ inches long and are fairly densely flowered. The female aments (near left) have smooth ovaries, but all other floral parts have long, silky, white hairs. Flowers are so hairy that the entire ament looks fuzzy.

Twigs are stout and usually thick, with wooly, white, or grayish hairs, especially on the newest growth. Lateral buds are of 1 size and are often wooly white or grayish. Leaf scars are somewhat V-shaped and contain 3 bundle scars.

Fruits ripen in early May or later in aments that look fuzzy. Ovaries are smooth, however, and split open to release the numerous tiny seeds that are attached to white fluff (pappi).

Heart-leaf willow does not attain a large enough size to develop a characteristic bark. Stems are brownish.

pussy willow

Salix discolor **Muhl.**

Family: Salicaceae

Pussy willow is 1 of the Midwest's most common shrub willows, and it is the only 1 prized as an ornamental. Its large, fuzzy, male flowers begin to appear in February, but they do not fully expand until early April. The white "pussy toes" add an interesting springtime dimension to an otherwise winter-like landscape.

Form and Size: It commonly grows to 12 feet but can become much taller. It usually has a single, low-branched trunk with numerous, upreaching, large branches. A large specimen is truly tree-like.

Habitat: Pussy willow is common in marshy, swampy ground, but it grows well on more upland sites. It requires full sun for best growth.

Wildlife Uses: This large willow is occasionally used by higher-nesting birds for nest sites, especially in wetland areas; such species include kingbirds, cedar waxwings, and green herons. The reader is referred to the *Salix* genus page 192.

Landscaping Value: Hardy to Zone 1, this is a wide-ranging willow that is commonly seen in domestic landscapes. It has to be given room to grow or pruned heavily to maintain its size. Only male plants are available, as the female flowers are completely different. Fall color of willows is less than spectacular, and they are susceptible to damage by numerous insects. There are several horticultural varieties, including 1 with pink male flowers. All require full sun for best flower production and growth.

Similar Species Distinctions:
—There are several native willows that attain the size of pussy willow. If out in early spring, one will notice only 3 willows in the Midwest that produce flowers before leaves. *Salix bebbiana* (Bebb's), *S. humilis* (prairie), and *S. discolor*. Bebb's is a large shrub like pussy willow, but the floral bracts (scales at the base of female flowers) are tan, not brown/black. Prairie willow has densely hairy twigs and buds.

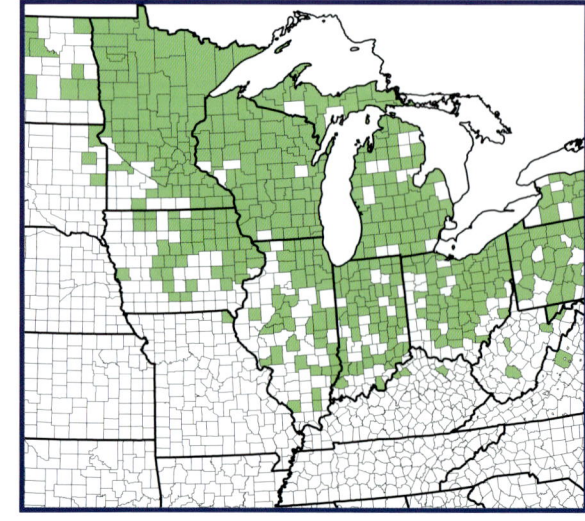

The alternate leaves are variable in shape and size, but they are usually narrowly elliptic, 3 to 4 inches long, and 1½ inches wide. They are dark green and shiny above, and whitened and sometimes hairy beneath. Leaf margins are entire, wavy, or slightly toothed. Stipules are common on summer growth.

Fruits ripen in mid- to late April (in Indiana) and are up to 3 inches in length. They are a loosely fruited cluster of capsules that split to release numerous tiny seeds that are attached to white fluff (pappi).

Flowering begins as early as late March (in Indiana) before the leaves emerge. Aments are about 1 inch long. Females (far left) are densely flowered and have a dark floral bract at the base of each hairy ovary. Ovaries are hairy. Male ament is at near left.

Twigs are brownish, rather stout for a willow, and usually have some degree of hairiness. Buds are of 2 sizes (flower buds larger), and the scales are brown. Leaf scars are somewhat V-shaped and contain 3 bundle scars.

Pussy willow bark is gray-brown with interlacing ridges. The size of the trunk of this species can be substantial, and with age, it may reach nearly 1 foot in diameter.

Missouri willow, diamond willow, heart-leaf willow

Salix eriocephala **Michx.**
Salix cordata **Muhl.**
Salix rigida **Muhl.**
Family: Salicaceae

Missouri willow was once commonly referred to as heartleaf, which is a confusing name when the mature leaves are considered. From a leaf standpoint, its closest shrub willow look-alike is blue-leaf willow (*S. myricoides*). Blueleaf tends to have more intensely white lower leaf surfaces, more hairy midribs, and wider leaf blades.

Form and Size: It can become large—up to 20 feet tall—but is generally 10 feet or less. It develops a full crown made up of many small branches, and it can form colonies through stem fragmentation.

Habitat: Missouri willow is found in wet areas along streams, in seepy sites, in road ditches, and near marshes. It requires full sun for best growth.

Wildlife Uses: Reader is referred to the *Salix* genus page 192.

Landscaping Value: Hardy to Zone 2, this large willow can be grown in an upland site, but plenty of room is necessary. Quite often, the tips of new leaves are bright red. It produces large fruit that is often reddish-tinged as well. Leaf backs can be quite glaucous white and are noticeable when the wind blows them. This is 1 of the willows that develops trunk cankers. People cut the stem, peel off the bark, and use them as walking sticks—commonly known as "diamond willow sticks." This species is not available from nurseries, but cuttings at least the diameter of a thumb will root.

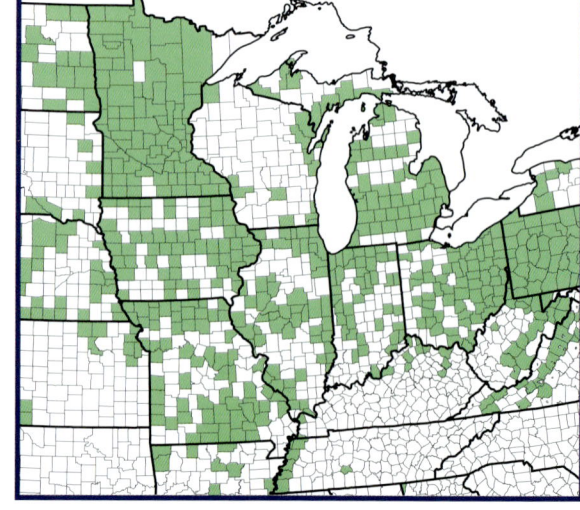

Similar Species Distinctions:
—**Blue-leaf willow** (*S. myricoides*) is mentioned above and most closely resembles Missouri willow. See species account for more comparisons.

The alternate leaves are narrowly oblong, 4 to 5 inches long, and about 1 inch wide. Margins are finely toothed; the leaf tip is pointed. The upper leaf surface is dull or shiny and usually smooth. The lower surface is slightly glaucous and lightly hairy all over. Stipules are usually present.

Flowers emerge just before or with the leaves in mid- to late April. Aments are large, up to 2 inches long, and moderately to densely flowered. Ovaries (near left) are smooth and often tinged with red. Male flower is at far left.

Fruits ripen in early May on large aments. The capsules are smooth, and they split to release the numerous tiny seeds that are attached to white fluff (pappi). Fruit is very similar to that of *S. myricoides*.

Twigs are reddish brown, slender, and usually covered to some degree with short hairs. Buds are all of 1 size and roughly the same color as the twig. Leaf scars are somewhat V-shaped and contain 3 bundle scars.

Bark is not well developed because of the small size attained by this species. Bark is gray-brown and somewhat interlacing. It can develop "diamonds" in the trunk and large stems (actually cankers) that are attractive when stems are peeled and made into walking sticks.

prairie willow

Salix humilis **Marsh.**

Salix humilis **Marsh. var.** *humilis*

Salix humilis **Marsh. var.** *tristis*
(Aiton) Griggs

Salix tristis Aiton

Family: Salicaceae

Prairie willow is 1 of a few of our native willows that flower before the leaves emerge; *S. discolor* and *S. bebbiana* are the others. Finding prairie willow in the spring is fairly easy—just look for a small (commonly) willow growing in an upland site that has flowers, no leaves, and hairy twigs. It is pictured on the opposite page with the fruiting capsules bursting. Seeds can be seen as dark specks mixed with the white fluff. Even at this stage, when fruits are almost past, the leaves are very small.

Form and Size: It can reach a height of 10 feet, but it is usually 3 to 4 feet tall. It can form small colonies by stem layering. Full sun is required.

Habitat: Prairie willow is usually found in drier sites than other willows, and it is common in dry openings, prairies, roadsides, and disturbed, sandy sites.

Wildlife Uses: Because of its upland nature, prairie willow supplies nesting sites for a different cadre of songbirds than do wetland-associated willow species, and it is used with some regularity. This willow also seems to be browsed more regularly by white-tailed deer than other willows. Reader is referred to the *Salix* genus page 192.

Landscaping Value: Hardy to Zone 4, this species can be of various sizes. It usually has a rounded form, densely hairy buds and twigs, and pretty, hairy female flowers and fruits. Prairie willow is a complex species with several varieties recognized by some taxonomists. Leaf size and form is often quite variable, and this species can hybridize with several other native shrub willows, complicating its identification further. This is 1 native willow frequently available from nurseries specializing in natives.

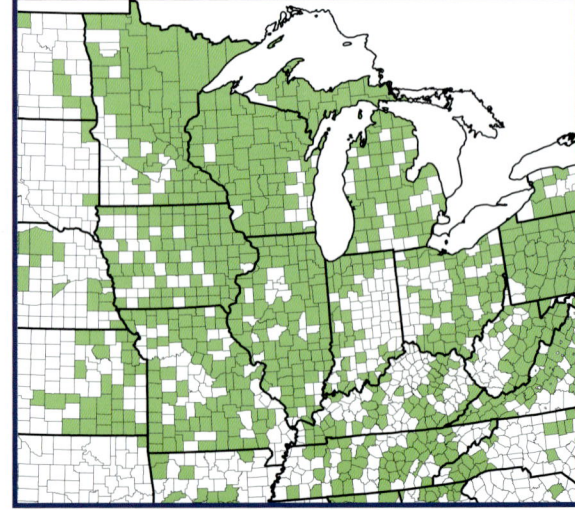

Similar Species Distinctions:
—**Pussy willow** (*S. discolor*) produces flowers early as does prairie willow, but they are much larger and have nearly black floral bracts. Its winter buds and twigs can be hairy, but the leaves are usually larger.

The alternate leaves are narrowly oblong, about 2 inches long, and about ½ inch wide. The upper surface is green, shiny, and smooth or hairy, with numerous impressed veins. The lower surface can be glaucous or glaucous/hairy. Leaf margins are usually wavy with a few rounded teeth. The leaf tip is abruptly pointed.

Prairie willow rarely attains a diameter that allows it to develop a characteristic bark. Larger stems are like most other willows—brownish.

Twigs are usually dark brown and densely covered with whitish hairs, especially on the newest growth. Buds are the same color as the twig and have a single bud scale that is also hairy. Buds vary little in size. Leaf scars are somewhat V-shaped and contain 3 bundle scars.

Flowers appear in early April in southern Indiana, before the leaves, in compact, relatively short aments. Ament length is variable, but it can be as much as 1 inch. Ovaries (near left) are quite hairy and in fairly densely flowered aments; male aments are pictured at far left.

Fruits ripen by late April and continue into May. The capsules have a pinkish cast and are covered with short, silky hairs. Capsules split open and release many tiny seeds that are attached to white fluff (pappi).

sandbar willow, longleaf willow

Salix interior **Rowlee**
Salix longifolia **Muhl.**
Salix exigua **Nutt.**
Family: Salicaceae

This is the only midwestern native shrub willow that spreads by root shoots. It suckers from the center of a colony outward. So the oldest, largest plants are in the middle; the youngest, smallest plants are along the edge of the linear or circular colony. This growth habit makes sandbar willow easy to spot from a distance. It also has the longest flowering period of any native willow. It can begin flowering in early April; this photo was taken on June 21 in west-central Indiana, and both empty fruiting capsules and new flowers can be seen.

Form and Size: It is a large shrub to small tree and can reach a height of 20 feet. It has a rather spindly growth form with slender branches that often die back. Suckering produces sizable colonies, especially when they grow in flood-prone areas.

Habitat: Sandbar willow is common throughout the Midwest in low, wet areas, including wet ditches, lake and pond margins, and river floodplains. Full sun is required for best growth.

Wildlife Uses: Wetland songbirds frequently use the unique, dome-shaped clones for nest sites and escape cover. Taller specimens frequently have witches'-brooms that supply unique nest-site opportunities. The reader is referred to the *Salix* genus page 192.

Landscaping Value: Hardy to Zone 2, this is a weedy, colonizing species that gives willows a bad reputation. In reality, none of the other native willows in the Midwest root sucker. This is certainly the least desirable willow species for a landscape (it is the ugly duckling of our willows), unless the goal is to stabilize soil along a wetland or waterway. It is adapted to flooding and grows quickly. It is unlikely to be sold by any midwestern nursery.

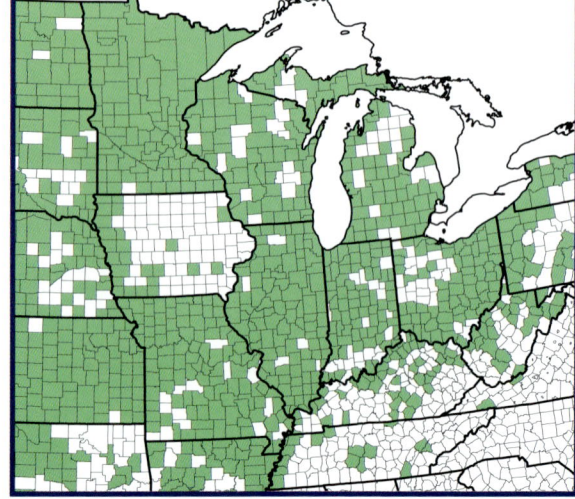

Similar Species Distinctions:
—**Black willow** (*S. nigra*) has similar leaves and similar growth form when young. Leaf backs are usually light green, never lightly glaucous.

The alternate leaves are elongate, linear, and about 5 inches long (⅓ inch wide). The upper surface is dark green, shiny, and mostly smooth. The lower surface is slightly paler, thinly glaucous, and sometimes hairy, giving the leaf a gray appearance. Leaf margins are slightly toothed. Stipules can be present.

Fruits ripen over a several month period, beginning in late April. The loosely fruited aments are 2 inches long and green. They are usually dangling down from the slender twigs. The capsules split open and release many tiny seeds that are attached to white fluff (pappi).

Flowering begins in early April, with the emergence of the leaves, and continues well into the summer. Aments are long and slender—about 2 inches long and only ¼ inch wide. Females (right) are loosely flowered and greenish, with sparsely hairy, short ovaries.

Twigs are slender, greenish yellow (especially on the new growth) to brown, and mostly smooth. Lateral buds have a single scale, are all the same size, and are the same color as the twig. Leaf scars are somewhat V-shaped and contain 3 bundle scars. Twig growth continues beyond the flowers (sylleptic growth), as seen to the left of this male ament.

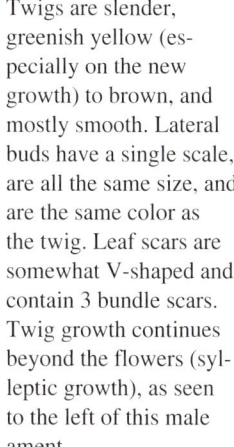

Mature bark is gray-brown and thin. It develops some non-continuous, interlacing ridges on older trunks.

shining willow

Salix lucida **Muhl.**

Family: Salicaceae

As its common name suggests, this willow does seem to shine because of its reflective upper leaf surfaces. A large male plant in flower is hard to miss in the spring (as shown here). Its large, 2-inch-long aments are loaded with bright yellow stamens that are prominent and showy—and very attractive to bees. In Iowa, it is listed as state threatened.

Form and Size: It is usually a large shrub and commonly reaches a height of about 10 feet in our region, but it can become nearly tree-sized and grow to 20 feet tall. It has an open, upright branching habit.

Habitat: Shining willow occurs only in wet sites in northern climates, in areas such as along marsh and pond edges, wet road ditches, and boggy sites in full sun.

Wildlife Uses: This species tends not to be as "limby" as many other willows and has lower cover value as a result. The reader is referred to the *Salix* genus page 192.

Landscaping Value: Hardy to Zone 1, this is a large, spreading shrub that has very striking leaves. The male plants are also pretty when in flower, and they attract myriads of bees and other pollinators. Stems are yellowish, shiny, and attractive. Most willows will grow in upland sites, if competition is not too great, and the authors have grown several in such a location with great success. Willows are a good source of soil stabilization along waterways, and planting them in such a location is beneficial. This willow is occasionally available through nurseries specializing in natives.

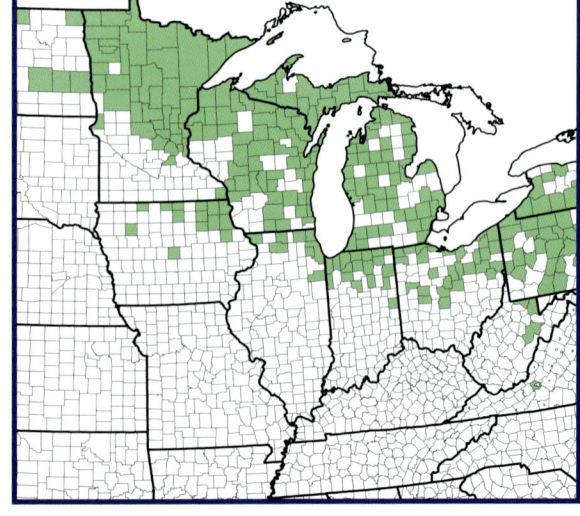

Similar Species Distinctions:
—**Autumn willow** (*S. serissima*) has similar leaves, but its leaf tips are much less elongated. It also lacks the glands that line the stipules, proximal blades (those that appear at the base of flowers), and petiole.

The alternate leaves are ovate-lanceolate and have a very elongated tip. The upper surface is dark green, shiny, and smooth at maturity. The lower surface is usually light green, and it may be either smooth or hairy. The leaf margin is finely toothed, and there are small glands near the leaf base. Stipules have gland-tipped, fine teeth.

Flowers appear in mid- to late April with the emergence of leaves. The aments are about 2 inches long and are moderately to densely flowered. Ovaries (right) are smooth and short; a male ament is pictured at left. Leaves on the flowering stalk have glandular, toothed margins.

Twigs are light brown and shiny (sometimes hairy). Lateral buds are all the same size; there is a single bud scale per bud that is usually the same color and texture (or nearly so) as the twig. Leaf scars are somewhat V-shaped and contain 3 bundle scars.

Fruits ripen in mid- to late May and are 2 inches long. They are loosely fruited with smooth, green capsules that split to release many tiny seeds that are attached to white fluff (pappi).

Mature bark is gray-brown, and with age, it develops very loose interlacing flat ridges.

blue-leaf willow, bayberry willow

Salix myricoides **Muhl.**
Salix glaucophylla **Bebb**
Salix glaucophylloides **Fern.**
Family: Salicaceae

Blue-leaf willow is aptly named, a factor to keep in mind when searching for this shrub. The leaves have a dark green upper surface. The underside, which is usually coated with a thick layer of white, glaucous bloom, gives the leaves a bluish appearance when they are blowing in the wind. It seems to be an optical illusion. This species is easily confused with *S. eriocephala*.

Form and Size: It is a moderate-sized shrub that reaches about 10 feet in height. It has a full, compact form with many small branches. It prefers full sun for best growth.

Habitat: Blue-leaf willow is common in the Upper Lake States and Canada. It occurs in open, sandy areas near Lake Michigan, on edges of ponds, and in fens and marshes. It can form colonies by stem fragmentation.

Wildlife Uses: Reader is referred to the *Salix* genus page 192.

Landscaping Value: Hardy to Zone 2, this northern willow is attractive when the lower leaf surface is exposed on a windy day. New leaves are commonly tinted red and stand out nicely against the lime-green existing foliage. This is usually a sizable shrub that would be a nice addition to any shoreline. This species is not available through native-plant nurseries.

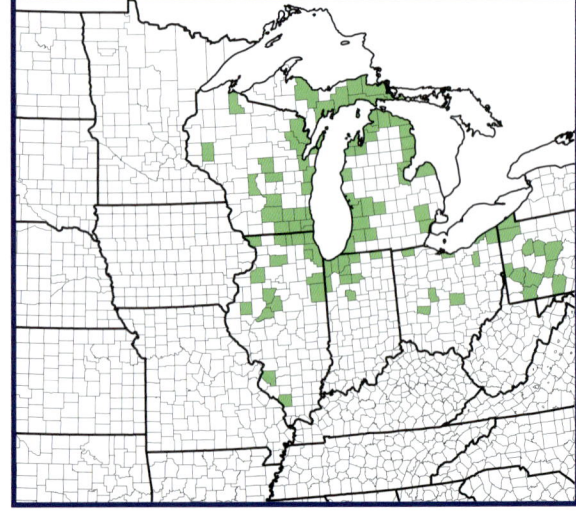

Similar Species Distinctions:
—**Missouri willow** (*S. eriocephala*) is so similar in every respect that distinguishing between the 2 is very difficult. *S. myricoides* leaves have fewer teeth per side and a dense, glaucous coating beneath.

The alternate leaves are narrowly oblong and elliptic to oblanceolate; they are about 3 inches long and ¾ inch wide. The upper surface is dark green, commonly shiny and smooth. The lower surface is usually coated with a heavy, glaucous bloom with hairs along the midrib (more hairs than *S. eriocephala*). Leaf margins are finely toothed, but they have less teeth than *S. eriocephala*. Stipules are usually present on vigorous summer growth. New growth is often tinted red.

Flowers appear as early as mid-April. Males (far left) open before the leaves; females (near left) open with the leaves. The aments are loosely flowered and about 1½ inches long. Ovaries are smooth and often tinted red; floral bracts are hairy all over. The aments are more loosely flowered than those of *S. eriocephala*.

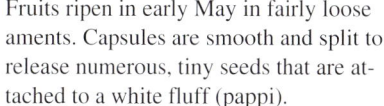

Fruits ripen in early May in fairly loose aments. Capsules are smooth and split to release numerous, tiny seeds that are attached to a white fluff (pappi).

Twigs are reddish brown and usually sparsely covered with short hairs. Buds are brown, often 2-toned, and have a single scale. Leaf scars are somewhat V-shaped and contain 3 bundle scars.

Bark does not develop any characteristic pattern, since blue-leaf willow remains of small diameter. Trunks are brownish, and similar to those of many other small shrub willows.

bog willow

Salix pedicellaris
Pursh.
Salix pedicellaris
Pursh. var. *hypoglauca*
Fern.
Family: Salicaceae

Bog willow is 1 of our easiest native willows to identify when in leaf. It is our only willow with a rounded leaf tip (usually). However, finding this diminutive species is the real challenge. It is almost exclusively associated with bogs. A common bog associate, leatherleaf, has leaves that are so similar in appearance that it is easy to confuse the 2. This willow is currently threatened in Iowa, and endangered in Ohio and Pennsylvania.

Form and Size: It is rarely more than 3 or 4 feet tall, but it can reach 6 feet in height. It is a small willow, with an erect growth form that is rather sparsely branched. It grows best in full sun.

Habitat: Bog willow is almost totally restricted to bogs and fens. Branches will trail through sphagnum hummocks, where they root and send new shoots upward, eventually forming small colonies.

Wildlife Uses: This diminutive willow sometimes provides nesting cover for songbird species that often nest in sphagnum hummocks in bogs, such as Nashville warblers and white-throated sparrows. The reader is referred to the *Salix* genus page 192.

Landscaping Value: Hardy to Zone 1, this willow prefers slightly acidic, moist sites. It is a beautiful, petite shrub that the authors have grown in 2 locations, both upland sites, with great success. One site is dark, prairie loam. The other is clay-loam mixed with sand and peat. Both locations have robust shrubs several feet tall, with deep green leaves. This species is not available through nurseries, but it can be propagated from stem cuttings.

Similar Species Distinctions:
—**Leatherleaf** (*Chamaedaphne calyculata*) is a common associate of bog willow and has very similar-looking leaves. The leaves, however, have silvery, circular dots on both sides that are easily seen without a hand lens.

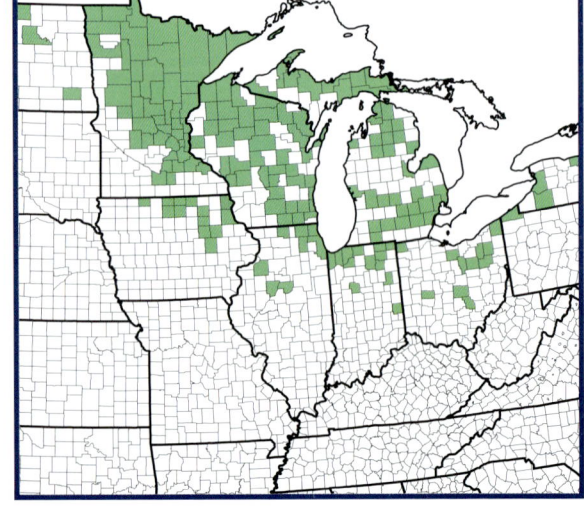

The alternate leaves are about 1½ inches long and ½ inch wide with entire margins. They are broadly elliptic to oblanceolate to obovate. The upper surface is dark green and dull; it is almost leathery and has a prominent, fine venation pattern. The lower surface is covered with a glaucous bloom and is smooth. There are rarely any stipules.

Flowers appear in late April into May with the emerging leaves. They are about 1 inch long in fairly loose aments. The smooth ovaries are attached to stalks (stipes) that are as long or longer than they are. Ovaries (near left) are often tinted red and have a single hairy floral bract at the base of each. Male ament is at far left.

Fruit ripens in late May and early June in 1 inch long aments. The ament, or catkin, is a head of smooth capsules that split open to release many tiny seeds that are attached to white fluff (pappi).

Bog willow is a small willow, and its stem diameter is never such that it develops any characteristic bark; it remains brown and smooth with slight fissuring.

Twigs are brownish, slender, and lightly covered with straight hairs. Lateral buds are of all the same size and are the same color as the twig. They have a single scale and are usually hairless. Leaf scars are somewhat V-shaped and contain 3 bundle scars.

meadow willow

Salix petiolaris **Smith**

Family: Salicaceae

Meadow willow is a northern species that is scattered in low, wet areas. It is a narrow-leaved willow that has several features to help it stand out among the willow crowd. If you find it in fruit, the aments are quite small, hairy, and loosely-fruited. The leaves are narrow, linear, and similar to several other native willows; however, when the leaves first appear, they are usually covered with long, silky hairs, as seen in this photograph. Meadow willow is listed as threatened in Ohio, and endangered in Pennsylvania.

Form and Size: Although it can become quite large (over 20 feet tall), it is usually only 6 to 7 feet. It is usually a moderate-sized willow with a full shape.

Habitat: Meadow willow is found locally in calcareous meadows and fens, marshes, and sandy, peaty areas. It forms loose colonies by stem fragmentation. It is often mistaken for silky willow (*S. sericea*), if meadow willow's leaves are hairy; however, leaf and fruit shape and size can distinguish between the 2. Full sun is required for best growth.

Wildlife Uses: Reader is referred to the *Salix* genus page 192.

Landscaping Value: Hardy to Zone 1, this is a tough willow accustomed to very cold temperatures. It is common in the upper Midwest, and it is a very attractive willow. Its leaves are pale beneath, and this feature is easily seen when the wind blows through them. This species would be a nice selection to plant in a wet site or even along a pond. It grows well in an upland site. Meadow willow is rarely available through nurseries, but it is easily propagated from stem cuttings.

Similar Species Distinctions:
—**Silky willow** (*S. sericea*) leaves are usually silky from flattened hairs beneath. New leaves of meadow willow are silky as well, and they have a similar shape, but mature leaves lose the hairs.

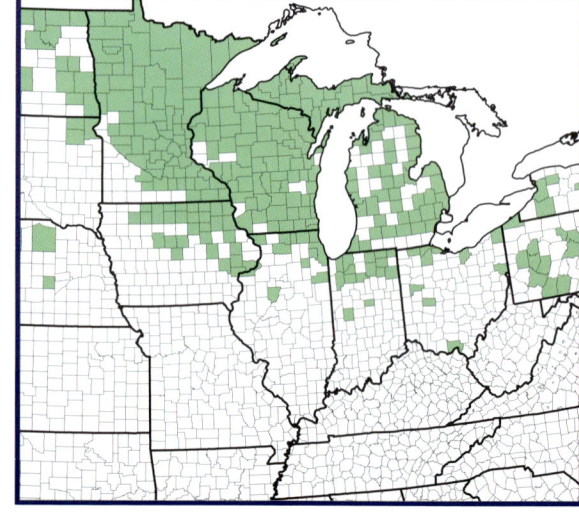

The alternate leaves are linear-lanceolate to ob-lanceolate, 2 to 3 inches long, and about ½ inch wide. The upper surface is green and smooth to hairy, while the lower surface is glaucous and varies from nearly smooth to quite hairy. Leaf margins are entire to finely toothed. Stipules are usually not present.

Flowers appear with the leaves from mid-April to early May. They are stout, fairly small (averaging about ¾ inch long) and loosely flowered. Ovaries (near left) are covered with flattened, silky hairs, and they have a hairy floral bract at the base of each. A male ament is pictured at far left.

Fruits begin to ripen in mid-May and continue into June. The aments are short, stout, hairy, and loosely-fruited. Each ovary is silky-hairy and splits to release many tiny seeds that are attached to white fluff (pappi).

Twigs are slender, dark brown to yellowish or reddish, and smooth to velvety-hairy. Lateral buds are twig-colored, multi-colored, or nearly black, with a single bud scale. Leaf scars are somewhat V-shaped and contain 3 bundle scars.

Meadow willow is not a large willow in our region, so there is no recognizable bark pattern. Large trunks are gray-brown with some fissuring.

silky willow

Salix sericea
Marsh.

Family:
Salicaceae

Silky willow is a
nice addition to an
urban landscape, for
it remains moder-
ate in size, does not
sucker, and has nice,
full form. The silky
white leaf-backs
shine when the
leaves blow in the

wind, adding a beautiful element to this species. Silky willow commonly hybridizes with other shrub willows, especially *S. eriocephala*.

Form and Size: It can reach a height of 13 feet or more, but it is generally shorter. Silky willow is often
found in colonies that are formed by stem fragmentation (stems break off, root where they fall, and grow
into a new plant).

Habitat: Silky willow can be found in a variety of habitats, but all are moist-to-wet, including bogs,
swampy woods, sandy swales, wet forest edges, roadsides, and pond edges. It grows best in full sun.

Wildlife Uses: Its limby structure makes it especially attractive as a nest site for wetland birds. Reader is
referred to the *Salix* genus page 192.

Landscaping Value: Hardy to Zone 3, silky willow is an attractive, medium-sized plant that has a neat,
compact appearance. Its leaves and flow-
ers are particularly pretty. The authors are
growing the species as an unpruned (at least
by us!) hedge, and the main problem is deer
browsing and Japanese beetles. All plants are
6 feet tall after 3 growing seasons. This species is occasionally available through native-
plant nurseries, but it is easily established
from stem cuttings.

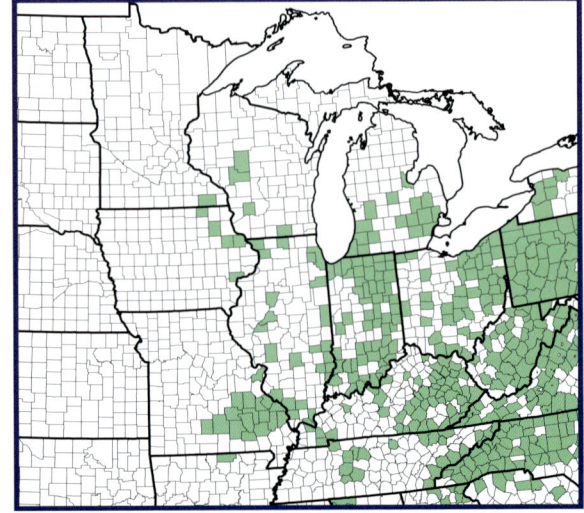

Similar Species Distinctions:
—**Meadow willow** (*S. petiolaris*) has
similar leaves and growth form. Its new
leaves are silvery underneath but are
usually more elongate.

The alternate leaves are about 3 inches long, ½ inch wide, and narrowly elliptic to elliptic. The margin is finely toothed, and the leaf tip is usually sharply pointed. The upper surface is green, dull, and smooth (can have a few hairs); while the lower surface is usually glaucous, it is often obscured by silky white hairs. There are usually no stipules, a feature that often helps to distinguish between this species and Missouri willow, *S. eriocephala*. Leaves turn black as they dry, which is quite unique to this species.

Flowering occurs from early to mid-April just before or with the emergence of the leaves. The authors have seen flowers on southern Indiana specimens as early as late March. The aments are about 1¼ inches long and are loosely to moderately flowered. The aments are silky in appearance, which is attributable to the hairs on the ovaries and the hairy floral bracts at the base of each ovary (near left); male ament is at far left.

Fruits ripen from mid-April to mid-May. The ament is comprised of hairy capsules that split to release many tiny seeds that are attached to silky, white fluff (pappi).

Silky willow does not achieve a diameter large enough for it to develop a characteristic bark. It is much like other willows, with brownish, fairly smooth bark.

Twigs are highly brittle at the connection to a branch; they are slender, brownish, and usually have some degree of hairiness, especially on the newer growth. Lateral buds are of 2 sizes along the twig, with the larger flower buds intermixed with the vegetative buds. Buds are the same color as the twigs, or sometimes 2-toned, and have a single scale. Leaf scars are somewhat V-shaped and contain 3 bundle scars.

autumn willow

Salix serissima (Bailey) Fernald

Family: Salicaceae

This willow has many characteristics that are similar to those of shining willow, *S. lucida*, causing confusion in identification. The lack of glandular stipules on the leaves of autumn willow is 1 of the most obvious differences. Another fairly obvious difference is the leaf shape, in particular, the leaf tip. *S. lucida* usually has an especially elongated tip, whereas that of *S. serissima* is pointed but not elongated. The species' ranges overlap, but because autumn willow flowers later, there is no hybridization between the 2. Autumn willow is listed as endangered in Illinois, and threatened in Indiana and Pennsylvania.

Form and Size: It can become large, to nearly 20 feet tall, but more commonly is only 5 or 6 feet in the Midwest. Like most willows, it grows quickly when it has little competition.

Habitat: Autumn willow tends to be found in or near bogs, fens, cedar swamps, and along shorelines of lakes and streams. It requires full sun for best growth.

Wildlife Uses: Reader is referred to the *Salix* genus page 192.

Landscaping Value: Hardy to Zone 1, this willow is beautiful, with slightly shiny leaves that have gland-tipped teeth along their margins. It is a very "neat" look for a willow leaf. The sizable, shiny fruits are often retained late into the summer. This is another willow the authors have planted in an upland site, where it continues to grow robustly. Stem cuttings easily root, which is the easiest means of acquiring most willows. This species could be used in a naturalized setting near a pond or a similar location. It has not been found in midwestern nurseries that sell natives, but it is propagated easily from stem cuttings.

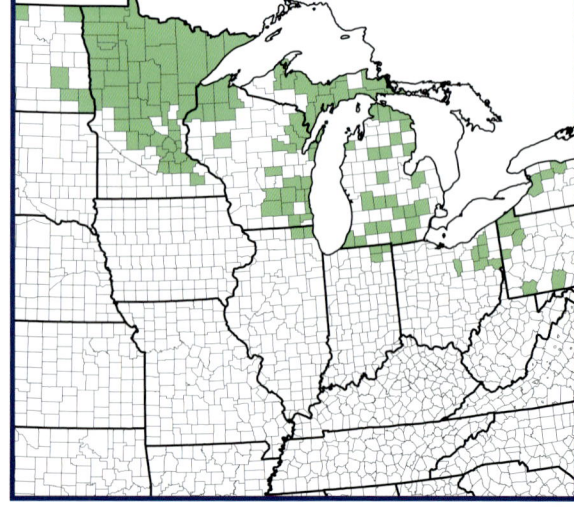

Similar Species Distinctions:
—**Shining willow** (*S. lucida*), as discussed earlier, has elongated leaf tips and obviously glandular stipules and proximal leaves.

The alternate leaves are about 3 inches long and ¾ inch wide. They are narrowly oblong, elliptic, or ovate with finely toothed margins. The leaf tips are pointed, but they are not elongated, as are those of *S. lucida*. The upper leaf surface is bright green, shiny, and smooth, while the lower surface is a bit paler, smooth, and shiny. Leaf petioles have paired glands, especially on new leaves.

Flowers appear with the emergence of the leaves in mid- to late May in northeastern Indiana, nearly a month later than those of *S. lucida*. These were photographed on May 14. The aments, or catkins, are about 1¼ inches long and somewhat loosely flowered. The ovary (near left) is smooth and has a tawny, hairy floral bract at its base; a male ament is at far left.

Fruits ripen slowly, and they are often seen in the fall. No other native willow retains its fruits as late as autumn willow. The fruit is a loosely fruited catkin with elongated, smooth capsules. The hairy floral bracts fall before the fruits ripen. Capsules split open to release many tiny seeds that are attached to white fluff (pappi).

In the Midwest, autumn willow does not attain a diameter large enough to produce any characteristic bark pattern. It is gray-brown and somewhat fissured.

Twigs are yellowish or tannish, often shiny and smooth. The lateral buds are of 1 size and darker-colored than the twig. Some of the buds pictured here are multi-colored. Buds have a single bud scale. Leaf scars are somewhat V-shaped and contain 3 bundle scars.

Spireas

Genus: *Spiraea*

In the world of horticulture, the genus *Spiraea* is no stranger. Numerous Asian species have been planted for several centuries in the United States, most commonly *S. prunifolia* Sied. & Zucc., bridalwreath spirea. The Midwest's 2 native spireas are much smaller, colonial shrubs usually associated with wetlands. Both make interesting landscape plants for several reasons. Their large, terminal inflorescences appear in the summer when most woody plants are past flowering. Additionally, their flowering period is quite prolonged, so a single plant can bloom for a month, moving from an initial terminal inflorescence to a series of axillary ones that bloom sequentially later.

Both species are easy to grow, thrive in full sun, and respond well to late winter pruning, which helps to maintain a full form. Interestingly, both species are fairly easy to purchase from nurseries that sell native plants, particularly those specializing in wetland restoration.

Both species of *Spiraea* occur in moist soils in openings where they have good cover values for songbird nesting, and occasionally protective cover, especially in *S. alba*. It is often part of good habitat structure for cottontails and woodcock. Seeds are sparingly used by birds and mammals. The genus has low browse preference by white-tailed deer; rabbits occasionally use bark and twigs.

Midwestern spireas have an interesting, upright, very straight growth form that is quite unique among native shrubs. Hardhack, *S. tomentosa*, is seen here in a wetland as the dominant shrub, although it very much resembles an herbaceous plant.

Spirea flowers all look similar, no matter what the species. *S. alba*, shown here, has 5 petals, long stamens, and a 5-lobed stigma, typical of them all. This meadow sweet picture illustrates the multiple, axillary inflorescences that develop over an extended period.

The simple, alternate leaves are similar between the 2 native species, but *S. tomentosa* has dense, whitish (sometimes cinnamon-colored) hairs on the lower leaf surface and over most of the plant.

meadow sweet, meadow spirea

Spiraea alba **DuRoi**

Family: Rosaceae

Meadow sweet is an attractive, colonial shrub that is always found in sunny, wet sites. However, the authors have grown it for several years in an upland site that receives only a half-day sun. Here it remains only waist-high and produces fewer flowers than normal. Flowering occurs over an extended period of several weeks, so the show seems to go on and on. It is currently listed as state endangered in Kentucky and Tennessee.

Form and Size: It can attain a height of 7 feet but is normally 3 to 4 feet tall. It is an erect, finely-branched, suckering shrub that forms colonies in low, moist areas.

Habitat: Meadow sweet is common in more northerly regions of the Midwest in low ground, such as the borders of lakes, wet prairies, and marsh meadows. It can tolerate partial shade but is a sun-loving species.

Wildlife Uses: Meadow sweet grows in wet openings where it occasionally forms loose colonies that, with other vegetation, produce good protective cover for cottontails and other early successional species. The upright growth form of this spirea produces an ideal nest-site for many associated songbirds, such as red-winged blackbirds, chipping and field sparrows, and indigo buntings. The species is lightly browsed by white-tailed deer, but it is far from preferred. Cottontails take bark in winter, if snow covers the ground. It produces a large quantity of seed, but recorded use by songbirds, pheasants, and small mammals is minimal.

Landscaping Value: Hardy to Zone 3, this slender, upright shrub has an unusual form that can only be found in spireas. Their rapid colonizing can easily be held in check by mowing. Beyond controlling suckers, this is an easily-grown species that requires little maintenance. Pruning stems back to 6 inches above ground in late winter promotes more dense growth. Japanese beetles do bother the plants, but not to a detrimental degree. This species is offered by several midwestern native nurseries.

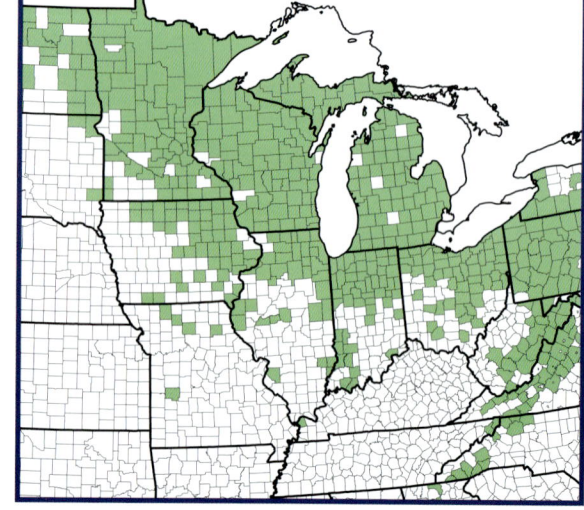

Similar Species Distinctions:
—**Hardhack** (*S. tomentosa*) has the same growth form, but its lower leaf surfaces are covered with dense hairs, and flowers are usually pink.

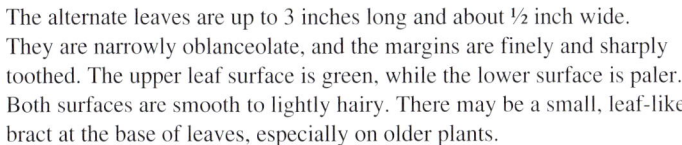

The alternate leaves are up to 3 inches long and about ½ inch wide. They are narrowly oblanceolate, and the margins are finely and sharply toothed. The upper leaf surface is green, while the lower surface is paler. Both surfaces are smooth to lightly hairy. There may be a small, leaf-like bract at the base of leaves, especially on older plants.

Fruits ripen in the fall in upright, persistent clusters that may be 12 to 15 inches tall. The individual fruits (follicles) have 5 segments, each with 2 to 3 tiny seeds.

Meadow sweet retains a small diameter, not much larger than a pencil, so there is no characteristic bark. With age, stems lose their thin outer, shreddy "bark" and become gray-brown and smooth.

Flowering begins in mid-June and can continue into September. Flowers are in large, terminal clusters (panicles) that are up to 4 inches tall, augmented with similar clusters from leaf axils to give a large "flower head." Individual perfect flowers are about ¼ inch in diameter, have 5 white petals (that sometimes have pink bases), and numerous long, white stamens. Complete flowering head can be seen on page 223.

Twigs are brownish to reddish brown with a thin layer of "bark" that splits and shreds with age. There are longitudinal ridges that are especially evident on vigorously growing stems. Lateral buds are plump and covered with numerous reddish brown scales that have silky, white hairs. There are often small, clustered buds at the nodes. Leaf scars have distinct, raised, triangular-shaped edges and contain a single bundle scar.

hardhack, steeplebush

Spiraea tomentosa **L.**

Family: Rosaceae

Hardhack is a spectacular native shrub with many outstanding features that are suitable for a native landscape planting. It has large, upright clusters of bright pink flowers that continue to be produced over an extended period of time. Nearly everything on the plant is densely covered with white or rusty-colored hairs. Its fruiting clusters, and sometimes the hairy leaves, overwinter, adding interest to a dull, cold landscape. The undersides of the leaves are always densely covered with whitish, brownish, yellowish, or rust-colored hairs.

Form and Size: Hardhack is found in low ground such as in acid bogs, on the borders of dune marshes, in old lake beds, and in wet prairies. It has a patchy distribution, but where found, can sometimes cover large areas.

Habitat: It is an erect, leafy shrub, generally 3 to 4 feet tall and colonial, suckering from the root system. It prefers full sun, and it seems to be adaptable to more upland sites.

Wildlife Uses: *Spiraea tomentosa* has essentially the same wildlife values as does *S. alba*. It does have a broader growth form, frequently having structure closer to the ground, which gives additional nesting cover for low-shrub nesting species, such as song sparrows and common yellow-throats, and overhead cover for ground-nesters. It is probably even less preferred as a deer browse species than meadow sweet.

Landscaping Value: Hardy from Zones 4–6, this spirea is probably restricted by climate, and it prefers the cooler, less humid regions of the upper Midwest. It is very hardy where it occurs, and it seems adaptable to upland sites. Everything on the plant from stems to flowers is hairy; the leaves are particularly attractive because of this feature. Leaves often overwinter on the plant, curl, and expose the velvety underside. Pruning in late winter to 6 inches or so encourages more dense growth. Hardhack can be purchased through some nurseries specializing in natives.

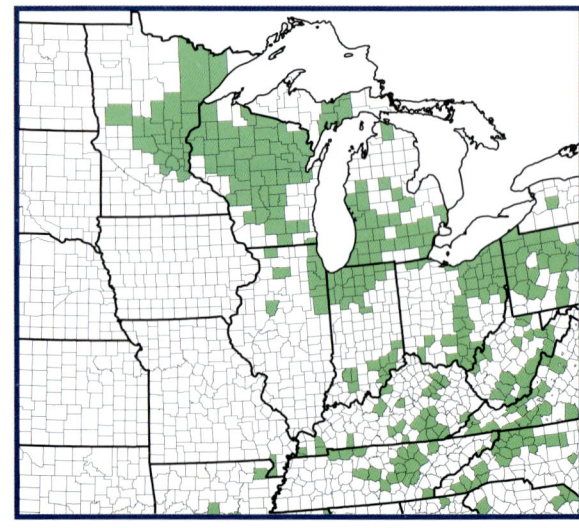

Similar Species Distinctions:
—**Meadow sweet** (*S. alba*) has similar leaves, but they are hairless beneath. Its flowers are usually white.

The alternate leaves are up to 3 inches long and ½ inch or more wide. They are ovate-lanceolate to narrowly oval and have coarsely or finely toothed margins. The upper surface is green and lightly hairy, while the underside is densely hairy and white, brownish, yellowish, or rusty in appearance.

Flowering occurs from mid-July into September. Flowers are in dense, hairy, upright terminal clusters (panicles) that are up to 6 inches tall augmented by axillary panicles that yield a fruiting head of 8 to 12 inches. Each flower is pink (rarely white) and about ⅛ inch in diameter. They have 5 pink petals and numerous, pink stamens.

Hardhack develops some roughness from raised lenticels on older stems, which become dark brown.

Fruits ripen in the fall in upright clusters of 8 to 12 inches. Each fruit contains 5 hairy follicles that split open and release the tiny seeds. Fruiting heads persist throughout the winter.

Twigs are slender, grayish-colored, and densely covered with whitish, matted hairs. After several years, the hairs slough off, leaving the older, brownish twigs and stems smooth. Thin, outer "skin" becomes shreddy and commonly peels away on older stems. Lateral buds are small, round, and covered with white hairs. Leaf scars have a distinct raised ridge, are triangular in shape, and contain a single bundle scar.

Snowbells

Genus: *Styrax*

There are few people familiar with these 2 uncommon shrubs, but 1 glimpse of either species in flower, and it is love at first sight. The beautiful, dangling, bell-like, snow-white flowers have few rivals in the wild, but they closely resemble those of the tree-sized *Halesia*s (they are in the same family).

Both American and large-leaf snowbell are very difficult to find in the horticultural trade, but instead several Asian species are sold, mainly in the Southeast, including *S. japonicus*. Most *Styrax* species are adapted to regions where winter temperatures do not drop below -10 degrees F. Both native species grow well in nearly full-day sun, although they are most often found in shaded woods.

While the flowers are striking, little is known about their use by wildlife. Fruits are capsule-like, containing 1 or more seeds that are likely used by wildlife, but use is rarely recorded. Browse value/use is unknown, although the authors have recorded heavy deer browsing on individuals planted in a rural arboretum. Growth form is such that neither species provides exceptional cover.

Leaves of the snowbells are simple and rather broad, with very irregular toothing, mainly on the terminal half. This is large-leaf snowbell (during a rain event), showing not only the variability of shape, but its pale yellow fall color.

The fruit of *Styrax* is hard, round, and nut-like, with a thin, hairy outer "shell" that splits when ripe. These are fruits of American snowbell, which are very difficult to distinguish from those of large-leaf.

The beautiful, white, fragrant flowers of the snowbells are a delight to see in May. Pictured is American snowbell.

Bark on *Styrax* is mostly smooth and gray, similar to that of a beech tree. There can be faint tannish-colored streaks running vertically along the trunk in older specimens, as seen on this American snowbell.

American snowbell, snowbell, storax

Styrax americana **Lam.**

Styrax americanum/americanus **Lam.**

Family: Styracaceae

This is 1 of the Midwest's little-known, native shrubs that has great potential as an ornamental. Its beautiful, large, dangling, white flowers are attractive in May and June, and flowers often appear again in early fall if the weather remains warm. Flowers, leaves, and twigs are similar in shape to the silverbells, *Halesia* spp., trees of the Appalachian Mountains. It is currently listed as state threatened in Illinois, and presumed extirpated in Ohio.

Form and Size: It can grow to a height of 15 feet but is generally shorter. It is finely-branched and has an open growth form. Sprouts form from the base of the plant, creating a multi-stemmed shrub.

Habitat: American snowbell is always associated with wet sites. It is found in wooded swamps, in wet woods, and along streams, where occasional flooding occurs. Typically, snowbell is found growing in the understory of swampy woods, where it experiences a great deal of shade.

Wildlife Uses: This species generally occurs in moist locations. Its cover value in such areas is average, and its growth form is not especially conducive to songbird use as nest-sites. The fruits are dry drupes with 1 or more seeds. They have been recorded as being heavily used by wild turkey—a preferred food. We would be surprised if they are not also regularly used by rodents, such as white-footed mice and gray squirrels, but we have not found such use recorded. There is no information on the browse use of the species, but we have recorded substantial browse by deer on a specimen in a rural arboretum.

Landscaping Value: Hardy to Zone 5, this shrub of shady, moist habitats can be a bit tricky to grow, but the authors have had success planting it in a dark "prairie" loam that is slightly acidic, in mostly full sun. It is particularly attractive in flower, and the fall color, a pale yellow, is nice. It has mostly smooth bark that resembles that of a beech tree. This species is rarely found in any native-plant nursery.

Similar Species Distinctions:
—**Big-leaf snowbell** (*S. grandifolius*) is found in drier, forested sites. Its leaves are generally larger (twice the size). Inflorescences are longer and contain up to 19 flowers, the petals of which are only slightly reflexed.

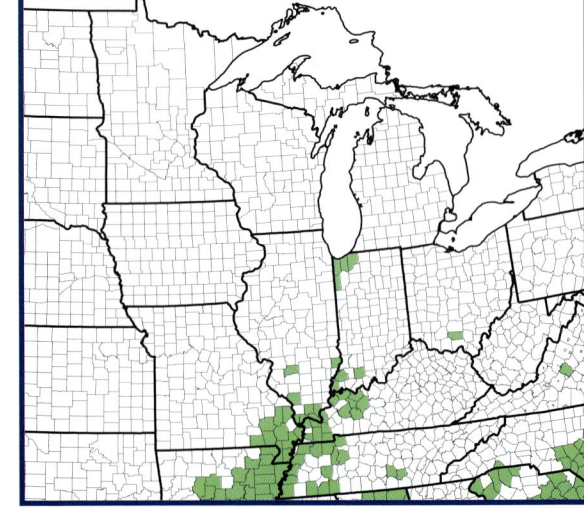

The alternate leaves are simple, up to 4 inches long, and narrowly elliptic. The upper surface is green and mostly smooth, while the lower surface is somewhat paler and mostly smooth. Leaf margins are obscurely toothed. The petioles are short and hairy.

Flowers appear the last of May and the first several weeks of June. They dangle in short, lateral clusters (racemes) that have 2 to 7 flowers. Flowers have 5 white, densely hairy petals that flare backwards from the cluster of stamens. Flowers are an inch or more across and fragrant.

Fruits are a dry drupe and ripen in early fall. Each fruit is about ⅓ inch in diameter and hard. When ripe, a single seed is revealed by the parting of 3 thin, hairy valves.

There is no true terminal bud. There is a pair of buds (pseudo-terminal) near the twig tip—both somewhat rounded. Buds have several fleshy "scales" but are considered naked. The largest is nearest the top. Both are dark tannish and densely hairy. Twigs are slender, dark tannish, and covered with short hairs. Lateral buds are paired or in 3's and diverge from the twig. They are considered naked, based on the type of fleshy scales they have. The upper, largest bud produces the flowering raceme. Leaf scars are U-shaped and deeply notched at the top, where a bud usually sits. There is a single row of "smiley face"-shaped bundle scars in the middle of the leaf scar. Pith color is green.

Mature bark is thin, medium gray, and develops vertical, broken, shallow, tan fissures with age.

big-leaf snowbell, large-leaf snowbell, storax

Styrax grandifolius/grandifolia **Aiton**

Family: Styracaceae

Big-leaf snowbell is at its northernmost range in southern Ohio, Indiana, and Illinois. It is a common shrub in the South, and it is usually distinguished from American snowbell by its larger leaves and bigger, more densely clustered flowers. It reportedly tolerates drier sites than *S. americana*. It is currently listed as state endangered in Illinois and Indiana, and presumed extirpated in Ohio.

Form and Size: Specimens in the northern portions of its range are 6 to 7 feet tall, but farther south, it can grow to 20 feet or more in height. In the forest understory, it is an open, fine-branched, large shrub.

Habitat: Big-leaf snowbell is found on wooded, dry upland sites. In the South, it occurs in wet woods and along stream banks, but it reportedly handles drier sites than does *S. americana*. It can form dense colonies through root suckers.

Wildlife Uses: This species often grows to small tree proportions, but in general, it has similar values for wildlife as *S. americana*. Cover value is unexceptional. Fruits are known to be used by the wood duck, but we would expect their use by wild turkey and woodland rodents as well. Browse value/use is unknown.

Landscaping Value: Hardy from Zones 6–9, this large shrub is not common in the Midwest. However, in the southern portion of our region, it could be utilized for the beauty of its flowers, which are present for a several week period in May. Its trunk has smooth, gray bark and is often covered with lighter patches of lichen. It prefers some shade, but it probably handles at least half-day sun. It is very tolerant of dry sites. Very little information is available, as it is virtually an unknown in the world of horticulture. No nurseries carry this species, but there are several Asian species planted in the East and Southeast.

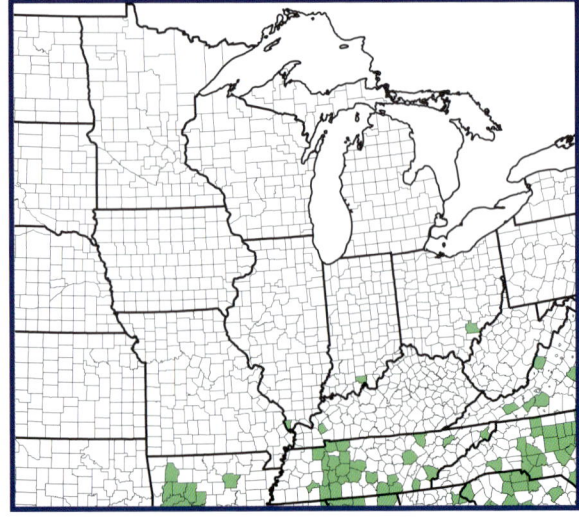

> **Similar Species Distinctions:**
> —**American snowbell** (*S. americana*) has many similar traits. It has fewer flowers per inflorescence with much more strongly reflexed petals. Leaves are usually much smaller.

The alternate leaves are simple and up to 8 inches long. They are obovate to broadly elliptic to ovate. The upper surface is green and smooth, while the lower surface is paler and hairy. Leaf margins are entire to few-toothed.

Fruits ripen in the fall on large clusters. The outer covering of the woody drupe does not split into 3's as it does in *S. americana*, but rather it usually remains intact. Each fruit is about ⅓ inch in diameter and hairy.

Flowers appear in late May in sizable racemes up to 6 inches long. There are more flowers (2 to 19) per raceme (but sometimes just a single flower), and a more drooping raceme than is found in *S. americana*. Flowers are up to an inch in diameter with 5 white, hairy petals that flex back slightly. There are 10 stamens.

There are no true terminal buds. Near the tip of the twig is a pair of round-topped, flattened, tan-colored, hairy buds. Buds are considered naked, because they have a unique kind of fleshy scale. Twigs are slender, tan-colored, and hairy. Lateral buds are stacked, round-topped, flattened, and hairy. The largest bud produces the raceme of flowers. Leaf scars are U-shaped, with a deep V-shaped notch in the top. In the center of the scar is a "smiley face"-shaped row of bundle scars. The pith is green.

In the Midwest, big-leaf snowbell is a small-diameter shrub that does not develop a characteristic bark pattern. Its trunk is dark brown-gray, thin, and commonly has white lichen patches scattered along it.

Blueberries, Farkleberry, Deerberry

Genus: *Vaccinium*

The Midwest is home to many species of *Vaccinium*. They are particularly common in the cooler, less humid regions of Minnesota, Wisconsin, and Michigan, where they often carpet the floor of an open woods or area that is managed with occasional prescribed burns. Blueberries are much sought after by wildlife and humans alike for their tasty fruit. Michigan is the home of many blueberry growers, and it reigns as the number 1 state in the United States for commercial blueberry production. Almost all commercially planted blueberries in the Midwest are cultivars of our largest native blueberry, *V. corymbosum*.

Blueberries, if given the right acidic soil and enough sun, provide year-round landscape appeal. It is hard to ignore a plant that adds so much—copious, whitish, bell-shaped flowers in spring, free summer food, spectacular fall leaf color, and green to pinkish red winter twigs. They provide all this with very little maintenance.

Distinguishing species of blueberries can be tricky at times, particularly if no flowers or fruit are available. There has been much splitting and disagreement by taxonomists over the years—too much to discuss here. For the most recent discussion on *Vaccinium* taxonomy, see *Flora of North America*, Vol. 8 (2009).

The most striking feature of the blueberries from a wildlife perspective is its fruit production, which can be quantitatively variable depending on species, location, and year. Most are juicy and highly palatable (preferred) to many songbirds, not limited to typical fruit-eaters like cedar waxwings and robins. Game birds, especially ruffed grouse and wild turkey, and a plethora of mammals, such as white-footed mice, squirrels, and black bears, also take fruits. The few species that do not have succulent fruit, like deerberry (*V. stamineum*), are frequently taken by white-tailed deer, along with associated foliage. Palatability of foliage and twigs, and thus browsing intensity, is quite variable, from highly preferred to seemingly spurned, both among species and within a species in different parts of its range. In general, the genus has very good cover value, largely because of its numerous small limbs and abundant foliage that supply protective wildlife cover, as well as nesting cover for birds. The suite of users varies somewhat, depending on species-specific size and habitat differences.

The tiny, urn- or bell-shaped flowers of blueberries and farkleberry are numerous and fairly long-lived. Given full sun the plants produce large amounts of flowers (and usually fruit if weather conditions are right). These flowers belong to farkleberry.

Twigs of most *Vaccinium* species, especially the first- and second-year growth, are attractive winter features in a landscape, varying from green to deep wine-reds. Deerberry twigs (near left) are usually wine-red, while highbush blueberry (far left) are commonly (but not always) greenish.

Some *Vaccinium*s are classified based on absence or presence of verrulose dots that cover the stems, as seen here on *V. angustifolium*.

The fall color of most *Vaccinium*s is just plain gorgeous. Similar to a sweetgum tree (*Liquidambar styraciflua*), there is just about every fall color imaginable. Seen here is highbush blueberry.

lowbush blueberry, early lowbush blueberry

Vaccinium angustifolium Aiton
Vaccinium pennsylvanicum Lam.
Family: Ericaceae

This blueberry is commonly associated with *V. pallidum* and black huckleberry (*Gaylussacia baccata*). Like all blueberries, it has spectacular fall color and produces delicious fruit in the summer. It is listed as state threatened in Iowa.

Form and Size: This is a small blueberry that can become 1½ feet tall but is usually shorter. It has a single stem and numerous fine branches and twigs.

Habitat: Lowbush blueberry is common in dry, sandy soil, such as that found in oak savannahs. It can sometimes be found in low areas with moist, acidic soils. This is a colonial species with clones produced by underground rhizomes that respond to disturbance. It is well adapted to fire, which is used as a management tool for blueberry crop production.

Wildlife Uses: This species is the most important commercial blueberry in the northeastern United States and Maritime Provinces of Canada, and our wildlife also recognize the tastiness of the small, sweet fruits. They are taken by many songbirds, game birds, and mammals, including black bears, whose productivity has been linked to the quantity of the lowbush blueberry crop. Cottontails and white-tailed deer regularly browse it, especially in late winter and spring, although its preference categories range from "preferred" to "rarely eaten." Its dense colonies and small stature provides excellent protective and nesting cover for ground-dwelling mammals and birds, from generalists like hermit thrushes to specialists like clay-colored sparrows.

Landscaping Value: Hardy from Zones 2–5, this is a cool-weather blueberry that thrives in full sun. Its tiny fruits are delightfully sweet, and its fall color is shades of deep orange-red. Twigs and branches are usually green, and they are easily seen in the winter. Several cultivars are available, but the wild plants are used commercially—apparently the growers do not care to "mess with perfection." 'Top Hat' is a cultivar recommended for patio containers.

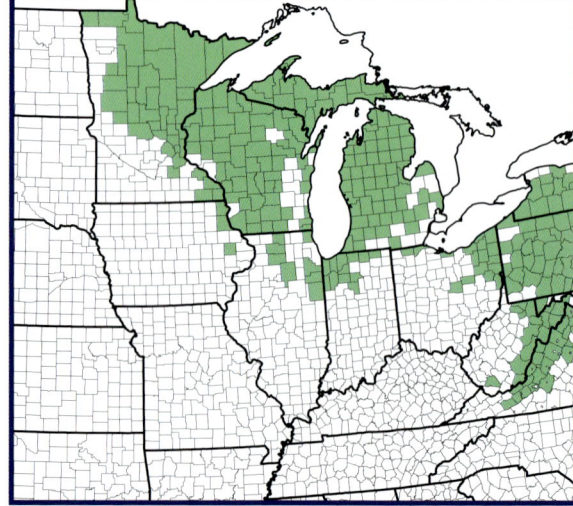

Similar Species Distinctions:
—Dryland blueberry (*V. pallidum*) is usually a bit taller and has fewer teeth (or none) on the leaf margins. Leaves are usually broader and whitish-colored beneath.

The alternate leaves are up to 1½ inches long and are lanceolate, oblong, or ovate. The leaf size is variable, being smaller on mature branches and larger on new, vigorous growth. The tip is sharp-pointed. Leaf margins are finely toothed; each tooth ends with a gland. The upper leaf surface is dark green and mostly smooth at maturity. The lower surface is somewhat paler and mostly smooth. The midrib on both surfaces can retain some hairs during the growing season.

Lowbush is a small blueberry, and it does not develop any characteristic bark. Its stems are about the diameter of a pencil, are brownish, and develop shallow fissuring with age.

Fruits ripen in July and are usually about ¼ to ½ inch in diameter. They are dark blue without a glaucous bloom, sweet, and fleshy.

Flowers can appear as early as early April, but they continue to be produced through most of the month. They are tiny, bell-shaped, perfect flowers that are about ⅓ inch long. Flowers are usually white, but they can be tinged with pink, and hang downward in small clusters.

There are no true terminal buds, but buds are of 2 kinds. The larger bud on the top of the twig (far left) is a flower bud; the smaller bud on the top is a vegetative bud. All buds are covered with rusty-brown, keeled scales. The keeled center is often elongated to form a long scale tip. Twigs are slender and usually some shade of green. If not green, they are reddish purple. New growth has scattered, short hairs and numerous tiny warts (verrulose texture). Leaf scars are narrow, V-shaped slits at each node.

farkleberry, sparkleberry

Vaccinium arboreum **Marsh.**

Family: Ericaceae

Farkleberry is a southern blueberry species that reaches the northernmost limits of its range in southern Indiana and Illinois. It has attractive, shiny leaves that display great fall color, especially when mixed with the various shades of its ripening fruits. Its thick, leathery leaves are retained longer than on most deciduous shrubs, and it has evergreen tendencies even in the Midwest.

Form and Size: It can grow to a height of 10 feet or more but is often shorter in the Midwest. It seems to thrive in a shady understory, but it tends to be a bit spindly when growing there. There is usually a single trunk and a small, open crown.

Habitat: Farkleberry is found growing on dry, acidic, well-drained, wooded ridges. A common associate is post oak, *Quercus stellata*, and several species of pines farther south in its range. It is usually found growing in fairly heavy shade, but it grows best with full sun.

Wildlife Uses: This species is frequently a lanky understory species that provides little in the way of cover; however, when it receives sufficient light to develop a fuller crown, it is selected occasionally as a nest site by songbirds, such as wood thrush. It is considered a preferred deer browse in some parts of its range, but in much of the Midwest, leaves on older plants are out of reach of deer. Its fruit is sweet but very dry and mealy; it seems to be 1 of the least preferred *Vaccinium* fruits, often lasting into the winter when they are taken by wintering songbirds and squirrels.

Landscaping Value: Hardy from Zones 6–9, farkleberry makes an interesting landscape plant. It has dark green, shiny, thick leaves in summer that turn dark red in the fall. The black fruit is persistent throughout much of the winter, and it has exfoliating, multi-colored bark. Old specimens are often gnarled and twisted. It is extremely heat and drought tolerant, but it requires acidic, well-drained soil. There are no known horticultural varieties of farkleberry, but it is occasionally offered through southern nurseries specializing in natives.

Similar Species Distinctions:
—Evergreen huckleberry (*Gaylussacia brachycera*) is a small (around 1 foot tall), colonial species found in woods in Kentucky and Tennessee. It has thick, evergreen, ovate leaves up to 1 inch long with round or sharp teeth along the margins. Its evergreen leaves are similar.

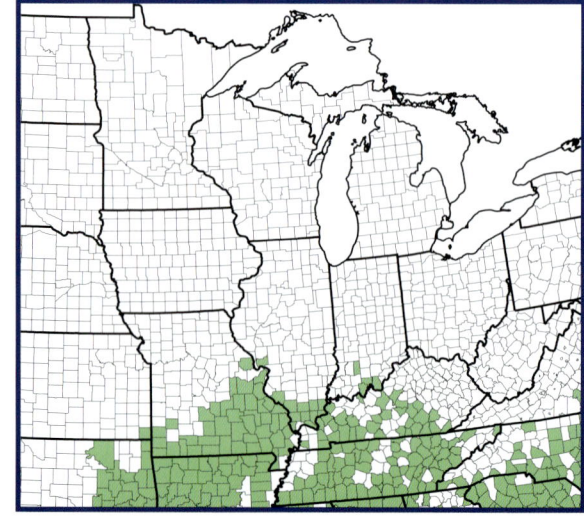

The alternate, thick leaves are up to 2½ inches long and have very short petioles. They vary from obovate to oval to nearly orbicular. The upper surface is dark green and shiny, with a hairy midrib. The lower surface is somewhat paler and hairy, especially along the veins. The leaf margin is curled under (revolute), entire, and glandular. In the Midwest, leaves are tardily deciduous. Farther south, they are evergreen.

Flowers appear from mid- to late May into June from leaf axils, or from a many-flowered, terminal cluster (racemes). They are white, broadly bell-shaped, and about ¼ inch long.

Fruits mature slowly during the fall, eventually turning black when fully ripe. They are about ⅓ inch in diameter, shiny, and dry. They have many seeds, and they are essentially non-edible for humans.

There are no true terminal buds, but the pseudoterminal and lateral buds are rounded at the base and pointed at the tip. Buds diverge from the twig and have brownish, somewhat loose-fitting scales. Twigs are somewhat slender, brownish gray, and mostly smooth. There are scattered hairs on the newest growth. Leaf scars are distinctly raised, and they contain a single bundle scar.

Mature bark is thin, scaly, and eventually exfoliates. When outer bark is shed, it reveals the reddish brown, inner bark.

highbush blueberry, swamp blueberry

Vaccinium corymbosum **L.**

Family: Ericaceae

This is "the" blueberry of commerce in the Midwest, from which many varieties have been developed. If one plants a mixture of varieties, fruit can be available for at least 6 weeks during July and August. Fall color of highbush is spectacular, as seen here in the early stages in September. Highbush is a pretty addition to a landscape, but acidic soil is required.

Form and Size: Usually it is 5 or 6 feet tall. It has a narrow base with many shoots rising from it, and a broad, domed crown. Acidic soil is a must, but it grows well in less-moist, upland sites. It requires full sun for best flower and fruit production, but it tolerates partial shade. It is also our largest species producing edible blueberries, sometimes reaching 10 feet in height.

Habitat: Highbush is a species mostly of sphagnum bogs, but it can be found in swampy woods and marshes, sandy places, oak woods, or boggy thickets. It requires acidic soils and grows best in full sun.

Wildlife Uses: Highbush blueberry is a large shrub that occurs in moist soils in openings, where it produces a good quantity of fruit. Users include many songbirds (e.g., scarlet tanager, thrushes, catbirds) and game birds. Many mammals use the fruits as well. Browsers, especially white-tailed deer, feed heavily on leaves in spring and to a lesser degree in summer. The large size and multiple limbs make the species an ideal nest site for mid-height nesters such as catbirds, robins, and cardinals. It does not form dense colonies, but its size and density alone provide protective cover for wildlife, especially birds.

Landscaping Value: Hardy to Zone 3, this blueberry is worth a try in any landscape where moist, acidic soil can be provided, since fruits are wonderful tasting and full of antioxidants. Spring flowers, blue summer fruit, fall color, and green or reddish twigs give year-round appeal. There are several hybrids available that are crosses between *V. corymbosum* and *V. angustifolium*. Many varieties offer smaller statue or larger fruit. Check to see which grow best in your area.

Similar Species Distinctions:
—Dryland blueberry (*V. pallidum*) can be confused with a small highbush. Blueberries can be difficult to identify. Site is often helpful, and knowing the local flora is a good way to rule out this and other potentially confusing species.

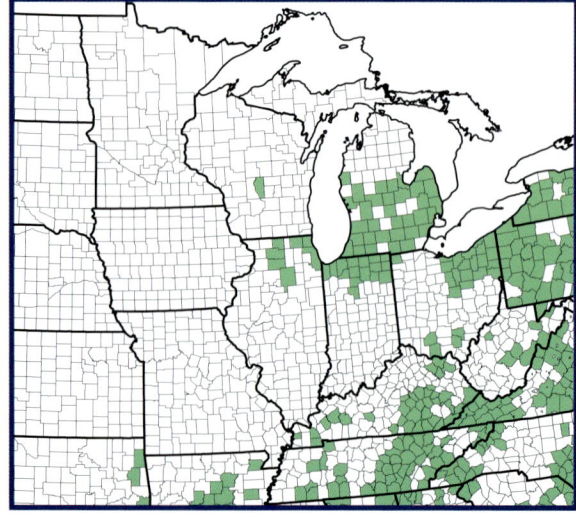

The alternate leaves are highly variable in shape, which led to several named natural varieties in the past. They can be lanceolate to ovate, obovate, elliptic-lanceolate, or oval. Leaves of vigorous growth are usually a much different shape than those on older twigs. The leaves are dark green on the upper surface and mostly smooth, except that hairs sometimes occur along the midrib. The lower surface is paler and somewhat hairy, especially along the veins. The leaf tip usually ends with a bristle-tipped gland. Leaf margins are entire to finely toothed.

Fruit begins to ripen in early July. Berries are up to ½ inch in diameter, blue-black, and usually covered with a glaucous bloom. They are sweet, fleshy, and tasty.

Flowers appear in late April through mid-May in short, terminal clusters. They are white, urn-shaped, perfect flowers that are about ⅓ inch long. The tips of the petals are curled back.

There are no true terminal buds, but buds are of 2 sizes. The larger, plump buds are flower buds (left), while the slender ones (right) are vegetative. Buds are covered with 6 or so rusty-brown scales that are keeled in the middle. The keel often extends beyond the scale margin, creating an elongated point. Twigs are fairly slender, reddish or greenish, and hairy, especially on the new growth. Leaf scars are narrowly V-shaped.

Mature bark is thin and flaky with a small, narrow pattern. Over time, the outer bark begins to fall, revealing the reddish brown inner bark.

Canada blueberry, velvetleaf blueberry, velvetleaf huckleberry

Vaccinium myrtilloides **Michx.**
Vaccinium canadense **Kalm**

Family: Ericaceae

Canada blueberry is just that—a blueberry of Canada. It thrives in the cooler, less humid climate of Canada and the upper Midwest. There are many species of blueberry in our region, but this 1 is certainly the hairiest. Its beautiful, densely hairy twigs and spectacular fall color—not to mention its edible fruit—make this species a great addition to home landscaping. It is currently listed as endangered in Indiana, and threatened in Iowa and Ohio.

Form and Size: It is only 1½ feet tall at its largest, and it suckers to produce a few-stemmed, finely-branched shrub. This is a colonial species, and it is often mixed with *V. angustifolium*.

Habitat: This blueberry is common in dry, sandy, or rocky soil in clearings and open woods and in tamarack bogs. It requires acidic soil, but it will grow in full sun to heavy shade.

Wildlife Uses: Canada blueberry often occurs with and is almost identical in growth form to *V. angustifolium*. Although they regularly grow together, *V. myrtilloides* is typically found in more moist sites, and thus, it may serve a slightly different wildlife clientele. Their values to and use by wildlife are almost identical; both may be killed in the northern parts of their ranges by overbrowsing by deer, rabbits, and moose if not covered by snow.

Landscaping Value: Hardy from Zones 2–6, this small blueberry is utilized by northern blueberry fans who pick the small, sweet fruits. This species loves acidic soil, both wet and dry, but it seems to be adaptable to droughty conditions. It has all the 4-season appeal needed: green or red, velvety winter twigs, beautiful spring flower clusters, blue, summertime fruit, and fall color extraordinaire. This species has not been seen by the authors in native-plant nurseries.

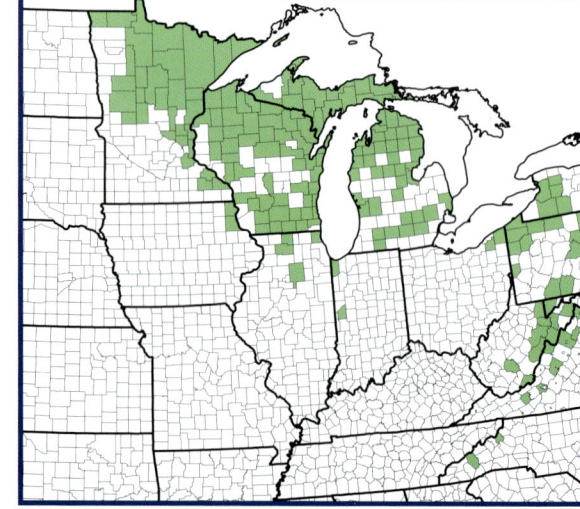

Similar Species Distinctions:
—**Lowbush blueberry** (*V. angustifolium*) is a common associate of the same stature. It has similar twigs and leaves, but it lacks the hairs on both. Its leaves have fine teeth along the margin.

The alternate leaves are oval to lanceolate and usually about 1 inch long. The upper leaf surface is green and densely hairy when young; only the veins retain hairs when the leaves are mature. The lower surface is paler and densely hairy. The margins are entire; the petioles are extremely short.

Flowers appear from late April to early May in small clusters (racemes). They are urn-shaped, whitish (often tinged with red), and about ¼ inch long.

Fruit ripens in early July. It is a fleshy, blue-black berry that may or may not have a glaucous bloom. Fruits are sweet and edible.

There are no true terminal buds. Buds are of 2 shapes, as seen here. The smaller, flattened buds are vegetative. The plump buds produce flowers. Buds are covered with about 6 rusty-brown scales that are keeled in the middle. The keeled "midrib" is often elongated into a point that extends beyond the scale margin. Scales are rarely hairy. Twigs are usually reddish or pinkish red and densely covered with stiff hairs. Leaf scars are distinctly raised and U-shaped, with a single bundle scar.

This is a small shrub that does not produce any characteristic bark. Stems are brown and somewhat scaly.

dryland blueberry, late low blueberry

Vaccinium pallidum **Aiton**
Vaccinium vacillans **Kalm ex. Torrey**
Family: Ericaceae

Dryland blueberry is usually found in large colonies mixed with *V. angustifolium* and black huckleberry. Dryland's whitish leaf backs help distinguish it from other native blueberries. It is pictured here in a common habitat near Lake Michigan—an oak savannah.

Form and Size: Dryland blueberry is a small shrub that is commonly 2 feet high or less. It usually has a single stem and few fine branches.

Habitat: It is found in dry, sandy soils in forests and oak savannahs and on forested, sandstone outcrops. It is usually found growing in forested areas where filtered light is available. Acidic soil is required for best growth.

Wildlife Uses: Dryland blueberry is a low-growing blueberry that usually occurs in large colonies, where it has cover value very much like those of *V. angustifolium* and *V. myrtilloides*. Its fruits are regularly used by wildlife, such as songbirds, game birds, and mammals; they are, however, produced over a long period of time (mid- to late summer) and thus provide extended availability to wildlife users, compared to most *Vaccinium*s. The leaves and twigs are sparingly taken by herbivores; it is not preferred by white-tailed deer, although leaves are occasionally taken in spring.

Landscaping Value: Hardy to Zone 4, this small blueberry has similar landscaping values as lowbush blueberry. It grows well in dry, sandy soils, with filtered light, but given more moisture and better quality soil, it thrives. Full sun is an option for this species, where fall color will be at its best. There are no cultivars listed, and it is rare to find this species in any nursery.

Similar Species Distinctions:
—**Lowbush blueberry** (*V. angustifolium*) usually has more narrow leaves with fine teeth along the margins, and usually it is not white beneath.
—**Highbush blueberry** (*V. corymbosum*) leaves can be similar, but the flowers, fruits, leaves, and plant itself are larger.

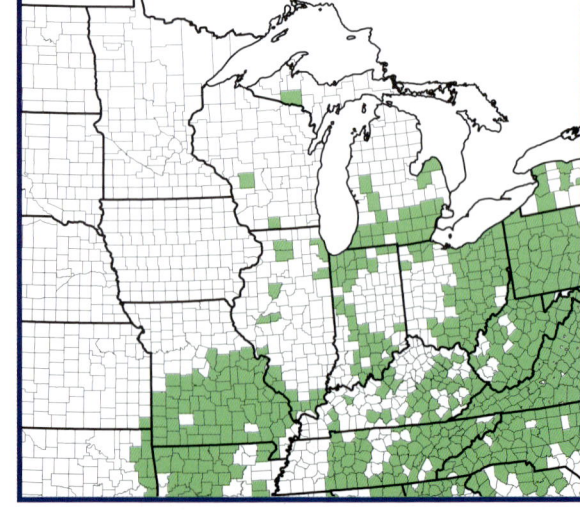

Fruits begin to ripen in early July and continue ripening into September. They are blue-black and usually covered with a glaucous bloom. The fruit, a berry, is about ⅓ inch in diameter, and it is sweet, fleshy, and edible.

The alternate leaves are about 2 inches long and lanceolate to ovate to broadly obovate in shape. The upper leaf surface is green and mostly smooth, except along the midrib, which is hairy. The lower surface is glaucous and mostly smooth. Leaf margins usually have some degree of very fine teeth. Leaf shape is variable, and leaves are especially large on vigorous growth.

Flowers appear in mid-to late April in the southern portions of its range. They are in short clusters (racemes), each flower being only ⅓ inch long. The flowers are creamy white-green, often blushed with pink.

Twigs are slender, reddish pink (sometimes green), and covered with many hairs. Leaf scars are distinctly raised, somewhat triangular in shape, and contain a single bundle scar.

There is no true terminal bud. The larger flower buds are plump and covered with 6 or more rusty-brown, slightly hairy scales. Scales can be slightly keeled along the middle. The smaller vegetative buds are more ovoid in shape.

This blueberry does not attain a diameter that allows any characteristic bark to form. Stems are usually greenish brown, with slight fissuring.

deerberry, squawberry

Vaccinium stamineum **L.**

Family: Ericaceae

Deerberry is by far the prettiest of our native blueberries when in flower. In mid-May, a profusion of large, white blooms greet the eye. Since deerberry is usually 4 to 5 feet tall and just about that big in diameter, there can be a tremendous number of flowers per shrub. It is worthy of ornamental status, but it is not available commercially.

Form and Size: This is a moderately-sized blueberry, reaching up to 6 feet in height. Although it suckers near the root collar, it usually has only a few, large trunks per shrub. It has a full, rounded crown comprised of many small twigs. A fully grown specimen (seen above) is commonly as wide as it is tall.

Habitat: It grows in acidic, dry, sandstone soils, and is found on wooded slopes, forest edges, and abandoned fields. It grows best in full sun, but without disturbance, it is quickly shaded by faster-growing species in these habitats.

Wildlife Uses: Deerberry typically grows in dry, acidic soils in open areas, glades, or forest edges. In those locations its large, dense form provides good nesting cover for shrub-nesting songbirds, but it does not form thickets and supplies little protective cover. Its fruits tend to be thicker-skinned than most blueberries and not as sweet. They are taken to a degree by songbirds and game birds, but as the name implies, they are often eaten, along with surrounding foliage, by white-tailed deer.

Landscaping Value: Hardy to Zone 5, this large blueberry has the biggest, most attractive flowers of any native *Vaccinium*. This is 1 native that the authors have had difficulties growing. The lack of acidic soils in west-central Indiana could be 1 reason, as this species requires just that. It is a tough, hardy, slow-growing species that is usually found in poor quality, dry soils. Leaves appear almost blue-green because of the glaucous leaf back and are attractive when blowing in the wind. This species is rarely found in native nurseries.

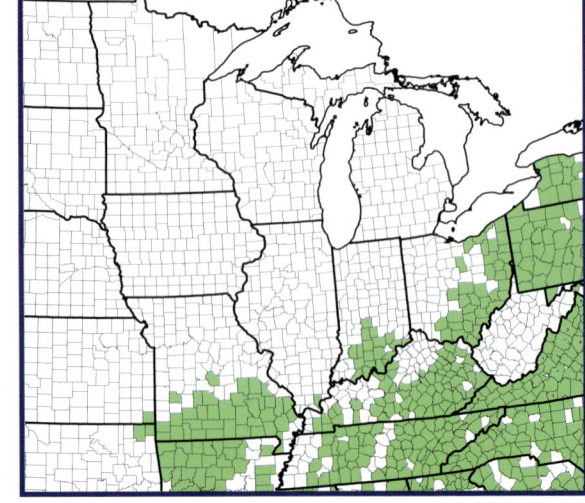

Similar Species Distinctions:
—A few *Vaccinium* species look similar, but the bluish leaves and entire margins help distinguish it.

The alternate leaves are up to 3 inches long and somewhat variable in shape, like those of most blueberries. They can be oval to oblong-lanceolate, oblong-cuneate, or oblong-obovate. The upper leaf surface is green and mostly smooth at maturity. The lower surface is green, but it may be very glaucous and hairy (rarely smooth). Leaf margins are somewhat wavy and lined with short hairs.

Fruit is a purplish, rounded berry when ripe. It is up to ½ inch in diameter, thick-skinned, and considered inedible.

Mature bark is thin and gray-brown, with a small, scaly pattern. The outer, scaly bark sloughs to reveal orange-brown inner bark.

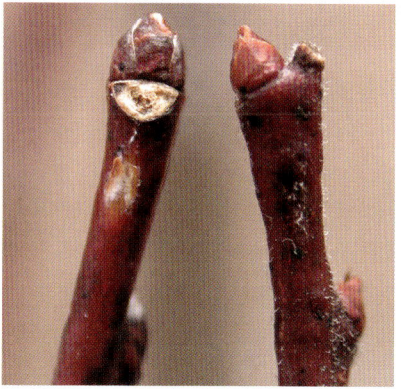

There is no true terminal bud. Buds are all 1 size; they are small, round at the base, pointed at the tip, and they diverge from the twig. There are 5 to 6 scales that vary in color, ranging from wine-red to rusty-red to reddish black. Scales are usually smooth. Twigs are slender, wine-red, and mostly smooth. Newer growth has a few, scattered, short hairs. Leaf scars are somewhat triangular-shaped and contain a single bundle scar. They do not have the small warty texture (verrulose) typical of many blueberries.

Flowers appear in May in a many-flowered cluster (raceme). Each flower is white, about ¼ inch long, and bell-shaped, with 5 spreading petals that are shorter than the stamens. The racemes have numerous, tiny, leaf-like bracts, 1 below each flower.

Indigobushes

Genus: *Amorpha*

Four species of *Amorpha* can be found in the Midwest, 2 of which are covered here. *A. nana* is a small plant of similar stature to *A. canescens*, but it has essentially hairless leaves and solitary inflorescences. It has a more western range, but it occurs in Minnesota and Iowa. *A. nitens*, mainly a species of the Deep South, has similar growth form to *A. fruticosa*, but it has shiny leaves and inconspicuous or no glands on the fruit.

The genus *Amorpha* is derived from the Greek word "amorphos," meaning deformed. This is a reference to the flower, which has only 1 oddly shaped petal that wraps around the 10 colorful stamens.

Several features that stand out about the members of this genus include the compound leaves and the slender, upright inflorescences. Another unusual feature is the blister-like glands that dot the tiny fruit pods. Some are actually filled with liquid and pop when pressed.

The 2 included species are so substantially different in form and habitat that generalizations relative to wildlife value are difficult. Neither is browsed to any great degree, as long as more palatable alternatives are available. Both produce large quantities of nutrient-rich seeds that are likely taken by birds and mammals, but use has been only occasionally recorded. The beautiful flowers of both attract many bees and wasps.

Leaves are always compound and similar in appearance to those of many legumes. The leaf of indigobush is often confused with that of *Robinia pseudoacacia*, black locust.

The flowers of all members of the genus *Amorpha* have a single petal that wraps around the long stamens like a blanket. This is a very unusual floral feature, especially for a legume. Pictured is the flower of leadplant.

Flowers of *Amorpha* are clustered together in slender, upright, mostly terminal inflorescences, making them hard to miss. Indigobush is pictured.

indigobush, false indigo

Amorpha fruticosa L.

Family: Fabaceae

Indigobush has to be 1 of our most striking native shrubs when in flower. Its royal purple spikes of flowers are accentuated by numerous, protruding stamens that show off their bright yellow-orange anthers. It is much larger than its cousin, leadplant, so a large shrub in flower is breathtaking.

Form and Size: Indigobush can become quite large—up to 12 feet in height. It is somewhat sparsely branched, but it has a full appearance during the growing season thanks to the fairly large, compound leaves. It requires full sun and moist, but well-drained soil; it is fast-growing.

Habitat: Indigobush prefers moisture and is nearly always found proximate to a lake, pond, or river. During flooding events, it can become covered with vegetative flood debris when close to a river or stream. There are numerous, recognized varieties within the large natural range of this species. In some states, aggressive progeny from planted individuals are problematic; some highlighted counties represent these escapes.

Wildlife Uses: When it gets sufficient sun, indigobush produces a multitude of flowers that attract bees, wasps, and butterflies. Such open grown specimens are occasionally dense enough for nesting by songbirds, such as willow flycatchers and yellow warblers. However, along riverbanks where they most typically grow in thickets, individuals are shaded by overstory trees and become rather spindly, providing little nesting cover. Red-winged blackbirds still occasionally use these shrubs as nest sites. Bobwhite quail have been recorded feeding on seeds, but production of nutritious seeds is so high that rodents and other birds almost surely utilize them. Foliage is considered unpalatable for white-tailed deer.

Landscaping Value: Hardy to Zone 4, indigobush is prettiest in late spring when flowering. It flowers profusely when grown in full sun, and it often continues sporadically until frost. It grows well on more upland sites and tolerates droughty conditions. In some eastern states, this species has been officially listed as a pest because of its abundant seedling production. There are several cultivars available, including 'Albiflora,' which has white flowers. Indigobush is available through nurseries specializing in natives.

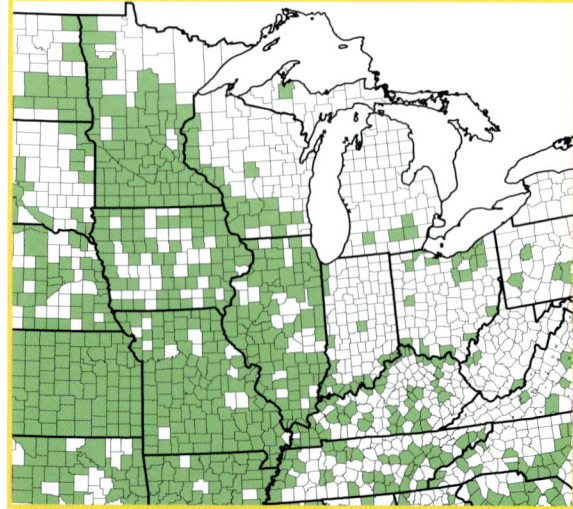

Similar Species Distinctions:
—**Black locust** (*Robinia pseudoacacia*), although tree-sized, has leaves that are very similar to indigobush. This shrub becomes large enough that it is often mistaken for a small black locust.

Leaves are alternate, compound, and up to 1 foot in length; they have up to 25 pairs of leaflets. Each leaflet is oval to oblong in shape, lightly hairy when new, and glandular-dotted beneath.

Flowers are in tall, upright, elongated spikes near the end of the branches. Indigobush blooms earlier than leadplant, mainly during May. Individual flowers are small, about ½ inch in length, with 1 royal purple petal, purple filaments, and yellow-orange anthers. The flowering period is quite long.

There are many fruits on large specimens. Each pod is about ⅓ inch in length and is more elongate and plump than those of leadplant. They are dark brown when ripe and covered with darker brown, somewhat fragrant, liquid-filled "blisters." Each pod contains 1 or 2 seeds.

Bark on the young indigobush usually has a distinct pattern of lenticels, which seem to be positioned in linear rows. The light-colored lenticels are very numerous on the greenish brown bark. Older bark develops thin, tan-colored fissures and is mostly light brown.

Twigs are brownish gray with numerous ridges and grooves running their entire length. Scattered lenticels are easily found. Lateral buds are usually stacked (superposed) above the leaf scars. The top rounded bud (flower bud) has 3 scales, the outer 2 of which appear loose; they are reddish brown and slightly hairy. The smaller, lower bud (vegetative bud) is more flattened and appears as a small tab. Leaf scars are raised, with a distinct lower ridge, and often have small appendages on either side that resemble small thorns.

leadplant

Amorpha canescens **Pursh.**

Family: Fabaceae

Leadplant is 1 of our more diminutive native shrubs, rarely reaching a height of 3 feet. Tall grass prairie is the usual haunt of this plant, and it tends to be associated with dry, sandy, gravelly soils. The Latin *canescens* refers to the abundant off-white or ashy-colored hairs that cover the plant; even the newly opening buds are covered.

Form and Size: Leadplant is a small shrub, rarely taller than 3 feet. Its form is upright and compact with a leafy appearance in the growing season. It has an open form with few short limbs, and it prefers full sun.

Habitat: It is most often found in a dry prairie setting as a scattered plant, but it can be found in more mesic sites. Establishment tends to be fairly slow, and its growth rate is moderate at best. Leadplant is a nitrogen fixer.

Wildlife Uses: Leadplant rarely grows densely enough to provide much wildlife cover, but in the typical prairie setting, it occasionally provides nest shelter for the ground nests of species like dickcissel and lark sparrow. It is considered a fairly good browse species for white-tailed deer, and cottontails browse stems in winter. Use of its seeds has not been recorded, but small mammals and granivorous birds almost certainly take them.

Landscaping Value: Hardy from Zones 2–6, it has a restricted range in the Midwest. This is a small shrub with gray-green leaves that contrast nicely to the normally green plants of a garden. The extended flowering period lends itself well to ornamental plantings, and it could be used in a rock garden or any dry site that provides a mostly sunny day. Leadplant is fairly easy to purchase through native nurseries and those specializing in restoration.

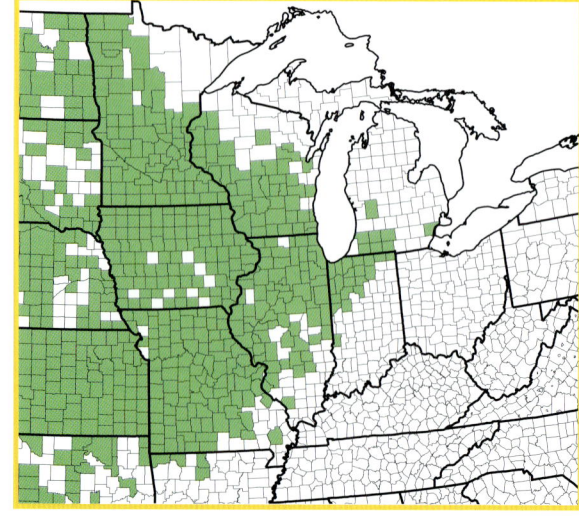

Similar Species Distinctions:
—**Dwarf false indigo** (*Amorpha nana*) is also found in prairies but has bright green, nearly hairless leaves. Each flower spike is single, versus the up to 15 terminal spikes of leadplant.

Leaves are alternate, compound, and vary in length from 2 to 4 inches. The leaflets per leaf are numerous—up to 49 pairs. Each leaflet is oval-shaped and dark green but appears gray because of the numerous hairs on both sides. Petioles are very short.

The lavender-blue flowers are abundant, occurring in showy, terminal, multiple spikes (up to 15). The flowering period is prolonged and can last through most of June and into July. Individual flowers are tiny, perhaps ½ inch in length, including the elongated stamens. Anthers are bright yellow and the filaments are purple, which adds to the overall appeal of this shrub when in-flower. Bees are particularly fond of leadplant flowers. One unusual feature of the flowers is the fact that they each have just 1 petal—most legumes have 5.

Twigs and stems are wooly-gray, especially during the growing season. Over time, the hairs fall and reveal a brown, smooth stem with tiny, plump, several-scaled, brown buds. Leaf scars are slightly raised and somewhat V-shaped. Older stems develop raised ridges, especially beneath leaf scars, which run along the length of the twig.

The small fruits are oddly-shaped, flattened pods, ¼ inch in length and clustered along the terminal spikes. Each hairy pod contains 1 or 2 tiny seeds. Fruit ripens in the fall, and individual pods tend to drop from the spikes before winter sets in, leaving a spindly, roughened "twig."

Leadplant remains a small shrub, and the bark that develops is thin and brown with no unique characteristics.

shrubby cinquefoil, shrubby five-finger

Dasiphora fruticosa (L.) Rydb.

Potentilla fruticosa L.

Family: Rosaceae

This is 1 of our most widely available ornamental shrubs. There are horticultural varieties that have different colors and larger flowers than our wild version. Shrubby cinquefoil is much more common in the more northerly regions of North America. The Latin *fruticosa* means shrubby, referring to the fact that this is the only woody cinquefoil species. This species is listed as threatened in Iowa and endangered in Pennsylvania.

Form and Size: Shrubby cinquefoil only reaches a height of 3 feet or so, and it is very bushy and upright in its growth. It has numerous delicate twigs that give it a full appearance.

Habitat: It is found in wet meadows, marsh and lake edges, fens, and calcareous bogs, growing in small (occasionally large) colonies. It is shade intolerant. This species has a very large natural range, and it is found across all of Canada and throughout much of the western United States. Individual plants are probably not long-lived.

Wildlife Uses: This is a species that occurs only in unique niches, usually water associated, in our region, often along streams. Song sparrows, yellow-throats, and other species that nest at the ground/small shrub interface use its dense structure as nest sites in these special areas. It is browsed lightly by white-tailed deer, and its seeds are sparingly used by songbirds and small mammals.

Landscaping Value: Hardy from Zones 1–6, this is a cool climate species that does not do well in the heat and humidity of the lower Midwest. Although it prefers calcareous soils, it is very adaptable. It handles droughty conditions once established in more upland sites. It has a long flowering period beginning in mid- to late May and continuing all summer long. Winter appearance is poor, but a light pruning before spring bud break helps rejuvenate the plant. Since the oldest stems produce more flowers, do not prune them all.

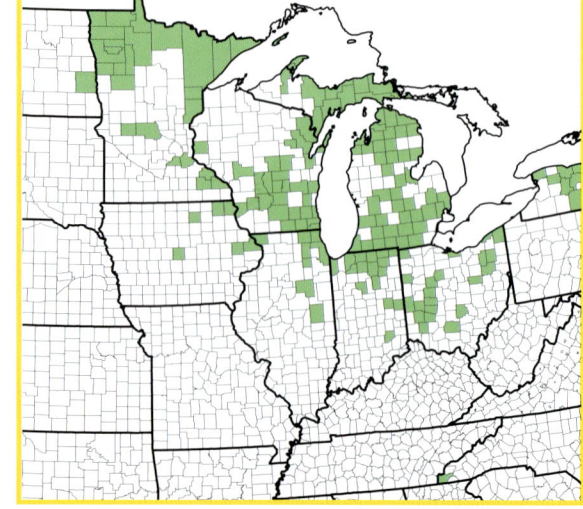

Similar Species Distinctions:
—No other shrub has leaves or overall form like that of shrubby cinquefoil. Overwintering fruit looks somewhat like those of members of the Asteraceae family, which may hamper identification for some.

Leaves are compound with 3 to 7 leaflets (usually 5—the terminal part of the leaf is usually interpreted as being 3 fused leaflets, but it appears to be 1 deeply-lobed leaflet), and are only about 1 inch in length. Leaves are light green and covered with long, white hairs on both top and bottom. Leaflet margins are entire. Brown, papery, forked stipules extend up the twig from the base of the petiole.

The fruits grow in terminal, leafy, hairy clusters and persist throughout the winter. Technically, each fruit is a head of achenes. When the fruit fully ripens in the fall, the seeds are released, and the sepals flex open and remain that way through the winter.

Flowers appear over an extended period of time, beginning in June and ending in early September. They are 5-petalled and about ¾ inch across. The petals are usually light yellow, but they can be orange or any color in between. Each flower has numerous stamens, which is a characteristic of the Rosaceae family. Flowers appear terminally, usually in clusters.

Terminal buds are difficult to find in the winter because of the persistent fruiting heads. Buds are covered by several loose-fitting, wavy, brownish, papery-thin stipules. The stipules are usually 2-toned and forked at the tip. There are often some silky, white hairs present on the scales. Twigs are very slender and reddish brown with scattered, whitish hairs. With age, the thin outer bark splits vertically and eventually exfoliates. Lateral buds are much like the terminal, but they develop from the top of slender short shoots that are covered with many loose scales. Leaf scars are tiny, half-moon shaped, and contain a single bundle scar.

Mature bark is thin, rusty-brown, and peeling to the point that it actually dangles from the stem near the base of the plant.

common hop-tree, hoptree, wafer-ash

Ptelea trifoliata **L.**
Ptelea trifoliata **L. var.** *mollis* **T. & G.**
Family: Rutaceae

Hoptree is 1 of our most common native shrubs. It is not exactly a showy species, but 1 of its best attributes has nothing to do with looks. Because it is a member of the Citrus family, its chemical properties are such that giant swallowtail butterflies (*Papilio cresphontes*) are drawn to it. They lay their eggs on the plants, and the butterfly larvae feed on its leaves. An easy way to find hoptree is to search during the winter months; since the fruit usually overwinters, plants are quite conspicuous at that time. It is listed as state threatened in Tennessee.

Form and Size: Hoptree can become quite large and will occasionally look like a small tree with a single, unbranched trunk. An 18-foot-tall specimen is not uncommon. Usually, however, plants are multi-stemmed at the base with irregular, rounded crowns.

Habitat: It can be found in a variety of habitats, but it prefers stream banks, lake shores, and riverbanks where it reaches its greatest size. It is also found on limestone bluffs and is shade tolerant. The variety *mollis* is common in the sand dunes along Lake Michigan.

Wildlife Uses: Hoptree occurs largely in open areas such as glades, forest edges, and fencerows where it develops a compact crown that is frequently chosen as a nest site by songbirds. Leaves are evidently little browsed by deer, and the unique, wafer-like fruits seem to be rarely, if ever, used by wildlife.

Landscaping Value: Hardy to Zone 3, this shrub is very adaptable to most sites and soil types, and it is equally at home in sun or shade. It does prefer moist, well-drained soil, however. There is nothing glaringly spectacular about hoptree, but it is quite unique. Its round-Band-Aid-shaped fruits persist most of the winter. Fall color is a feeble attempt at yellow. There are several cultivars, including 'Aurea,' which has yellow new leaves that transition to light green at maturity. Hoptree is rarely offered through any nurseries.

Similar Species Distinctions:
—**Poison-ivy** (*Toxicodendron radicans*) leaves are so similar that they are often confused with those of hoptree; and poison-ivy can grow shrub-like. Be sure to check for differences in twigs, buds, and fruits before touching.

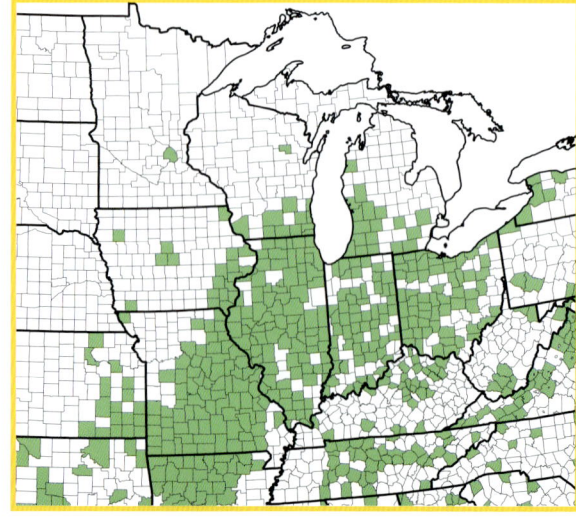

The compound leaves are alternate, about 3 inches long, and trifoliate (rarely 5 leaflets). The terminal leaflet is larger than those on either side. Leaves are grass green, smooth, and hairless at maturity. Those of variety *mollis* are thicker and densely hairy beneath. Leaves are dotted with translucent glands. Leaflet margins are entire or slightly toothed and wavy. The petiole is nearly as long as the leaflets. Crushed leaves have a pleasant odor.

Flowers appear in mid-May in greenish, terminal clusters after the leaves are mostly developed. There are perfect and single-sexed flowers mixed in the clusters. Clusters are large, often 4 or 5 inches across. Some say the flowers are ill-scented (a skunk-like smell), while others claim just the opposite.

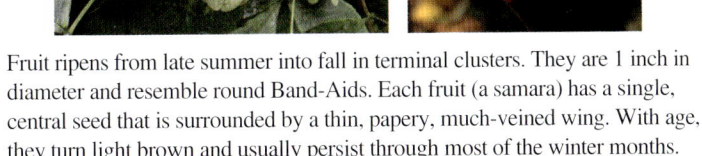

Fruit ripens from late summer into fall in terminal clusters. They are 1 inch in diameter and resemble round Band-Aids. Each fruit (a samara) has a single, central seed that is surrounded by a thin, papery, much-veined wing. With age, they turn light brown and usually persist through most of the winter months.

Twigs are medium brown and mostly smooth, except for the variety *mollis*, which has persistent, dense hairs for a year or more. Twigs are dotted with tan-colored lenticels. Scratched twigs give off a strong odor that is often likened to that of freshly pulled carrots. Leaf scars are horseshoe-shaped and contain 3 bundle scars. Lateral buds are positioned inside the leaf scar. A common winter feature of the twigs is sticky, white froth masses that are deposited by the twomarked treehopper (*Enchenopa binotata*). It is placed there in the fall as protection after the female oviposits her eggs inside the twig.

Mature bark is dark brown and thin with a roughness caused principally by raised lenticels. Younger bark is smooth, dark reddish brown, and dotted with tan-colored lenticels.

There is no true terminal bud, but buds are almost conical. They are covered with dense, tan-colored hairs, are small, and largely sunken into the twig.

Sumacs

Genus: *Rhus*

Probably because of their commonness, sumacs have, for the most part, been generally ignored by those who landscape. William Cullina, an avid *Rhus* fan, and author of *Native Trees, Shrubs, and Vines* (2002) championed the taxon by creating an acronym to help clean up their image. SUMACS, he says, stands for "sexy, undemanding, mellifluous, appropriate, colorful shrubs." In his book, he then defends and elaborates on each adjective. Truly, they are spectacular in the fall, and their colonial nature can give an area a great deal of colorful panache.

With all the "toxic" members of this family now quarantined into their own genus, *Toxicodendron*, there are 4 *Rhus* species in the Midwest and eastern United States. A hybrid between *R. glabra* and *R. hirta* (*typhina*) known as *R.* x *borealis* Greene is fairly common where the 2 ranges overlap.

Members of this genus are tough, hardy, easy to grow and maintain, and fairly easy to purchase at nurseries. They should be considered in any landscaping scheme.

This genus has species that are among the most aesthetically pleasing in the Midwest, species that also have multiple and unique wildlife values. In general, plants have stout, upright limbs (except for *R. aromatica*) that give little opportunity for placement of nests by songbirds; the individual stems are typically clonal and thereby offer protection from predators for ground-dwelling birds and mammals. The fruits tend to be high in energy (oily), but they are not preferred, usually being utilized late in the winter. Stems, bark, and twigs are also used by several herbivores, differently for different species, but all use tends to climax in the winter.

There is probably not any other group of shrubs that can topple the "Kings of Color" from their throne. Every shade of fall color is seen on these plants, if they are given enough sun during the day. Here smooth sumac grows strikingly red against the dead foxtail fruiting heads.

All the midwestern sumacs have compound leaves and alternate branching. Winged sumac, shown here, has an expanded rachis (central portion of the leaf), which makes identification easy.

Sumac fruiting heads are comprised of small, oily, hairy seeds in an upright, usually densely-packed head. Fragrant sumac, seen here, has the smallest fruiting head of our natives.

Most sumacs are clonal by nature, and they form colonies via underground rhizomes. Seen here is staghorn sumac claiming an old homesite.

fragrant sumac, aromatic sumac, squaw-bush

Rhus aromatica **Aiton**

Rhus canadensis **Marshall**

Family: Anacardiaceae

This is 1 of our least common sumacs, but it is readily available in many nurseries. Its many attributes include unusual, silky soft, trifoliate leaves, gorgeous fall color, attractive flowers and fruit, and overwintering "catkins." On the dunes of northwestern Indiana and northeastern Illinois, a densely hairy form is found that is often designated as *R. aromatica* var. *arenaria* (Greene) Fernald. This variety is listed as state rare in Indiana and presumed extirpated in Ohio; *R. aromatic* var. *aromatic* is not state listed.

Form and Size: Fragrant sumac can grow to 6 feet high and nearly that broad, but when growing on unstable dunes, it seems to spread along the sand surface in a diffuse manner, and it only reaches a height of a foot or 2.

Habitat: Although common in the dunes area, it is also found on rocky bluffs of streams and ravines and in open woods. It prefers full sun but handles some shade.

Wildlife Uses: Fragrant sumac differs from other sumacs in having a lower growth form with some limbs typically touching the ground. That characteristic makes this species a much better cover plant than other members of the genus, supplying both winter cover and some nesting cover for ground/shrub nesting birds such as brown thrasher. As with other sumacs, the fruits are not highly preferred and last into late fall when they are typically taken by birds and mammals. Unlike other sumacs, these fruits can be proximate to the ground and easily available to a larger spectrum of users, including bobwhites. Rabbits take bark when winter conditions are severe, an emergency food role shared by all sumacs.

Landscaping Value: Hardy to Zone 3, fragrant sumac makes a great addition to a landscape. It prefers slightly acidic soil, but it is adaptable as long as the soil is well-drained. Flowers are not the main attraction, but rather the unique leaf shape, attractive fruit clusters, and brilliant fall color. Full sun is required for the best leaf color. A popular cultivar 'Gro-low' is no taller than 2 feet and spreads like a carpet across the ground. Its only drawback is all the leaves it collects over the dormant season.

Similar Species Distinctions:
—**Poison-ivy** (*Toxicodendron radicans*) leaves are very similar, but twigs are tan-colored and fruit is a yellowish, berry-like drupe; it is normally a vine.

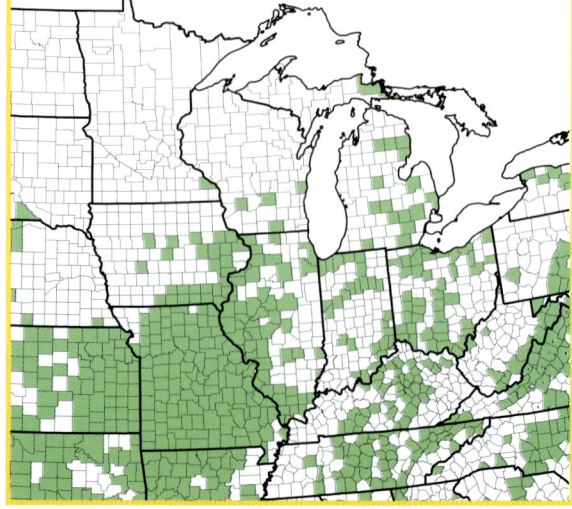

The compound leaves are alternate, trifoliate, and up to 4½ inches in length. The terminal leaflet is the largest, and all have irregularly toothed or lobed margins. Leaves are green and somewhat hairy above, paler and hairy beneath. Variety *arenaria* is particularly densely hairy, especially on the new leaves. Petioles are about 1 inch long and usually hairy.

Flowers appear in late April or later in upright spikes. They are usually separate sex, but there can be a mix in any given cluster. Some flowers are perfect as well. Each tiny flower has 5 pale, yellowish green petals. Flowers appear before or with the leaves. Males are shown on the left; females on the right.

Fruit ripens in late summer into the fall in upright dense clusters. Each fruit is less than ¼ inch in diameter, red, and densely hairy. A large quantity of fruit is rarely produced, and unlike most other sumacs, it is usually present for only a brief time in the fall.

There is no true terminal bud, but the overwintering, catkin-like flower buds appear singly or in small clusters and are about ¾ inch long. They have numerous, overlapping, dark brown scales that are fringed with whitish-colored hairs. Twigs are slender, tannish, and malodorous when scraped or cut. There is usually some degree of hairiness, particularly on the variety *arenaria*. Twigs have a soft, silky feel because of the hairs, and there are scattered, rust-colored lenticels present. Leaf scars are circular and distinctly raised, but lateral buds are almost nonexistent. If visible, they are tan-colored and mostly sunken into the twig.

Fragrant sumac usually does not attain a diameter large enough to develop much of a characteristic bark pattern, but it is thin, brownish, and somewhat roughened from fissuring. Younger plants retain the rust-colored lenticels that are present on the twigs.

dwarf sumac, shining sumac, winged sumac

Rhus copallinum/copallina **L.**

Family: Anacardiaceae

Dwarf sumac is really not a small shrub, as the name implies. It is frequently 6 feet tall, but it can become much taller. There are few native trees or shrubs that rival a sumac in terms of fall color—the plant's exposure to full sun helps intensify the color. All sumacs have a nectar-producing disk in each flower, which helps attract swarms of bees in the summer.

Form and Size: Dwarf sumac can grow quite tall—18 feet or more. It is a clonal species that colonizes an area from an ever-spreading root system. Like most sumacs, it prefers full sun. It is a very adaptable species and can be used in almost any situation where a mass planting is desirable. There are several horticultural varieties of smaller stature.

Habitat: Dwarf sumac occurs in sandy, gravelly, open habitats, on ridges or bluffs, and in moist settings near lakes and bogs.

Wildlife Uses: This is a shrub that pioneers old-fields and tends to be clonal, but not to the degree that *R. glabra* and *R. hirta* are. Its cover values, therefore, are minimal. It produces a fruiting head that tends not to be as durable as those of congeners, often drooping by midwinter, with fewer fruits available to the winter songbirds and game birds that utilize them. The twigs of dwarf sumac are heavily browsed by white-tailed deer in winter (including those with fruiting heads), even when other green forage is available. Rabbits take bark from stems in late winter when snow covers the ground, but not to the degree they take *R. glabra*.

Landscaping Value: Hardy to Zone 4, this is a sizable shrub that becomes larger as one moves south through its range. It makes a nice addition to a more naturalized setting, but make sure there is room for a control mechanism, such as a mower, as the suckers can take over a large area. It is very low maintenance and not particular about where it grows, except it does not do well in poorly-drained sites. Full sun gives the best fall color and fruit production. Dwarf sumac is available through native-plant nurseries.

Similar Species Distinctions:
—**Smooth sumac** (*R. glabra*) has a similar number of leaflets, but white leaf backs and no winged midrib.
—**Staghorn sumac** (*R. hirta*) has larger leaves and twigs covered with long, dense, brown hairs.

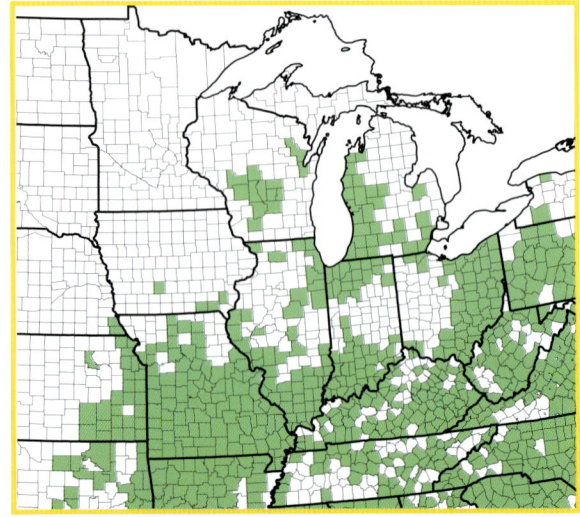

The compound leaves are alternate, about 1 foot long, but are commonly half that length on older specimens. There are 7 to 15 or more ovate or oblong-lanceolate-shaped leaflets per leaf. Leaves are dark green and glossy above, lime green and hairy beneath. The rachis is winged between each pair of leaflets. Leaflet margins are usually entire.

Fruit is a tiny drupe in dense, terminal clusters that ripen in early fall. Each fruit is only ⅛ inch across and is red, hairy, and glandular. By winter, and even into the spring, the fruiting heads droop or become crooked, a characteristic of this sumac.

Flowers appear in July and August in terminal clusters that are about 6 inches tall. The flowers are usually all a single sex on a given plant and have 5 greenish yellow petals.

There are no true terminal buds. All buds are tannish, covered with appressed hairs, and rounded. They are partially sunken into the twig and are set down into the horseshoe-shaped leaf scar. The leaf scars contain a single row of bundle scars that follow the curve of the scar. Twigs are stout, pale-colored, and covered with short hairs. There are numerous, obvious, rust-colored lenticels scattered all along the twigs. During the growing season, milky sap oozes slowly from any cut twig. The pith is usually light brown and is substantial.

Bark is thin, tan to brown, and roughened with sizable lenticels that have expanded with age. Older specimens develop some peeling bark near the base.

smooth sumac, common sumac

***Rhus glabra* L.**

Family: Anacardiaceae

This is a common sumac in the Midwest, and it is easily distinguished by the white leaf backs and white glaucous coating on the twigs. A wonderful winter feature of smooth sumac (also of staghorn) is its large, deep red, upright clusters of fruit that persist throughout the winter. Large colonies, when in fruit, brighten any cold, snowy day.

Form and Size: It can reach heights of 20 feet but is more often 6 to 8 feet tall. Its spreading root system allows it to form colonies quickly. Individual stems are short-lived, but new shoots are always emerging. Like all sumacs, it prefers full sun, where its fall color is best developed.

Habitat: Smooth sumac occurs in all of the lower 48 states, so across its natural range it occurs in many habitats. It is found from sites with dry, sandy, gravelly soil to moist settings to those with heavy, clay soils. It is common along fencerows, on roadsides, and in abandoned areas.

Wildlife Use: Smooth sumac is clonal and occasionally grows densely enough that it provides escape cover for ground dwelling birds and mammals, about the only cover value this species supplies. It produces very large fruiting heads; in most cases, these heads last well into the winter when use increases as wintering songbirds (such as eastern bluebirds, hermit thrushes, and robins), game birds, and some mammals begin to utilize them. In late winter, if snow is deep, cottontails begin to eat bark of smooth sumac (seemingly suddenly so), occasionally chewing through moderately sized stems, and then proceeding to debark the fallen stem and eating any fruiting head attached thereto.

Landscaping Value: Hardy to Zone 3, it is very adaptable, thriving just about anywhere. Full sun brings out the best fall color and the most fruit. A hybrid between smooth and staghorn sumacs, *R.* x *borealis* Greene, is found where their ranges overlap. Growth of most sumacs is fast, and individual stems are short-lived. Smooth sumac is found in midwestern nurseries specializing in natives.

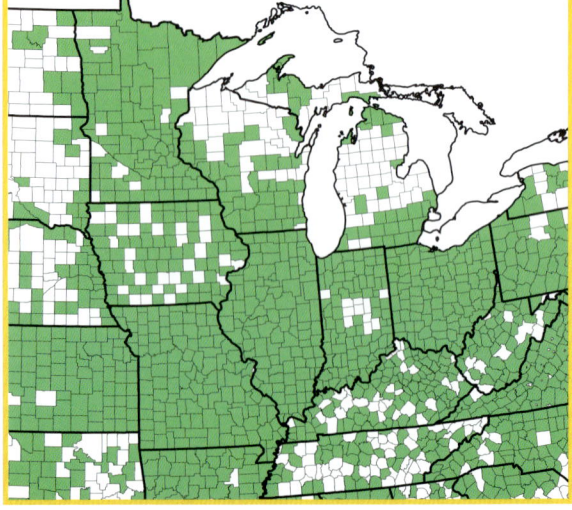

Similar Species Distinctions:
—Dwarf sumac (*R. copallinum*) has similar leaves, but they have winged rachises. Twigs are dotted with cinnamon-colored lenticels.

The compound leaves are alternate and up to 1½ feet long with 15 to 25 leaflets each. The leaflets are oblong-ovate to lanceolate-oblong with coarsely toothed margins. The upper surface is dark green and smooth; the lower surface is glaucous and smooth. The rachis is smooth and reddish, with a glaucous coating.

Flowers are in large, terminal clusters that are up to 10 inches tall. They develop in June and July, are greenish yellow, and usually have single-sex flowers on a given plant.

Mature bark is thin, grayish brown, and develops fissures and a rough texture from elongated lenticels.

The tiny fruit is only ⅛ inch in diameter, but it is in a dense, terminal, upright cluster that can be 8 inches long. The fruit, a drupe, is deep red, hairy, and oily. The fruiting heads remain upright and persist throughout the winter months.

There are no true terminal buds. Buds are more prominent than those of dwarf sumac and diverge from the twig. They are tannish-colored and densely hairy. Their shape is similar to an upside down, pointed ice cream cone. Twigs are very stout, greenish when new, but wine-red when mature. There is a glaucous coating over the twig, and it has scattered, tiny lenticels that are the same color as the twig. Broken twigs or leaves exude a milky, sticky sap. The pith is substantial and darker in color than in our other sumacs. Leaf scars are large and horseshoe-shaped; they contain a single row of bundle scars.

staghorn sumac

Rhus hirta (L.) Sudw.
Rhus typhina L.
Family: Anacardiaceae

This is our largest native sumac, sometimes growing to 30 feet in height. Like smooth sumac, its upright, deep red fruiting heads overwinter well, and give the cold landscape a bright splash of color. This species has a most interesting winter feature—its twigs and young branches are covered so densely with long, brown hairs that they very much resemble deer antlers in velvet, hence, the common name.

Form and Size: It can become 30 feet tall, but it is more commonly 20 feet or less. It forms open colonies from its spreading root system and is fast-growing. Its large size and extensive colonies can create what appears to be a small patch of woods.

Habitat: Staghorn sumac occurs on rocky slopes, sandy barrens, dry forest edges, and lake shores. It prefers full sun and will display quite dramatic fall color when grown there.

Wildlife Uses: Staghorn sumac has wildlife values almost identical to those of smooth sumac. A singular difference is perhaps a result of this species occurring at a more northerly latitude than smooth sumac; the very stout twigs of this species are browsed rather heavily by moose and to a lesser degree by white-tailed deer in the northern reaches of our region. Deer rarely browse either staghorn or smooth sumac twigs at more southerly latitudes.

Landscaping Value: Hardy to Zone 4, this species is particularly interesting in the winter with its antler-like twigs. It adapts well to almost any site except poorly drained, and it is sometimes so aggressive that it is considered a weed or an invasive by some. Plant this where it receives full sun, but make sure it is controllable. Mowing over suckers keeps it in check. Several cultivars are available, including 'Laciniata,' which has divided leaflets, giving it a fern-like appearance. Several midwestern native-plant nurseries sell staghorn.

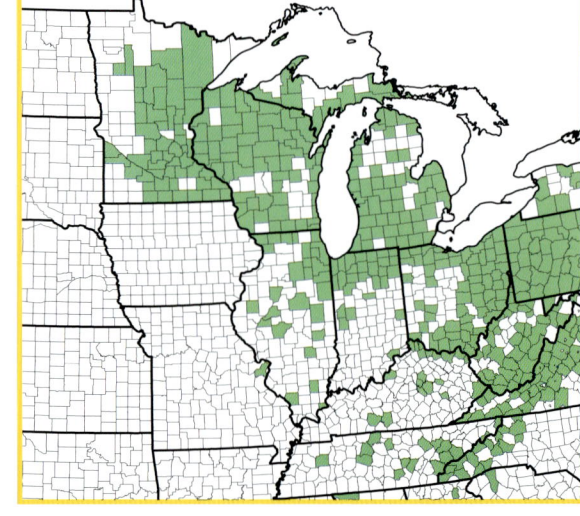

Similar Species Distinctions:
—**Ailanthus** (*Ailanthus altissima*) has similar but much larger leaves. It is an exotic tree commonly found in the Midwest. Just look for long, brown hairs on the twig. Ailanthus has smooth, brownish twigs.

The compound leaves are alternate, over 1 foot in length, and largest on the newest twigs. There are anywhere from 11 to 25 leaflets that are oblong to linear-oblong in shape. The upper leaf surface is dark green with a hairy midrib and veins. The lower surface is glaucous-white. Leaflet margins are coarsely or finely toothed. The rachis and petiole are hairy.

Flowers appear in June and July in large, dense, terminal clusters that are up to 1 foot tall. They are greenish yellow, and there are usually single-sex flowers on separate plants.

Fruits mature in late summer in upright, dense clusters. The individual fruit is a drupe only ⅛ inch across and deep red, hairy, and oily. The fruiting head is densely hairy and up to 1 foot tall. They persist throughout the winter months and well into the spring.

There is no true terminal bud. Lateral buds are similar to those of smooth sumac—tan-colored, densely hairy, and conical-shaped. They diverge from the twig. Twigs are very stout, medium brown, and densely hairy. Older twigs lose their hairs and remain brown. Pith thickness and color is similar to that of smooth sumac, and broken twigs and leaves exude milky sap. Leaf scars are horseshoe-shaped and contain 3 groups of bundle scars.

Mature bark is thin, gray, and somewhat rough from raised, elongated, rust-colored lenticels.

Roses

Genus: *Rosa*

Humans love roses, as evidenced by how many domesticated varieties can be found throughout the world. They are planted and admired for the beautiful, scented flowers, which differ greatly from our natives. It is easy to spot a native rose by its flower—it has only 5 petals. In addition, the natives typically have pink petals, pretty but not spectacular.

Identifying roses can be tough, but like the willows, one has to be willing to accept variability within a species, as there is a great deal. Hybridization also occurs between several of the native roses, which compounds the sometimes difficult task of identification. Taxonomists will probably never be able to define some species to everyone's satisfaction; the many complex reproductive strategies that roses possess causes the substantial variation encountered.

There are at least 7 native roses in the Midwest. Three are not covered in this book, including *Rosa acicularis* Lindl. which is found in the Lake States, west-central United States, and most of Canada. *Rosa arkansana* Porter, which mainly occurs in the Central Plains States, is found in the Lake States, Illinois, Minnesota, Iowa, and Missouri. *Rosa woodsii* Lindl. is a western species that is found in Minnesota, Iowa, and throughout much of the Great Plains.

Roses have many gardening fans, but 1 of the biggest challenges facing them is the introduction of the Japanese beetle. During a severe attack, all leaves on a plant can become skeletonized, and every flower petal can be eaten.

This genus has substantial wildlife values that encompass both food and cover. All species produce fruits called "hips" that wildlife regularly use, but the predominant user-group is predicated on the size of the hips, which may be too large for regular consumption by small songbirds. Indeed, the small hips of the exotic multiflora rose, which are very attractive to even small songbirds, are 1 of the features that has exacerbated its spread. All hips contain sizeable seeds that are very attractive to small rodents. The growth form of roses, along with their generally prickly nature, makes the roses important cover plants, both for protective cover and nesting sites for songbirds. Most are shrubby, but 2 (*R. setigera* and *R. multiflora*) have climbing tendencies as well, which expand their cover values substantially.

Leaves of roses are compound and usually have 7 to 9 leaflets. The leaf base has flared stipules that usually have either a unique shape or some type of glands. The leaf of meadow rose is shown.

Roses are famous for their prickles, and some are pricklier than others. Some species typically have stout, paired prickles at the nodes, as seen on this swamp rose. The number of much finer stem bristles can vary greatly on a given plant.

The native roses are not nearly as showy as those tinkered with by horticulturalists, for they have just 5 pink (rarely white) petals. Shown here is *Rosa palustris*, swamp rose.

meadow rose, smooth rose

Rosa blanda Aiton.

Family: Rosaceae

Roses can be difficult to identify because of overlapping characteristics in similar species. As with willows, one has to be willing to accept variability within a species. Of our native roses, this species most closely resembles *Rosa carolina*, which can be a common associate. It is listed as threatened in Ohio.

Form and Size: Meadow rose can grow to 6 feet in height, but it is most often 3 feet or less. It spreads by underground rhizomes and forms clumps that are interconnected.

Habitat: In the northern portions of the Midwest, meadow rose is found in almost any habitat, from prairies to open woods to roadsides. Along the shores of Lake Michigan, it is found in stable dunes. It is normally found growing in full sun, but it handles partial shade.

Wildlife Uses: Meadow rose is found mostly in the northern portion of our region and, as with most roses, occurs in open woods and openings. Its stems may become rather long and upright if supported by other vegetation, and in such situations, it yields important protective and nesting cover. While it reproduces through rhizomes as do most roses, it does not typically form dense thickets, which add to the cover value of several other species. The hips are too large to be attractive to most songbirds, although they are used by ruffed grouse, wild turkeys, and mammals. Seeds are removed from hips and eaten by white-footed mice.

Landscaping Value: Hardy from Zones 2–6, this species probably has a restricted southern range based on environment, particularly humid summer conditions. It seems to be adapted to just about any soil type and moisture regime, and all but the shadiest sites. It is noted for its nice fall fruit display and red leaf color, but it also has interesting white glaucous coating over the pinkish red twigs. Japanese beetles can ruin a plant and its flowers if given the chance. Without competition, this species becomes aggressive and suckers profusely. It is occasionally offered in native-plant nurseries in the Midwest.

Similar Species Distinctions:
—**Wild rose** (*R. woodsii*) has a more western range but occurs in our region. It hybridizes freely with *R. blanda*, creating confusing hybrids.
—**Pasture rose** (*R. carolina*) has similar growth form and many overlapping features, often making them difficult to differentiate. It usually has more stem prickles and a hairy ovary.

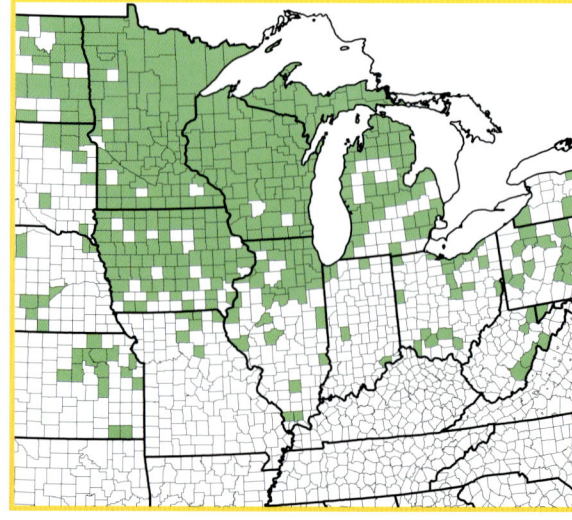

The compound leaves are alternate and about 4 inches long with 5 to 7 leaflets. Each leaflet has a finely toothed margin. The upper surface is green, dull, and smooth. The lower surface is pale and hairy. The petiole is densely hairy. The stipule is broad, entire, somewhat wavy, and glandular (sometimes hairy).

Fruit ripens in the fall and is a shiny, red hip about ⅔ inch long. Both the fruit and its glandular calyx lobes sit erect and persist throughout the winter.

The pretty pink flowers appear in early June in a small cluster or singly. They are perfect and up to 2¾ inches across, with notched, wavy petals and numerous, yellow stamens. The calyx lobes are glandular, while the outside of the ovary is (usually) smooth.

Terminal buds are small, ovoid, and covered with a few reddish, smooth scales. Twigs are slender and pinkish red with a white, glaucous coating over much of the surface. There are a varying number of prickles along the twig, sometimes almost none, sometimes densely scattered. Lateral buds are similar to the terminal bud, only smaller; they diverge away from the twig. Leaf scars are a thin line above each node; each 1 contains 3 bundle scars.

This rose does not develop much diameter, so there is not much character to its bark. The stems become darker red and develop slight fissuring near the base. Prickles may be present on stems, but not on this specimen. Older stems can develop a white, almost waxy coating, as seen here.

pasture rose

***Rosa carolina* L.**

Family: Rosaceae

The extremely variable characteristics of pasture rose are generally considered a function of the environment in which the plant grows. However, it hybridizes with *R. palustris* and many other native roses, which creates another level of complexity. The flowers of our native roses are nearly always 5-petaled and a delicate pink color. Individuals of *R. carolina* about 1 foot tall, pictured here, are commonly found.

Form and Size: It is rarely over 3 feet tall, and it is more commonly 1 foot or so. It is few-branched from a single stem that arises from underground rhizomes.

Habitat: Because pasture rose is such a common species, its habitats vary, including upland woods and forest edges, dunes, roadsides, prairies, abandoned fields, and thickets. It grows best in full sun.

Wildlife Uses: A diminutive and upright, clonal rose, this species produces substantial cover for rabbits and quail, as well as other ground-dwelling, openland species. Its short form is especially attractive to several shrubland nesters such as field sparrows, indigo buntings, and, if the situation is right, yellow-breasted chats. The hips are smaller than those of most congeners and taken by some songbirds. The major users, however, remain game birds, such as bobwhite, and mammals, from fox squirrels to white-tailed deer. Leaves are also browsed by deer, and bark from stems is taken by cottontails if snow covers other resources.

Landscaping Value: Hardy to Zone 4, this sun-loving rose is petite and quite pretty when in flower. Even though it is found in a variety of habitats, it prefers well-drained sites. The fruit is usually bristly and bright red, and fall color is shades of yellow, orange, and red. There is a variety known as *R. carolina* var. *alba* that has white flowers. This easy-to-grow species is available at some midwestern nurseries specializing in natives.

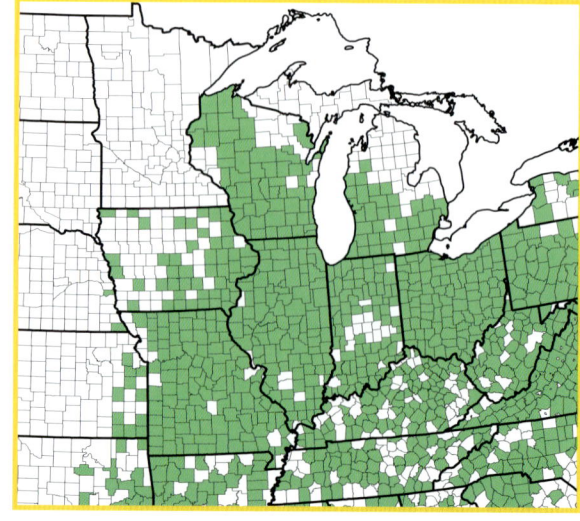

Similar Species Distinctions:
—**Meadow rose** (*R. blanda*) has many similar features, but look for less thorns along the lower stem, smooth hypanthium (outer portion of the ovary), and more white, glaucous bloom along stem of meadow rose.

The compound, alternate leaves have 5 to 7 leaflets that are about 2½ inches long. The leaflet margins are sharply or coarsely toothed. Leaf surface is green and mostly smooth, while the lower surface is paler and may be smooth, hairy, or hairy only along the midrib. The leaf petiole and rachis are hairy and often glandular-prickly. Leaf stipules have stalked glands along the entire or toothed margins.

Flowers occur from early June to early September and are a delicate pink with 5 petals and many yellow stamens. The 3-inch-wide, perfect flowers usually appear singly at the top of the plant. The ovary surface and pedicel are usually covered to some degree with stalked glands. The glandular sepals reflex after flowering and gradually break apart as the fall progresses.

The new twigs are deep pinkish red and mostly smooth. There can be patches of glaucous (white) bloom, especially on new growth. Twigs quickly develop prickles to varying degrees. There is usually a pair of straight prickles at the nodes. Twigs can be densely covered with various-sized prickles, particularly on vigorously growing specimens. The prickles can be straight or recurved and slender or thick. Many prickles are deciduous after the first year. Lateral buds are similar to the terminal, except smaller. Leaf scars are a fine line that wraps half-way around the twig; they contain 3 bundle scars.

The diameter of a large pasture rose is about as big as a pencil, so there is no diagnostic bark. The stems are green to red with a varying amount of prickles.

Terminal buds are few-scaled, mostly smooth, and deep pinkish red.

Fruits (hips) ripen in the fall; they are bright red and up to ⅔ inch long. They can be smooth or glandular when ripe and often persist into the winter months.

swamp rose

Rosa palustris Marshall

Family: Rosaceae

Swamp rose is appropriately named, as it is never found far from wet sites. However, the authors have grown a single specimen on an upland site for a number of years with great success. This is 1 of our largest native roses, and its fall color is beautiful. Like many roses, it has a tendency to hybridize principally with *R. blanda* and *R. carolina*.

Form and Habitat: Swamp rose grows to 6 feet tall or more and is much branched. It prefers full sun but will handle partial shade. Seen here is swamp rose in December in southern Indiana that is loaded with fruit.

Habitat: Swamp rose is common in poorly drained edges of swamps, marshes, streams, and low roadsides.

Wildlife Uses: As the name implies, swamp rose grows in wetland areas, often forming dense thickets of upright stems that provide substantial escape cover in these water-edge habitats. The stems can grow to substantial height (almost 10 feet) and thus supply nesting habitat to some higher nesting songbirds, like red-winged blackbirds, catbirds, and yellow-billed cuckoos. The hips are among the largest of our roses and thus have limited bird use. Raccoons and other mammals eat the fruits, and rodents frequently open hips in the winter and eat seeds. As they do with most roses, white-tailed deer browse leaves from swamp rose with regularity.

Landscaping Value: Hardy to Zone 4, this rose is adapted to wet, poorly-drained sites, but it grows well in fertile, well-drained soils. It has unusual, nearly purple mature bark with tan-colored thorns. The glandular fruits usually overwinter, and the combination purple twigs and red fruits catches the eye. Like many members of the Rosaceae family, this is a highly preferred species of Japanese beetles. Swamp rose is easily found in nurseries in the Midwest that specialize in natives.

Similar Species Distinctions:
—This is the only large, native rose essentially restricted to wet sites. It hybridizes with several other natives, in particular *R. blanda*, which is known as *R.* x *palustriformis* Rydb.

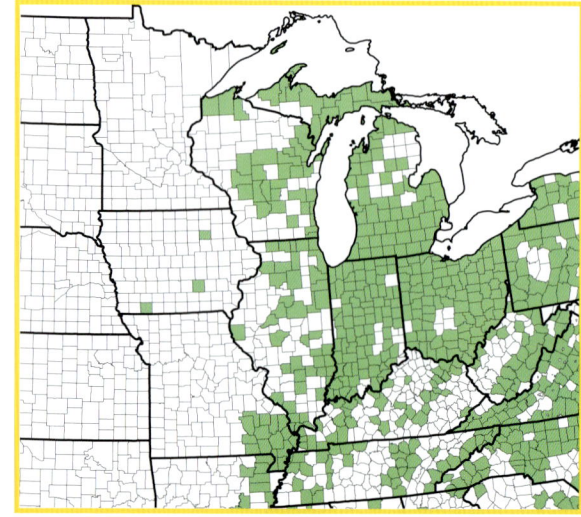

The compound, alternate leaves normally have 7 leaflets (sometimes 5 or 9) that are up to 3 inches long. The leaflet margins are finely toothed. Leaves are green, mostly smooth above, and slightly paler beneath. There can be hairs along the midrib beneath. The petiole and rachis are usually hairy and sometimes prickly. Stipules are narrow with glandular, hairy margins.

Fruits (hips) ripen in the fall and are red and glandular. They are about ½ inch long and remain on the plants through the winter months. The glandular sepals reflex after flowering and usually persist throughout much of the winter.

Flowering begins in late June and continues through July. Flowers are pinkish, perfect, 5-petaled, and up to 2 inches across. The ovary surface and pedicel are covered to some degree with stalked glands that persist into the winter.

The terminal bud is few-scaled and a deep wine-red. The margin of each scale is lined with white hairs.

Trunk diameter is rarely more than 1½ inch, but it is enough for the shrub to develop some light-colored fissuring along the stem that contrasts with the dark, wine-red or purple outer bark. Prickles thicken and straighten with age, turn beige, and develop an oval-shaped base.

New twigs are a deep wine-red and smooth. There is a pair of straight or recurved prickles at the nodes. Lateral buds are similar to the terminal. Leaf scars are a thin line that wraps halfway around the twig and contains 3 bundle scars. Older twigs are usually covered to some degree with fairly stout, straight or recurved prickles.

climbing prairie rose, prairie rose

***Rosa setigera* Michx.**

Family: Rosaceae

This is our only native, climbing rose, and many cultivars have been developed from it. It has several characteristics that are similar to the common, exotic multiflora rose (*R. multiflora*). Both have long, arching branches; both have small fruit. But that is where the similarities stop. Leaves, flower color, prickle size, and stipule shape are some of the differentiating characteristics. It is seen here in 1 of its most common habitats—a forest edge, using a fallen tree to support its long branches.

Form and Size: Its long, arching stems can reach to 13 feet or longer, and they spread over existing shrubs or over tree limbs for support. A fully grown plant appears large and spreading.

Habitat: This species occurs along fencerows and roadsides, along forest edges, and in thickets. It seems to be quite content with partial sun and produces fruit prolifically where found.

Wildlife Uses: When growing in the open, prairie rose can form dense shrubs up to 4 feet tall, similar to several other roses. However, if its location is on roadsides, fencerows, and forest edges, it assumes a climbing nature and climbs onto other shrubs and trees. In doing so, it provides very good escape cover and nesting cover for a wider variety of songbirds than do most shrub roses. The fruits are small enough to be attractive to larger songbirds (e.g., catbirds, wood thrushes, robins) and to gallinaceous game birds. Deer also eat the fruits and browse leaves through much of the summer.

Landscaping Value: Hardy to Zone 3, this climbing rose needs room to grow, perhaps over a trellis or fence. Stems can be cut back to promote branching. They grow well in partial shade but handle nearly full sun. This species does not seem too particular about soil type, but it probably does not handle poorly-drained sites. The authors have not seen this native rose offered in any midwestern nursery.

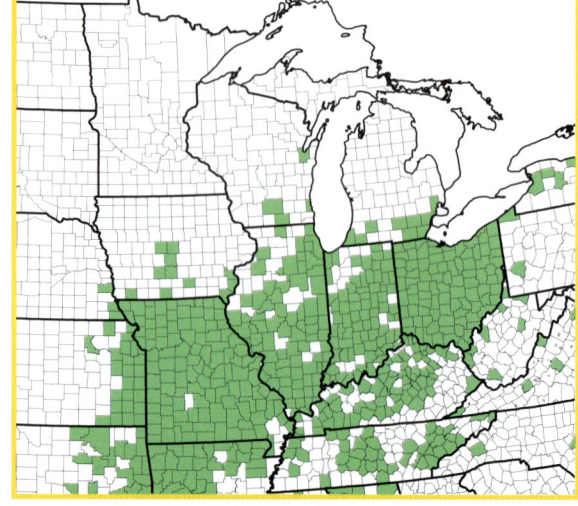

Similar Species Distinctions:
—The only native rose that climbs. The biggest difference between this native and the exotic and very common multiflora rose is the flower color. Multiflora has white flowers, which natives rarely do.

The compound, alternate leaves commonly have only 3 leaflets, especially on the older growth (sometimes 5 on the new growth). Leaves are up to 4 inches long and are dark green and glossy with impressed veins. Leaflet margins are finely, singly or doubly toothed. Upper surface is smooth, while the lower surface is paler and usually smooth. Stipules are narrow and entire, with glandular or hairy margins. The petiole and rachis are usually prickly.

Terminal buds are ovoid with 5 to 6 reddish, smooth scales.

Flowering begins in mid-June and continues through July. The pink (mostly), 5-petaled flowers are up to 3 inches across, and are unisex—very unusual for a rose. The ovary surface and pedicel are glandular. Sepals are glandular and reflex back during flowering.

The diameter of this species is rarely over 1 inch, but fissuring occurs on the largest stems to expose the beige-colored inner stem. Straight or recurved prickles, which are also beige, remain scattered along the stem.

Twigs are smooth and greenish or reddish, with numerous tiny, white lenticels. The beige prickles are paired near the nodes and usually recurved. They can also be scattered along the twig singly. Lateral buds are similar to the terminal but much smaller (those seen here are expanding in the spring). Leaf scars are half-circles at each node that contain 3 bundle scars.

Blackberries, Raspberries, Dewberries

Genus: *Rubus*

Blackberries, raspberries, and dewberries—oh my! This is a large, complex genus that most people collectively term "briers" or "brambles." If you want a good challenge, this genus is for you.

Because of the difficult nature of the genus with its complex reproductive strategies, which includes hybridization and genetic heterogeneity, we have concluded that it would be best to follow Voss (1985) and take a practical approach in covering the species of the Midwest. Anyone interested in a recent, in-depth discussion of *Rubus* is directed to *Trees and Shrubs of Minnesota* (Smith 2008) and *Flora of North America*, Vol. 9 (in press). In both works, taxonomists who specialize in *Rubus* cover the most recent genetic data and theories on species distinctions.

There are several midwestern *Rubus* species used as ornamentals, 1 of which is *R. odoratus*, purple-flowering raspberry. It has large, maple-like leaves and large, rose-like flowers. Most brier patches found in a yard or near a human dwelling are in a neglected, overgrown area where birds are responsible for its establishment. These patches need not be eliminated and can provide a wonderful, annual supply of tasty, summer berries with just a little maintenance. Fall color, particularly of the blackberries, can be phenomenal if the plant is growing in full sun.

One would be hard-pressed to name a genus of shrub that has greater overall wildlife value than *Rubus*. In general, it can be said that they provide outstanding cover, highly preferred fruits, and substantial browse—often all in a single species. Taxonomically, the genus is difficult, so our descriptions will focus mainly on the wildlife values of: 1) the highbush blackberries, 2) the raspberries, and 3) the dewberries. To avoid redundancy we may refer the reader to accounts of congeners that have almost identical values.

Blackberries tend to be robust, with long canes and dense growth that provides very good cover. Raspberries are usually less robust, providing less cover than blackberries, yet a considerable amount, especially when growing with (and supporting) grass and forbs. Dewberries tend to be prostrate, creeping plants that provide minimal cover.

The stems of briers are known as canes, and each cane lives just 2 growing seasons. Our natives produce flowers and fruits on the second-year cane, which are known as the floricanes. The first year, vegetative canes are known as primocanes, and they do not produce flowers or fruits. Seen here is a black raspberry cane below a blackberry cane, showing not only the color difference, but the shape of each.

Leaves of most *Rubus* are compound, but there are often a different number of leaflets per leaf depending on the age of the cane in question and the location of the leaf on the cane. Shown are leaves of *R. idaeus* L. var. *strigosus*.

The tasty, juicy summer fruit of the genus is worth the pain of collecting it; it is impossible to remain scratch-free. However, they are some of the Midwest's best wild fruits, free for the taking.

common blackberry, Allegheny blackberry

***Rubus alleghaniensis* T.C. Porter**

Family: Rosaceae

Common blackberry is just that—common; it is found throughout the Midwest in disturbed areas. Blackberries, like all the "briers," develop brier patches that are impenetrable to mammals of much size, but they are the perfect habitat for shrub-nesting birds. Humans, wild mammals, and birds are fond of the juicy berries that are produced in the summer. A human might receive a good scolding from a catbird or brown thrasher that is not only nesting in the patch in which the person may be picking, but also protecting its food source.

Form and Size: Its new canes are erect, but eventually the tips curve toward the ground. An individual plant can produce 10-foot-long canes, and a patch of blackberries can fill in a large area if left undisturbed.

Habitat: Common blackberry occurs in many habitats, including fencerows, roadsides, open woods, forest edges, and old-fields. It grows best in full sun; each cane lives 2 years.

Wildlife Uses: This is likely 1 of the most valuable shrubs for wildlife. Cover value is superb, with canes not only upright and often growing very densely, but also often interspersed with herbaceous growth that adds to the cover for ground-dwelling species. Many species of birds nest in the tangles; it provides escape cover for others. The fruits are eaten by almost all species of birds and a range of mammals as well—including foxes, coyotes, and deer. Deer also browse the leaves substantially in spite of prickles, but they really relish the lush foliage of new canes as they emerge in the spring.

Landscaping Value: Hardy to Zone 3, this is a ubiquitous blackberry of the Midwest and northeastern United States. Canes reproduce quickly and are short-lived, but they can fill in an area in short order if not controlled. If flowers are your desired landscape objective, cut the flowering canes to the ground once flowering is past. The non-flowering canes will be what flowers next spring. If you want fruit and fall color, delay cutting those canes until the winter months. If attracting wildlife is your goal, leave the patch alone. Fall color can be spectacular. There are horticultural varieties that are thornless and readily available.

Similar Species Distinctions:
—Too many to cover here. See authorities listed on genus page 278.

The compound, alternate leaves have mostly 5 leaflets (sometimes 3) and are up to 8 inches long; terminal leaflet may approach 5 inches in length. The leaf has a long, prickly petiole. The upper leaf surface is dark green and sometimes lightly hairy, while the lower surface is somewhat paler and hairy, with short, hooked prickles running along the mid-veins. Leaflet margins are finely, sharply toothed. Leaves on the primocanes (first-year stems) may be different (5-leaflets) from those of the floricanes (the second-year flower and fruit producing stems), which may be simple or trifoliate. There is a pair of stipules at the base of the petiole.

Fruit is an aggregate drupe that is fleshy and black when ripe in July. It is up to 1 inch in length and many-seeded. Unlike a raspberry, a blackberry's receptacle does not separate from the fruit when picked.

Flowers are produced in profusion in mid- to late May on lateral branches of the previous year's canes. They have 5 white, crinkled, strap-like petals and numerous stamens. The pedicels are usually densely hairy or glandular-prickly.

Twigs are thick and strongly angled, with a deep, wine-reddish or greenish color. There are scattered, sharp, straight or curved prickles along the entire stem (cane). Canes have substantial whitish pith and die after the second growing season—after flowering and fruiting is over. Lateral buds diverge strongly from the cane and have numerous scales.

Blackberry canes are ridged, which is easily seen in this cross section.

northern dewberry

Rubus flagellaris Willd.

Family: Rosaceae

Like so many species of *Rubus*, this 1 tends to be highly variable in its traits. Swink and Wilhelm (1994) in *Plants of the Chicago Region* recognize a separate species—*R. enslenii*—as being distinct from *R. flagellaris;* it has "solitary flowers, weak subherbaceous stems with weak or absent prickles." Voss (1985) in *Michigan Flora* lumps them together. Either way, this is another creeping *Rubus* that is common throughout much of the Midwest.

Form and Size: Northern dewberry has a prostrate or low-arching form that spreads near the ground, often in large colonies. Individual stems can reach 9 to 10 feet in length.

Habitat: It prefers full sun but tolerates partial shade. It is found in nearly all soil types, but it is common in degraded soils of old-fields, open woods, stable dunes, and roadsides. It is usually found in full sun.

Wildlife Uses: Northern dewberry is a trailing, vine-like species. As such, it does not supply the vertical cover of the blackberries and raspberries. It still does, however, provide cover for ground nesters, such as common yellow-throat and eastern towhee. The late summer fruits are not produced as prolifically as in highbush blackberries and black raspberries, but they are juicy and readily used by wildlife, especially small terrestrial rodents that can easily access them. Cottontails browse leaves in summer and stems in winter.

Landscaping Value: Hardy to Zone 3, this dewberry creeps along the ground, and without much competition, it can nearly cover large areas of disturbed soil. Allowed to grow, it is a nice ground cover that is particularly lovely in the fall with its multi-colored leaves. It has large flowers for a *Rubus* that sit upright on the stems. Cleaning leaves away from the plethora of stems could prove challenging and painful if it were used as a ground cover in an urban setting. This species is not available from nurseries.

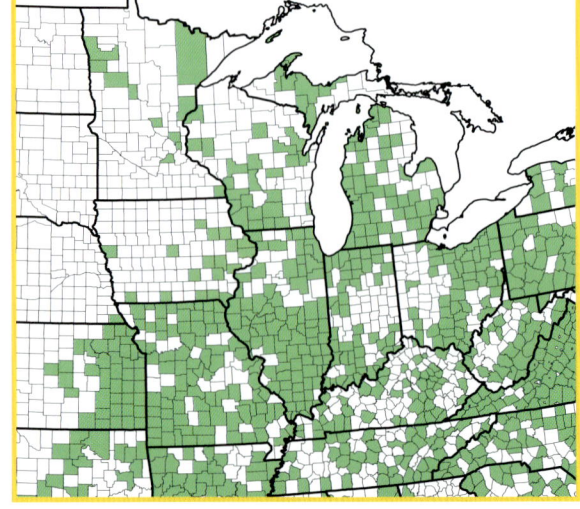

> **Similar Species Distinctions:**
> —There are several species of dewberry in the Midwest. We cover 2 here, and the leaves and stems of *Rubus hispidus* are discussed within those accounts.

The compound, alternate leaves are usually trifoliate (rarely 5) on young and old canes. All the leaflets are roughly the same size, their margins being sharply, doubly toothed. Leaflets are widest below the middle and are green on the top and bottom. The amount of hair on top and bottom is variable, and there are sometimes glandular hairs beneath. The petiole and lower midrib have scattered prickles. There is a pair of leafy stipules at the base of each petiole. Typical fall color is developing on these leaves.

Flowering occurs from early May into June on upright branchlets. The flowers have 5 white, crinkled petals and numerous stamens; flowers are 1 inch or more in diameter. Flowers are single or in terminal clusters of 2 or 3. Sepals are hairy and do not reflex back.

Twigs are slender and pinkish red or green, and they have scattered, slender, straight or curved prickles as well as glandular hairs or bristles. Older stems are smooth or have scattered, weak, recurved prickles. This species has stronger, more broadly-based prickles than *R. hispidus*, swamp dewberry.

Fruits ripen as early as late June. They are small, aggregate drupes about ⅔ inch long. When ripe, fruit is black and juicy.

swamp dewberry

***Rubus hispidus* L.**
Family: Rosaceae

As the name implies, swamp dewberry is always found in moist habitats, mostly in the northern portions of the Midwest. It is fairly easy to recognize by its creeping habit, extremely fine, bristly stems, and small, shiny leaves. Fruits are much smaller than the native, large blackberries, but they are still edible—although a bit sour.

Form and Size: It is prostrate, with trailing to low-arching stems that root at the tip. It is colonial, and usually found carpeting an area.

Habitat: Swamp dewberry is mainly found on the borders of lakes and marshes, at the base of wooded slopes, in tamarack bogs, and in low black and pin oak woods. It probably prefers acidic soil and partial shade. It tolerates poorly-drained, wet sites and full shade.

Wildlife Uses: Although swamp dewberry differs from northern dewberry in having bristles rather than prickles, its value to wildlife is almost identical—differing largely in the habitat occupied. As the name suggests, this species occurs in wet places—edges of streams and lakes and moist bottomlands—and thus has a somewhat different cadre of users that are served in the same manner.

Landscaping Value: Hardy to Zone 3, this dewberry makes a nice, delicate-looking ground cover. It prefers moist, well-to-poorly-drained soils that are acidic. Leaves are small, more rounded than those of other native *Rubus* species, and shiny. It has tiny fruits in late summer and nice fall color. Swamp dewberry is not available through nurseries.

Similar Species Distinctions:
—**Northern dewberry** (*R. flagellaris*) has smooth stems except for rather tough, slightly curved prickles instead of straight bristles. Leaflet tips are pointed.

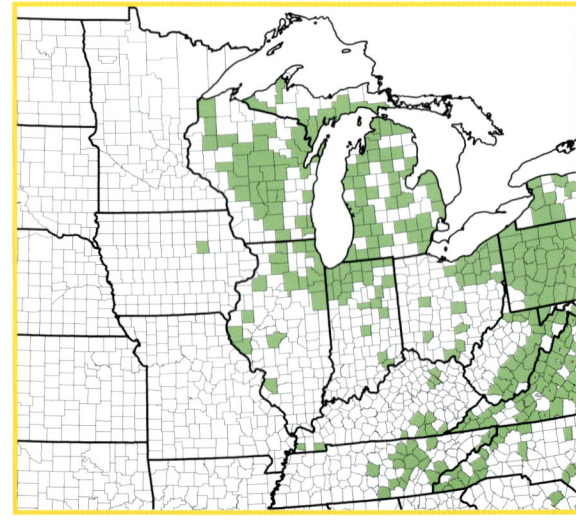

The compound, alternate leaves are trifoliate (rarely 5) and up to 2 inches long. They are semi-evergreen and firm to the touch. They are dark green, shiny, smooth above, and somewhat paler and mostly smooth beneath (there can be hairs running along the veins). Leaflets are mostly the same size, and the margins are doubly-toothed. Leaf petioles are smooth, hairy, or bristly.

Flowers first appear in early June and can continue to occur until late August. They are about ¾ inch in diameter and have 5 white petals and numerous stamens. There are 2 to 6 flowers in terminal clusters. The pedicels are hairy and sometimes bristly. Sepals are densely hairy and reflex at flowering.

Twigs are very slender and green. They are usually covered with straight bristles that are less than ⅛ inch long. Sometimes bristles are not obvious and/or the twigs may have short, glandular hairs.

Fruit begins to ripen in August and is small—only ⅔ inch long when ripe. It is reddish purple, juicy, and sour.

common red raspberry

Rubus idaeus L. var. *strigosus* (Michx.) Maxim.
Rubus idaeus L. ssp. *strigosus* (Michx.) Focke
Rubus strigosus Michx.
Family: Rosaceae

Common red raspberry has had a confusing list of synonyms over the years, as taxonomists have disagreed over what the correct classification should be (and they probably still do). At any rate, red raspberry is a cool-weather plant that is extremely common in the northern Lake States, where it forms sizable brier patches, as seen here. It is easily recognized by its extremely bristly canes.

Form and Size: Canes can grow to 6 feet or more in length, and they are usually more erect than other native briers of the Midwest. It forms patches that spread from underground shoots, and it thrives in full sun to partial shade.

Habitat: Red raspberry is not particular about where it grows, and it is found in wet areas (such as swampy woods, old tamarack bogs, and marshes), roadsides, open upland woods, pine stands, and old-fields. It seems to prefer moist, rich soil.

Wildlife Uses: Red raspberry is a species that occurs in the northern part of our region, becoming 1 of the most ubiquitous shrubs with the increase in latitude. It is more upright in growth form than *R. occidentalis* and forms dense colonies through rhizomes. It has many of the same cover values as the highbush blackberries, although it has bristles rather than their stout prickles. The fruits are produced in July in the southern part of its range and taken by many birds and mammals. Leaves are browsed, often rather heavily, by white-tailed deer and moose.

Landscaping Value: Hardy to Zone 2, this is a cool-weather plant that is too weedy and common to be considered as an ornamental. It aggressively encroaches into off-limit areas in a domesticated landscape (the authors have some in a native shrub plot). Flowers are not showy, and the fruits are less palatable than those of black raspberry. Leaves are more interesting that most *Rubus* because of their white leaf undersurfaces. A cultivated variety 'Aureus' has bright yellow leaves.

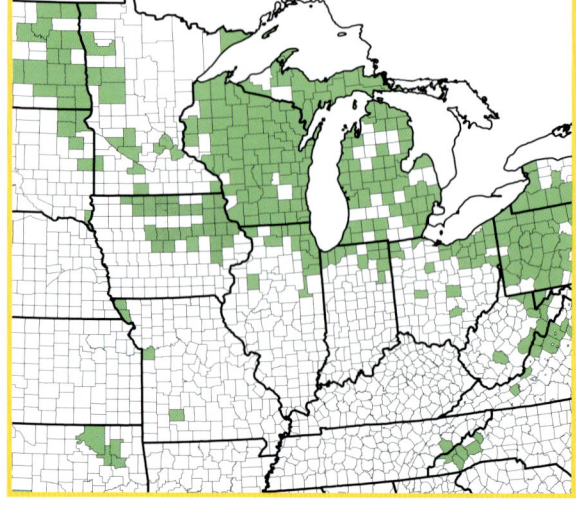

> **Similar Species Distinctions:**
> —**Black raspberry** (*R. occidentalis*) has less bristly canes that are coated with a white glaucous bloom. Ripe fruits are black, tasty, and edible.

The compound, alternate leaves are of various shapes. The second year flowering canes (floricanes) have trifoliate leaves, while the first year primocanes produce leaves with 3 to 5 leaflets. Sometimes the terminal leaflet is 3-lobed or actually divided. Leaflets are long-tipped with irregularly toothed margins. The upper surface is green and smooth, while the lower surface is white with dense hairs. The color contrast between upper and lower leaf surfaces is striking.

The small flowers appear in May and June. They are in flattened clusters of 2 to 5 flowers. The flowers are about ⅓ inch in diameter and have 5 tiny, white petals and numerous stamens. Sepals are glandular and slightly hairy; they spread at flowering. Pedicels have glandular bristles.

The canes are medium brown and usually densely covered with bristles and curved prickles. Canes tend to remain more upright than most other species of briers; the tip does not curve and root in the ground. Lateral buds diverge from the twig and are reddish brown with loose-fitting scales.

Red raspberry fruit is an aggregate drupe that ripens in July and is about ⅓ inch in diameter. Like all raspberries, the receptacle remains behind when the berry is pulled from the plant, leaving the picked fruit with a hollow center. It is red, juicy, and edible when ripe.

black raspberry, common blackcap raspberry

Rubus occidentalis **L.**

Family: Rosaceae

This common *Rubus* is found throughout the Midwest, often growing alongside the also prevalent *R. allegheniensis*. Together, the 2 provide much relished food for many fruit-eating birds and mammals. Colorful canes are distinctive in winter.

Form and Size: Canes arch as they elongate and can reach 6 feet in length or longer until their tips reach the ground, where they root. With a little time, sizable brier patches are created.

Habitat: Black raspberry can be found in a variety of wet-to-dry habitats—in open woods and forest edges, clearings, old-fields, borders of streams and lakes, and fencerows. It grows best in moist, partially shaded sites.

Wildlife Uses: This widely distributed raspberry is the perfect complement for any of the highbush blackberries from a wildlife perspective. In Indiana, for example, black raspberry produces fruit in June, followed by blackberry, which typically fruits in July. Thus, having both in an area gives a prolonged period of fruit availability for wildlife users, which are similar for both. From a cover perspective, black raspberry does not usually grow as densely as blackberries, depending on tip-rooting for spread, and canes are more arching and remain closer to the ground. This supplies even better cover for small ground-dwellers and for low nesting songbirds. Browsing by deer is not as heavy as it is for blackberries.

Landscaping Value: Hardy to Zone 3, this species is most important for the great tasting food it provides. It does have unusually-colored canes and 2-toned leaves, but most people just find briers to be a pain—literally. Black raspberry is not particular about where it grows, but it prefers moist, well-drained soils and half-day sun. There are several varieties on the market that produce larger fruit and are disease resistant.

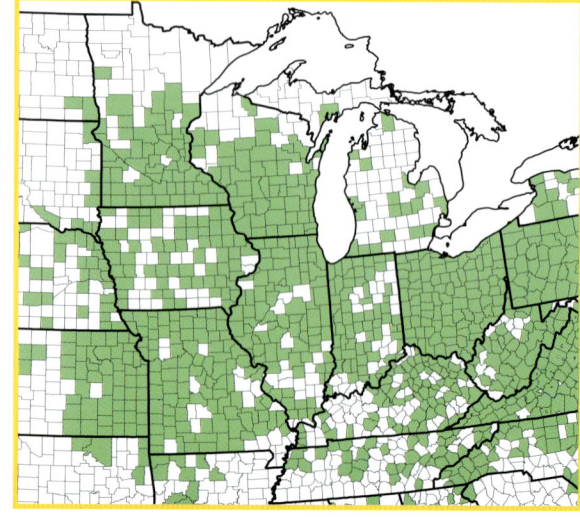

Similar Species Distinctions:
—No other native brier has canes the color of black raspberry.

The compound, alternate leaves are usually trifoliate, but the new primocanes (canes of this year) can have leaves with 5 leaflets. The margins are sharply, doubly-toothed on leaves up to 10 inches long. Half of that length, however, is from the long, prickly petiole. The upper leaf surface is green and smooth; the lower surface is white and thinly hairy with fine prickles running along the midrib.

The fruit is an aggregate drupe that ripens from mid-June to mid-July. When ripe it is black, juicy, and very tasty. Fruit size varies from ⅓ inch diameter to nearly 1 inch, depending on genetics, site, and weather conditions.

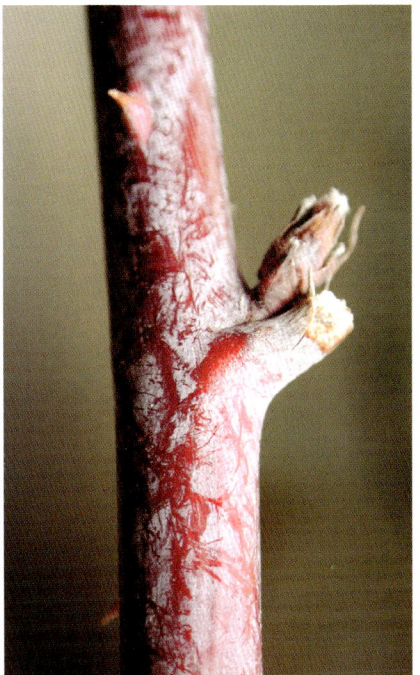

Flowers appear from early May into June. They are perfect and have 5 short, white petals and numerous stamens. They are about ½ inch in diameter. Sepals are hairy, white, and spread during flowering. Pedicels are prickly.

The new canes are greenish red, eventually turning a bright, wine-red color. Canes usually are coated with a glaucous bloom that easily rubs off with the touch of a finger. The bloom gradually weathers away to reveal the bright wine-red beneath. Canes have scattered, fine, stiff prickles running their entire length. Lateral buds diverge from the twig and have numerous, loose-fitting, reddish brown scales. There are no terminal buds since cane tips bend and root at the tip.

Pennsylvania blackberry, leafy-flowered blackberry, yankee blackberry

Rubus pensilvanicus/pensylvanicus **Poiret.**
Rubus frondosus **Bigelow**
Family: Rosaceae

For this species the authors follow Edward Voss (1985) in *Michigan Flora* (and his *Rubus pensil-vanicus* "complex") because of the difficulties of identification caused by a great deal of variability among individuals. *Rubus* specialists over the last century have not agreed (and still do not) on classification of this complex group. Swink and Wilhelm (1994) in *Plants of the Chicago Region* cite "segregates" of this species as was done by earlier researchers. One somewhat constant feature of the complex is a loose, almost umbel-like inflorescence, rather than the typical racemose flowering structure.

Form and Size: Like most blackberries, this species also develops long, arching canes that can reach 7 feet or more. They do not root at the tip, as do those of the raspberries. Underground rhizomes create dense brier patches.

Habitat: This species is common throughout much of the Midwest, and it is found in essentially the same type of sites as *Rubus allegheniensis*. It requires full sun for best growth, but it is not particular as to soil type.

Wildlife Uses: This species is essentially identical in wildlife value to *R. allegheniensis*. This could have been the species to which Aldo Leopold was referring in his classic *Sand County Almanac* when he spoke of going from 1 patch of "red lanterns" to another in his fall rabbit hunting. Red lanterns were patches of blackberry (either this species or *R. allegheniensis*), the leaves of which had turned their beautiful fall color of wine-red, which were ideal rabbit havens.

Landscaping Value: Hardy to Zone 3, this blackberry has the same qualities as *R. allegheniensis*, with the exception of more upright branching. Fruits are edible, fall color is splendid, and canes have sharp prickles that tear the skin. This species is not available commercially.

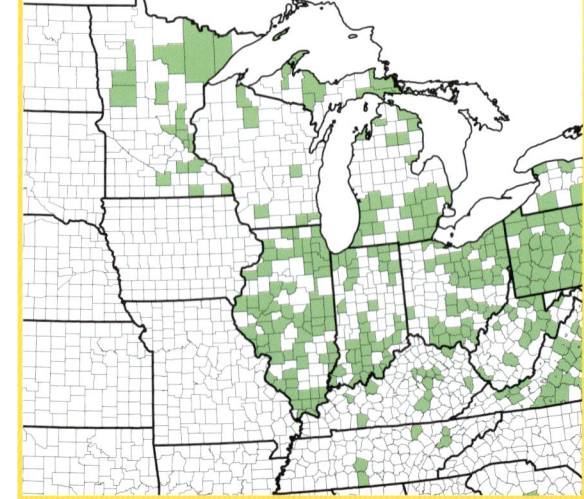

> **Similar Species Distinctions:**
> —Several blackberries look very similar and are difficult to tell apart, especially without flowers and fruit. We do not attempt to distinguish among them here.

The alternate, compound leaves are up to 4 inches long and nearly as wide. Leaves on the new canes (primocanes) have 5 leaflets, while leaves on the old canes (floricanes) have 3. The upper leaf surface is green and lightly hairy. The lower surface is paler and hairy with small prickles along the central vein. Leaflet margins are irregularly toothed. Petioles are hairy and lined with tiny prickles.

Twigs, or canes, are greenish or reddish with continuous grooves (sometimes none). Canes can be strongly or weakly armed with broad-based, straight or recurved prickles. Canes may have gland-tipped hairs as well. Lateral buds have numerous, loose scales that are the same color as the cane. Buds diverge from the cane. Cane tips usually die back during winter months, so there is no terminal bud.

The flowers begin to open in late May in more flat-topped inflorescences than those of other blackberries. The flowers are about 1 inch in diameter and have 5 white petals. Pedicels are glandless, or nearly so, and densely hairy.

Fruits begin to ripen in early July. The fruit is a drupe (an aggregate of druplets) that is shiny black when fully ripe. Each fruit can be up to ¾ inch long; it is juicy and sweet.

poison sumac, swamp sumac, poison elder

Toxicodendron vernix (L.) **Kuntze**

Rhus vernix **L.**

Family: Anacardiaceae

It is claimed that poison sumac is more toxic than the native vine poison-ivy (*T. radicans*). Most parts (not the hairs and pollen) of the plant contain toxidendrol, an oil that causes an allergic skin reaction in many people. Fortunately, most people do not wander through the boggy habitats where this species occurs. The toxic nature of this plant is unfortunate, because it is quite beautiful in many respects. Its deep green, shiny leaflets stand out against the bright red rachis.

Form and Size: It can become quite large, 15 feet or taller. Its crown seems sparse, because leaves develop from the new growth on the tips of its scant branches.

Habitat: Poison sumac is a shrub of the bogs, swamps, and seepy, poorly drained areas. Like all sumacs, it spreads by underground suckers to form loose, clonal colonies. Poison sumac will tolerate partial shade, but it prefers full sun.

Wildlife Uses: Poison sumac is limited to very wet areas where it occasionally reaches small tree proportions with a small crown; it is at times chosen as a nest site by songbirds, but its structure is not conducive to its use. Its whitish fruits are taken by songbirds, such as eastern bluebirds and robins, but frequently some remain on the shrub into the winter, suggesting that preference is not high.

Landscaping Value: Hardy to Zone 4, this very toxic plant should never be planted where humans or pets might contact it.

Similar Species Distinctions:
—**Black ash** (*Fraxinus nigra*) is a tree found in the same habitats as poison sumac and has similar leaves, but all ashes have opposite branching.

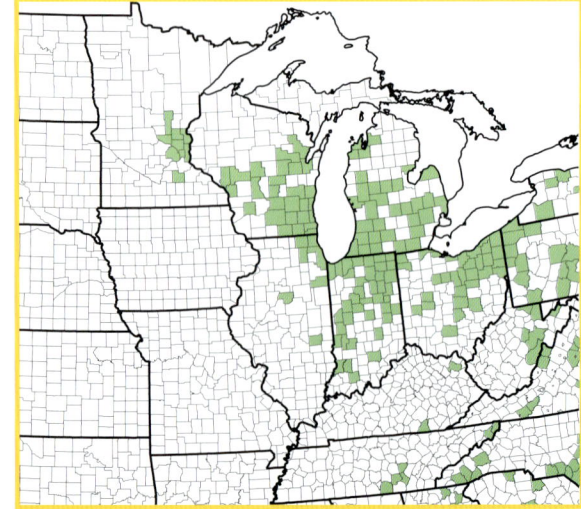

The compound leaves are alternate, 1 foot or less in length, with 7 to 13 leaflets. The leaflets are of variable shape and always appear to extend upward from their attachment point. Leaves are dark green, shiny, and smooth above, pale yellow-green and mostly smooth (at maturity) below. The rachis is usually bright red and smooth. Leaflet margins are entire or slightly wavy.

Fruits ripen in late summer and hang downward in drooping, open clusters. The individual fruits are about ⅛ inch in diameter, yellowish, and shiny. Fruit commonly persists into the winter months.

Flowers appear from the new growth in open, elongated clusters (panicles) from the leaf axils. Individual flowers are tiny, greenish yellow, and usually single sex on a given plant. Flowering occurs in June.

The terminal bud, although not a true terminal in the technical sense, is dark brown, short, and blunt-tipped. The 2 largest scales are almost fleshy. Buds have some scattered hairs. Twigs are fairly stout, pale-colored, smooth, and dotted with lenticels. The lenticels are the same color as the twig. Leaf scars are sizable and shield-shaped, and they contain numerous bundle scars that are located along the margin of the scar. Lateral buds are small, rounded, and few-scaled.

Mature bark is smooth, thin, and dotted with rust-colored lenticels. The color is usually very pale gray or tannish gray.

prickly-ash, toothache tree

Zanthoxylum americanum **Mill.**
Xanthoxylum americanum **Mill.**
Family: Rutaceae

Prickly-ash is a tall, suckering shrub that has several unique attributes. It is in the Citrus family, as is hoptree (*Ptelea trifolia*), and contains certain chemical properties that attract the Midwest's largest butterfly, the giant swallowtail (*Papilio cresphontes*). Adult females lay their eggs on both shrubs in order to provide food (the leaves) for their growing larvae (known as "orange dogs"). A butterfly enthusiast can easily attract this beautiful butterfly to his or her property, simply by planting either shrub.

Form and Size: This can be a large shrub to small tree, depending on location in the United States. The largest known specimen is 28 feet tall, but it rarely grows more than 10 feet. It freely suckers from the roots and forms colonies. It has a much-branched, upright form and is very prickly all over.

Habitat: Prickly-ash prefers moist sites in partially shaded woods, and it is usually scattered in its distribution. It is most common in low, wet woods, along streams, and in old tamarack bogs, but it can also tolerate drier sites and is found on rocky, wooded slopes. In some regions of the Midwest, it is somewhat of a pest and invades abandoned fields and grazed pastures. Although it is generally found growing in partial shade, it tolerates full sun.

Wildlife Uses: Prickly-ash's principal wildlife value is as a protective cover plant, especially in bottomlands where it occasionally forms seemingly impenetrable thickets through root suckering. In these habitats, it is also frequently used as a nest site by songbirds, especially wood thrushes and robins. Its fruits are small enough for use by even the smallest songbirds, but they seem to be little used. Red-eyed vireos and bobwhite have been recorded feeding on them; others likely do as well. The shrub is rarely browsed by white-tailed deer.

Landscaping Value: Hardy to Zone 3, prickly-ash is just that—prickly. If you do not care for plants with spines and prickles, this is not a shrub for you. It is insect and disease-free, and it attracts the largest butterfly in the Midwest. Its caterpillars do so little eating that damage is barely noticeable. The prettiest feature on prickly-ash is its ripe, shiny, black fruits when they pop out of their reddish, fleshy cases. Twigs are an unusual purplish color and contrast nicely with its rusty-red winter buds. This species is likely unavailable through any nursery in the Midwest.

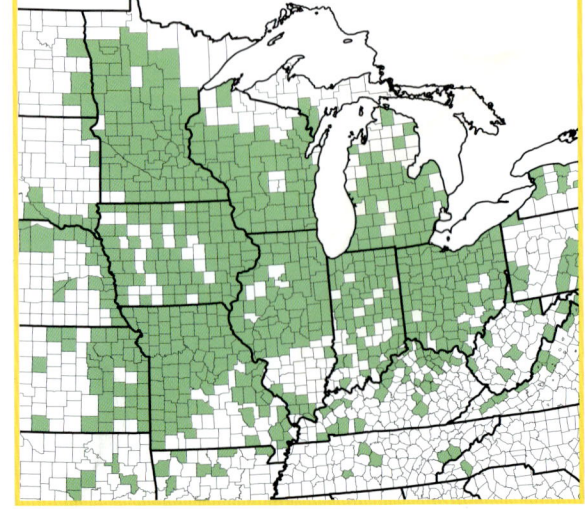

Similar Species Distinctions:
—**Ash** (*Fraxinus* spp.) trees have similar leaves but opposite branching. They do not have the fine prickles running along the bottom of the rachis like the prickly-ash does.

The alternate leaves are compound and up to 11 inches long, but they are usually shorter. They have 5 to 11 leaflets that are opposite each other along the rachis—a feature that creates an ash-like appearance. Leaflets have a finely-toothed margin with a yellow gland between each tooth. The upper leaf surface is green and mostly smooth; the lower surface is paler and somewhat glaucous with hairs mainly along the veins. Leaflets held up to the light reveal a "dotted" translucence. The rachis and petiole have small prickles along them. Crushed leaves give off a unique smell, resembling citrus.

Flowers appear before the leaves, as early as the second week of April, in clusters (cymes) along last year's twigs. This species usually has single-sex flowers on a plant, but bisexual flowers can be found. They are tiny, about ¼ inch long, and have 5 greenish petals (there are no sepals). The females (near left), with their rather long style, appear beaked.

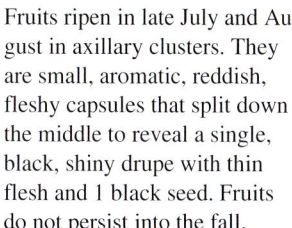

Fruits ripen in late July and August in axillary clusters. They are small, aromatic, reddish, fleshy capsules that split down the middle to reveal a single, black, shiny drupe with thin flesh and 1 black seed. Fruits do not persist into the fall.

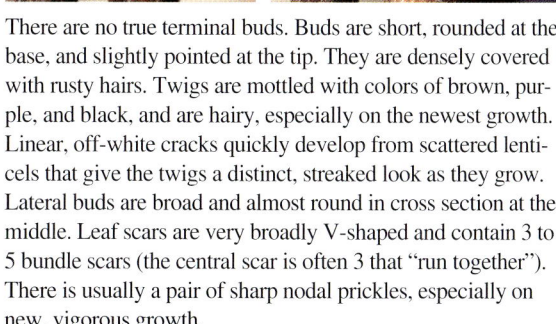

There are no true terminal buds. Buds are short, rounded at the base, and slightly pointed at the tip. They are densely covered with rusty hairs. Twigs are mottled with colors of brown, purple, and black, and are hairy, especially on the newest growth. Linear, off-white cracks quickly develop from scattered lenticels that give the twigs a distinct, streaked look as they grow. Lateral buds are broad and almost round in cross section at the middle. Leaf scars are very broadly V-shaped and contain 3 to 5 bundle scars (the central scar is often 3 that "run together"). There is usually a pair of sharp nodal prickles, especially on new, vigorous growth.

In the Midwest, prickly-ash generally does not become large enough to develop much bark character. Mature bark is smooth and grayish or brownish, with visible streaking running the length of the trunk. This specimen is "stained" with gray lichen patches.

NATIVE VINES

crossvine

Bignonia capreolata **L.**

Anisostichus capreolata **(L.) Bureau**

Family: Bignoniaceae

Crossvine is by far 1 of the showiest native vines in the Midwest, and its unusual semi-evergreen leaves warrant a second look. Its common name refers to the cross-shaped pith of the twig.

Form and Size: Crossvine can grow to a height of 80 feet into the tops of forest trees, particularly in the southern states where the growing season is longer. It can grow quickly and live fairly long. When growing to such great heights, only the leaves clinging to tree bark are seen. Vines can grow to a diameter of 6 inches or more at the base.

Habitat: The common habitat for crossvine is wooded, bottomland floodplains where moisture is nearly always present. It can be found growing along wooded streamsides as well. It prefers full sun but handles partial shade.

Wildlife Uses: Crossvine occurs in the southern portions of our region, usually in bottomland habitats. It can climb high in trees as well as along fences and onto low shrubs. Thus, it supplies nesting cover for low-nesting species, such as cardinals and catbirds, as well as higher nesters, such as gray squirrels. While vines climb high, their diameter does not grow large enough to deform the trees on which they grow. The unique flowers bloom early and supply important hummingbird food at that time; seeds in the resulting pods seem little-used by wildlife. Foliage and twigs are favorite browse for swamp rabbits (endangered species in Indiana) and are taken regularly by white-tailed deer.

Landscaping Value: Hardy to Zone 5 (barely), it can be killed back to the ground during cold winters in this zone. Although found in moist sites in the wild, it will handle drier uplands. Its flowering period is a month or more, and it can be grown as a ground cover or up a trellis or pole. Given free range it can become a bit pesty. There are numerous cultivars available, including 'Tangerine Beauty.'

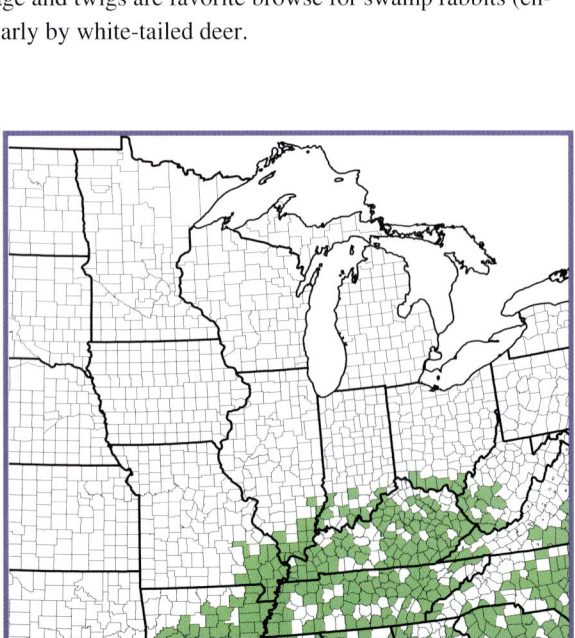

Similar Species Distinctions:
—**Trumpet creeper** (*Campsis radicans*) has similar flowers that are usually orange and are borne terminally. Pods are not flat.

The evergreen leaves are actually compound and are divided into matched pairs of leaflets. Each pair of leaflets has a branched tendril that ends in adhesive disks. Leaflets are oblong to ovate, dark green, and mostly smooth above and beneath. At the base of each petiole is a stipule. Leaflet margins are entire. Winter leaves are often tinted red, especially underneath.

The 5-lobed, tubular flowers are in axillary clusters from late April through May. They are up to 2 inches long, reddish orange on the outside, and yellowish inside. Their floral scent has been described as that of chocolate or beef bouillon.

The flat, linear, dangling capsules ripen in the fall and are up to 7 inches long. When ripe, they split along the sides to release many thin seeds. Each seed has large, flat, papery wings on each side.

Mature bark is light to dark brown, rough, and scaly with vertical exfoliation.

Winters twigs are greenish to reddish with small slender buds of the same color. Pith is cream-colored and cross-shaped.

Cranberries

Genus: *Vaccinium*

Cranberries are to the blueberries a bit like cousins. They have similar genetic makeup, but they really do not look much alike. Over the years, taxonomists have disagreed on how they should be classified. Some suggest a separate genus, *Oxycoccus*, while others leave well enough alone and stick with *Vaccinium*. The current thinking, based on molecular data, suggests that the cranberries should remain within the genus *Vaccinium*. For more information on this sometimes complex taxon, see *Flora of North America*, Vol. 8 (2009). This compendium separates the cranberries into their own section the author calls *Oxycoccus*.

Cranberries are small vines that creep along sphagnum hummocks in bogs and other wet habitats in cool regions of the United States and Canada. They retain their leaves in the winter, although they turn deep red when cold weather arrives. Large cranberry, *V. macrocarpon*, is grown commercially (think Ocean Spray commercials) in several states, including Wisconsin, and in Canada. It has actually escaped from cultivation in the Pacific Northwest.

Determining cranberries to species is not always the easiest task. Subtle differences in their tiny leaves and flowering stalks are the major ways. Cranberries have small bracts on the flowering stalk (pedicel) that are unique to species. Large cranberry has green bracts, while small cranberry, *V. oxycoccos*, has red bracts. The leaves of small cranberry are more linear in outline because the leaf margins are curled under. Fruits of both overlap in size, but large cranberry generally has larger fruit.

Large cranberry, seen here, has green, leafy bracts on the pedicel of each flower. The color, length, and location of these bracts are usually significant in determining species. Small cranberry's bracts are nearly always red and < ⅛ inch long. They are usually less than halfway up the pedicel. Large cranberry's bracts are > ⅛ inch long. They are usually more than halfway up the pedicel; they often cradle the flowers, as seen here.

Small cranberry leaves (pictured left) are more linear or acute in outline and have leaf margins that are curled under. Large cranberry (pictured right) leaves are broader without as strongly curled margins.

The cranberry leaves turn deep red when cooler weather sets in. Once temperatures return to between 40 and 50 degrees F, they begin photosynthesizing, and the green color returns.

large cranberry, cranberry

Vaccinium macrocarpon **Aiton**
Oxycoccus macrocarpus **(Aiton) Pers.**
Family: Ericaceae

Large cranberry is a common, creeping vine in sphagnum bogs. This species is easy to mistake for our other native cranberry (*V. oxycoccos*), with which it is often found growing. The 2 are commonly mixed with other shrubby bog specialists and mosses, mainly sphagnum. It is listed as endangered in Illinois, and threatened in Tennessee.

Form and Size: Large cranberry is a tiny, trailing vine that may grow to several feet in length, usually mixed with, and rooting in, wet sphagnum in bogs.

Habitat: It is usually found in open bog mats in full sun, but it tolerates partial shade. This species grows well when planted in more upland sites, but it requires acidic soil.

Wildlife Uses: Large cranberry, as well as small cranberry, usually grows complexed with sphagnum in bog habitats. In such situations, it provides some cover for hummock nesting songbirds, such as Nashville warblers. It provides little other cover value. It produces prodigious fruits, given the diminutive size of the plant, but quantities are limited, and they are used sparingly by larger birds and rodents like chipmunks and red squirrels. They appear to be browsed little if any by herbivores.

Landscaping Value: Hardy from Zones 2–6, this vine is capable of growing in a domesticated environment given the acidic soil required for growth. It grows slowly and makes a nice evergreen ground cover that is red during winter months. Flowers are tiny, but showy, and the fruits are edible. If not picked, they remain on the stems throughout the winter. This species is available through nurseries simply because of its commercial value.

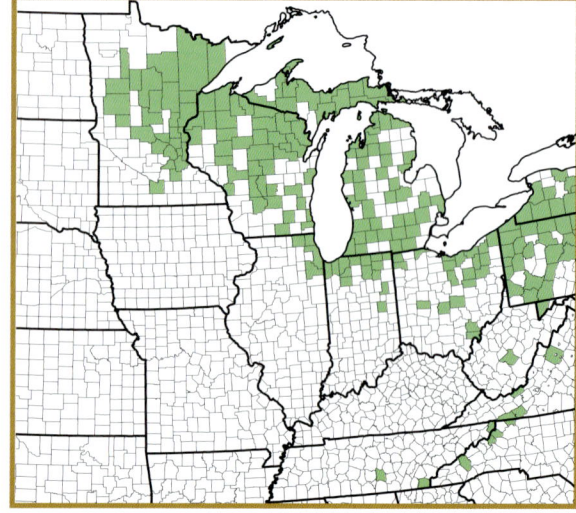

Similar Species Distinctions:
—**Small cranberry** (*V. oxycoccos*) has leaves with curled margins, smaller pedicel bracts that are red, and smaller fruits. More details are mentioned on both species' accounts.

The alternate leaves are evergreen and oblong-elliptic. They are green and smooth above, and covered with a heavy, glaucous bloom beneath. The petiole is very short, and the margin is thickened and revolute-entire. The leaf tip is usually more rounded than those of small cranberry.

Flowers appear in mid- to late June, at least several weeks later than those of small cranberry. They are nodding, pale pink flowers on hairy pedicels. The flowers are about ⅓ inch long, have completely reflexed petals, and 8 yellow-orange, clustered stamens. The long pedicels have a pair of leafy bracts that are usually greater than ⅛ inch long. The bracts are located above the middle of the pedicel. Pedicels develop from leaf axils several inches below the tip of the stems.

Fruits begin ripening in August and continue into the fall. The fruit is an edible berry that is from ⅓ to ⅔ inch in diameter. They are fleshy, red when ripe, and persist through the winter months.

The terminal bud is absolutely round and has numerous scales (green or red, depending on the time of year) with whitened margins. Lateral buds are extremely tiny and often not present. Twigs are very slender, greenish, and lightly covered with short hairs, especially on the new growth.

small cranberry

Vaccinium oxycoccos **L.**
Oxycoccus oxycoccos **(L.) MacM.**
Oxycoccus quadripetalus **Gilib.**
Family: Ericaceae

Small cranberry is quite rare in the Midwest, but it has a very large range across all of Canada and the northeastern United States. It is listed as threatened in Indiana and Ohio, and endangered in Illinois. This species occurs in bogs and produces smaller fruit than large cranberry, *V. macrocarpon*, but they are just as edible. Small cranberry creeps along sphagnum mats and almost gets lost in all the other vegetation.

Form and Size: Small cranberry is a trailing, delicate, creeping vine that crawls along vegetation in sphagnum bogs, usually alongside large cranberry. It prefers full sun for best flower and fruit production, but it handles partial shade.

Habitat: This species grows in acidic, moist conditions such as bog mats and hummocks, cedar swamps, or fens.

Wildlife Uses: This species often grows mixed with large cranberry in bog habitats; it has essentially identical wildlife values.

Landscaping Value: Hardy from Zones 2–5, this species is probably limited by climate, and it grows where there is less summer heat and humidity. Adequate habitat is also a limiting factor. This delicate vine has the same attributes as large cranberry, and growing it is not difficult, as long as proper soil acidity is provided. Fruits are a bit smaller but just as tasty (if you like cranberries). The authors know of no cultivars, and they have not seen it offered in any nurseries.

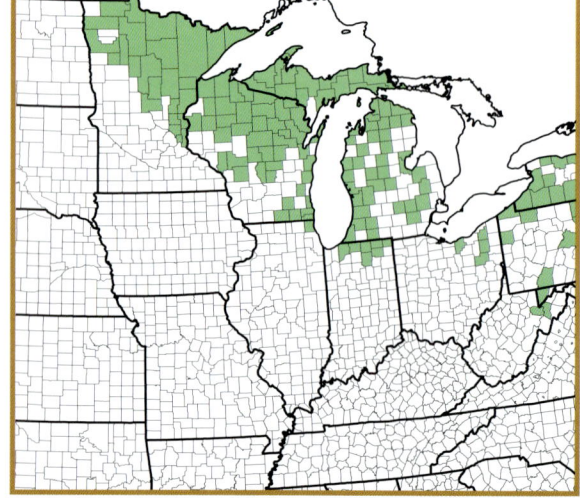

> **Similar Species Distinctions:**
> **—Large cranberry** (*V. macrocarpon*) has larger everything. See its species account for differences.

The alternate, evergreen leaves are up to ⅓ inch long with revolute, entire margins that give the leaf a pointed look. The upper surface is green and smooth, while the lower surface is glaucous and smooth. During the winter, leaves often turn dark red, and they remain that color until warm, spring temperatures arrive.

Flowers appear in early to mid-June. They are on long, hairy pedicels that are nearly terminal on the stem. Flowers have 4 pink petals, which are strongly reflexed, and 8 yellow-orange, clustered stamens. There is a pair of small, leafy, red (usually) bracts up to ⅛ inch long located at or below the middle of the pedicel.

Buds are extremely small and covered with green or red, overlapping scales. Twigs are very slender, reddish, and covered with short hairs, especially on the new growth. Lateral buds are tiny and rarely present.

Fruits begin to ripen to red in the fall, and they are up to ⅓ inch in diameter. They are juicy, edible berries that are sour to the taste. Fruit often persists through the winter months.

Honeysuckles

Genus: *Lonicera*

The midwestern honeysuckles in this group seem to be hybrids between shrubs and vines. They do not do much of what vines usually do, but rather they spread out and over their own stems or on top of other ground vegetation. If they do climb something, it is a small diameter tree or shrub, and they loosely twine around it, since they produce no tendrils. Individual stems of 10 feet in length have been measured on *Lonicera dioica* in Missouri (Kurz 1997). Apparently it is not necessary for them to climb, but they can, given an opportunity. They have been described as twining shrubs, climbing or scrambling vines, trailing or climbing vines, and many other unique classifications.

There are (at least) 5 species of these trailing shrubby vines in the Midwest. *Lonicera flava* is more of a southern species found in Kentucky, Tennessee, and Missouri, and it is listed as state endangered in Illinois. *Lonicera hirsuta* has a more northerly range in the upper half of Minnesota, Wisconsin, and Michigan, and it is easy to identify with its hairy leaves and stems.

These honeysuckles are very attractive when in flower and fruit, and they will quickly adapt to climbing a fence or trellis. The authors planted *L. reticulata* for growth on a wooden trellis. It grew so well that it broke the trellis from its own weight after a period of 7 to 8 years.

Our native *Lonicera* vines have generally lower wildlife value than exotic *Lonicera* shrubs and vines. When growing in relatively full sunlight they do frequently climb onto shrubs and obstructions and provide songbird nesting habitat in the process. All produce fruits that are small enough to attract even small songbirds, such as warblers, vireos, and thrushes, and they are generally quickly taken. Forage value of all is usually low; the exotic Japanese honeysuckle is by far the most preferred by white-tailed deer and cottontails.

The form of these honeysuckles is rather sprawling, as seen in this photo of grape honeysuckle. Here it is growing on a ridge that overlooks a stream, and the individual stems are seen with the water as a backdrop. Eventually, the stems will grow on top of, or up, whatever they encounter, but usually not up anything much over a few inches in diameter.

Leaf shape on a plant is different not only along any given stem, but also between flowering stems and vegetative stems. The terminal pair of leaves is fused to form a collar, but those below may be somewhat fused or just single and opposite each other. Grape honeysuckle, seen in this photograph, has at least the upper 4 pairs of leaves fused.

Flowers of the vining honeysuckles are tubular, clustered, and 1 inch or more in length. The color can be variable on some species, and it ranges from a pale yellow to magenta. Hummingbirds are attracted to the nectar-laden flowers. Seen here are flowers of *Lonicera dioica*.

limber honeysuckle, red honey-suckle

Lonicera dioica **L.**

Family: Caprifoliaceae

This rather uncommon native honeysuckle is sometimes referred to as a climbing shrub. It is quite unobtrusive and only flamboyant when in flower. Unlike the many introduced honeysuckles, this is never aggressive and is well behaved. There are several varieties based on regional variation of floral hairs and glands. It is state endangered in Illinois and Kentucky.

Form and Size: Limber honeysuckle is commonly seen as a sprawling shrub, but it is also found masquerading as a vine that loosely climbs to 10 feet. It has no tendrils or rootlets to use as climbing aids, and it does not wrap tightly like bittersweet. It is slow-growing, at least until it becomes established, and is short-lived.

Habitat: This species is found in a number of habitats including dunes, tamarack bogs, rocky slopes of streams, wooded openings, or forest edges, but it prefers moist, friable soil of slightly acidic sites. It handles partial shade but will grow in full sun.

Wildlife Uses: Limber honeysuckle shares some of the same wildlife values of trumpet honeysuckle but differs in several substantial ways. It often grows in woodlands under at least partial shade, where it flowers sparsely and often sets no fruit. When it does grow in the open or forest edge, it can flower prolifically, although flowers are not as attractive to hummingbirds as those of *L. sempervirens*. Fruits are regularly taken by birds and mammals. This species occasionally climbs over other woody plants, but it is often nearly a self-supporting shrub. Thus, it is most frequently used for nesting by small shrub species, such as indigo buntings or chestnut-sided warblers. It is only rarely browsed by white-tailed deer.

Landscaping Value: If a compact native vine is desired, this is a good choice for any landscaping. It has beautiful tubular flowers of many shades from yellow to deep red, and it has deep burgundy fall color. It prefers slightly acidic soil but can exist in moisture regimes from wet to dry. It also handles full sun to partial shade. It is a delicate plant that is somewhat susceptible to leaf blight and powdery mildew, but it is worth the effort to establish it.

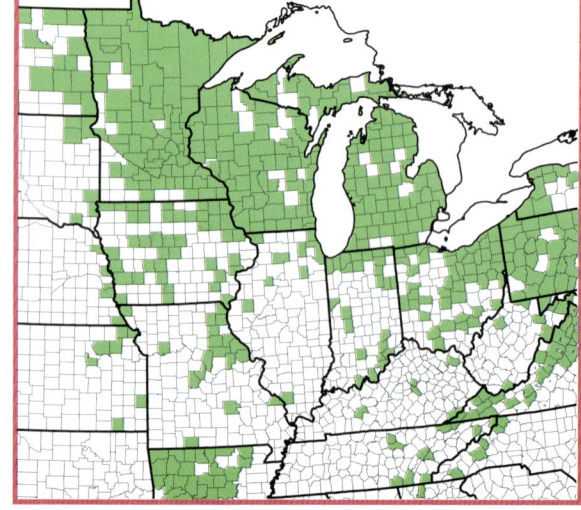

Similar Species Distinctions:
—**Hairy honeysuckle** (*L. hirsuta*) has similar leaves but hairs on both leaves and stems. No whitish coating on leaves.

The opposite leaves are oblong, with most pairs united at the base. The uppermost pair of leaves on the flowering stems is disk-like. Non-flowering stems often produce more elliptic-shaped leaves without petioles. Leaf margins are entire, and the tips are usually pointed. The upper surface is usually dark green and smooth, while the lower surface is coated with a whitish bloom.

Flowers are in terminal clusters and begin to appear in May. They are tubular, about 1 inch long, with a variety of colors from yellowish to orange to reddish purple. The stamens are long and protruding, and the throat is often hairy.

The orange-red, fleshy berries ripen in late summer in terminal clusters. They are about ¼ inch in diameter.

Bark on mature plants is brownish gray, thin, and fibrous. It peels into thin strips as the plant matures.

New growth is often covered with a dense, whitish bloom that is easily rubbed off. One-year-old twigs are light brownish gray and smooth. Lateral buds are small, diverging away from the twigs, and covered with loose scales the same color as the twig. Pith is hollow and white.

grape honeysuckle

Lonicera reticulata **Raf.**
Lonicera prolifera **(Kirchner) Rehder**
Family: Caprifoliaceae

Classifying this honeysuckle is difficult, as it could be considered a shrub or a vine. Since it has a sprawling nature, we are calling it a vine, but we use the term very loosely. Many people do not realize there are native honeysuckles in the Midwest, but there are several. This is possibly the Midwest's most common, but it is listed as endangered in Kentucky, and possibly extirpated in Tennessee.

Form and Size: Form is somewhat variable. It will grow supported on its own, appearing very "shrub-like," but long, sprawling stems stretch outward until they reach a large plant; these long stems then spread out and over that plant. It will grow up onto a large shrub or small tree and seem vine-like as well, even though it has no tendrils or rootlets and does not usually twine around anything. Some plants can develop stems 10 feet in length, but those that remain mostly shrubby grow to a height of around 3 to 4 feet.

Habitat: Grape honeysuckle is found in open woods and on rocky bluffs. Some disturbance is necessary to provide adequate light for the plants to prosper. They prefer well-drained soil, but they tolerate heavy clay and rocky sites. It can grow very quickly and is probably short-lived.

Wildlife Uses: Grape honeysuckle has almost identical wildlife values to those of *L. dioica*; where it receives sufficient light, it is intermediate between *L. dioica* and *L. sempervirens* in climbing propensity, and thus serves a series of songbird species that prefer to nest at varying heights.

Landscaping Value: This is a very adaptable species, and it tolerates all but poorly drained, acidic soils. It is hardy to Zone 4 and is a prolific bloomer. It has a long flowering period, but since the flowers are pale yellow, they are not as showy as those of some honeysuckles. It quickly adapts to a trellis, can be grown in full sun to partial shade, and is attractive with its bluish white leaves and red fruits in the fall. Powdery mildew can be a problem.

Similar Species Distinctions:
—**Yellow honeysuckle** (*L. flava*) has little-to-no whitish bloom on upper surface of leaves. Its leaves are less round and upper pair of leaves are longer than broad.
—**Limber honeysuckle** (*L. dioica*) leaves are similar to those of yellow honeysuckle, but it commonly has deep red flowers.

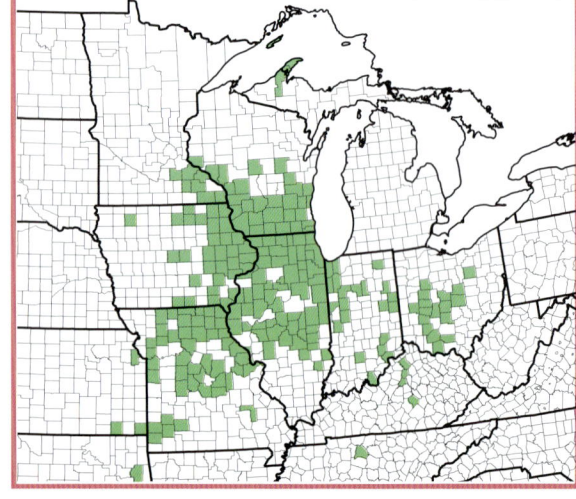

The opposite leaves are of 2 types. The upper leaves of flowering stems are fused together, forming a collar, and below those, they are joined at the base. Leaves of non-flowering stems are on short petioles and opposite or whorled. Leaves are nearly round and pale green with a whitish coating on both top and bottom, particularly the bottom. The lower leaf surface is usually lightly hairy. Margins are entire and leaf tip is blunt.

Flowering begins in May and continues into June after leaves are fully developed. They are in terminal clusters, each about 1 inch in length, tubular, and pale yellow. Flowers are hairless with long, protruding stamens.

Twigs are slender, brownish, and smooth. Lateral buds project outward at 90 degree angles from the twig and are covered with loose, pale-colored scales. Older twigs begin to shed their outer "skin" in a thin, loose layer. Pith is hollow and white.

Bark on mature plants is thin, light brown, and shreds into strips. The stem diameter is never large enough that much bark develops.

Fruit is a round, red berry that ripens in mid- to late summer and is produced in terminal clusters. Each berry is about ¼ inch in diameter.

trumpet honeysuckle, scarlet/ coral honeysuckle

Lonicera sempervirens L.

Family: Caprifoliaceae

This beautiful vine has been cultivated so long that its exact natural range in the Midwest is uncertain. Its specific Latin name *sempervirens* means "always living," which is a reference to evergreen leaves. That is a bit of a misnomer, because even in the South, it is only semi-evergreen.

Form and Size: This twining honeysuckle is commonly found in small- to medium-sized trees and may reach a height of 30 to 40 feet. It produces short, clustered rootlets at the nodes, which helps its attachment to its support. It does not develop a large trunk, at least in the Midwest.

Habitat: Trumpet honeysuckle is usually found growing up trees along forest edges in moist, well-drained soils that have neutral to slightly acidic pH. It grows best in full sun, but it handles a great deal of shade. All records north of Kentucky likely represent individuals escaped from cultivation.

Wildlife Uses: The beautiful, tubular flowers of trumpet honeysuckle are especially attractive to hummingbirds; the resulting fruits are readily taken by songbirds, game birds, and small mammals. It is not a preferred browse species for deer or cottontails. Compared to our other native, vining honeysuckles, this species is much more of a climber, and plants frequently grow densely, although individuals are scattered. They are often used for nesting by songbirds, especially mockingbirds, brown thrashers, cardinals, and towhees.

Landscaping Value: Hardy to Zone 4, this is a great vine for trellises or for large fences and walls. It has an amazingly long flowering period; it blooms from spring to fall. Leaves are tardily deciduous, and there is no real fall color. It has escaped from cultivation, but it does not appear too aggressive. There are at least a dozen cultivars, and it is readily available through some midwestern nurseries, especially those in our southern region.

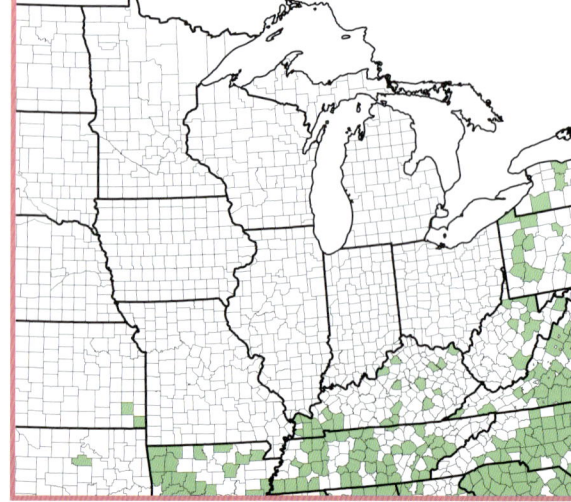

Similar Species Distinctions:
—**Limber honeysuckle** (*L. dioica*) has similar leaves, but they are much narrower and usually have no petioles. Flowers can be reddish, but they are more commonly yellowish.

The variable leaves are usually ovate-elliptic but are different on flowering and nonflowering stems. They are up to 2½ inches long. On flowering stems, the upper pair is connected to form a cone, and many of the leaf pairs below that are attached at the base (see page 312). Leaves of nonflowering stems are not connected, and have short petioles. Leaf tips are rounded or notched and margins are entire. The upper leaf surface is dark green and smooth, while the lower surface is covered with a white, glaucous bloom. The petioles are ¼ inch long.

Fruits ripen in late summer and fall. The orange-red berries are ¼ inch across and are slightly translucent.

Flowers first appear in April. They are in terminal clusters, red, and slender. Each flower is up to 2 inches long and has a yellow throat. Stamens and the stigma are slightly exerted.

Twigs are slender and have a glaucous bloom when young. Winter twigs are tan-colored and smooth with vertical "cracks" in the skin. Pith is white and hollow in the middle. Lateral buds are small and covered with a few loose, tan-colored scales. Buds are at right angles to the stem. Leaf scars are shaped like a "smiley face."

Bark is brownish and shreddy. It eventually splits and peels away from the trunk to reveal tan-colored inner bark.

trumpet creeper, trumpet vine

Campsis radicans (**L.**) **Seem.**

Family: Bignoniaceae

This native vine probably has the most extended flowering period of any in the Midwest. Its large, showy flowers are present all summer long and attract the attention of humans, hummingbirds, and insects alike.

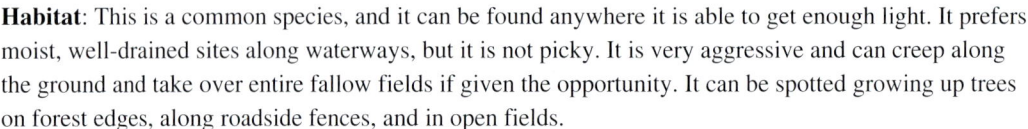

Form and Size: Trumpet creeper climbs by twining or by rootlets that attach to tree bark or other substrates. It grows quickly, and it can reach heights of 50 feet into treetops. The species is fairly long-lived, especially once it is associated with a large tree in an undisturbed area.

Habitat: This is a common species, and it can be found anywhere it is able to get enough light. It prefers moist, well-drained sites along waterways, but it is not picky. It is very aggressive and can creep along the ground and take over entire fallow fields if given the opportunity. It can be spotted growing up trees on forest edges, along roadside fences, and in open fields.

Wildlife Uses: Trumpet creeper's most frequently mentioned wildlife value is the attractiveness of its beautiful orange-red flowers to hummingbirds; however, the species has major cover value as well. Although it can climb high into trees, it is more frequently a runner than a climber, and it often envelopes fences and other obstructions in its path. When this happens it forms excellent and regularly used nesting sites for many birds—e.g., mockingbirds, brown thrashers, and blue grosbeaks. When it exploits fallow fields, it can be a pest to humans while supplying cover for cottontails, bobwhites, and other ground-dwelling birds and mammals. Leaves are regularly browsed by white-tailed deer and cottontails.

Landscaping Value: This is 1 tough native vine that can handle heat, drought, and just about any soil type. It is hardy to Zone 4. If planted, it should be in an area where it is easily contained, as it can become almost uncontrollable. It is particularly attractive grown on a pole. Give it full sun for best flower production. Fall color is yellow. Fruiting pods are sizable and can create a bit of litter. Several horticultural varieties are available, including 'Flava' and 'Crimson Trumpet.'

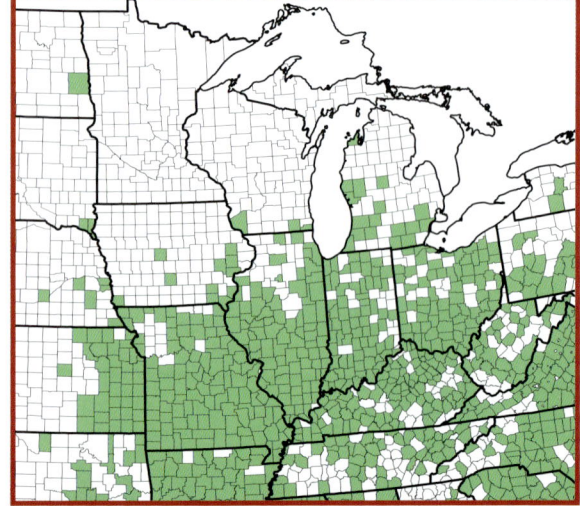

Similar Species Distinctions:
—Opposite branching vines are rare, so look for that first.
—**Wisteria** (*Wisteria* spp.) leaves are similar, but branching is alternate. Pods are similar, but flattened.
—**Pepper vine** (*Ampelopsis arborea*) leaves are twice to thrice compound, and they have alternate branching. Fruit is a black berry.

Leaves are opposite, compound, and have 7 to 13 leaflets per leaf. Each leaflet has an elongated tip and toothy margins. Leaves are dark green, shiny, and mostly smooth above, paler green and lightly hairy beneath. They can be up to 15 inches in length but are usually shorter.

Flowers are produced from June through September in terminal clusters. They are about 3 inches long, reddish orange, and trumpet-shaped with 5 lobes at the opening. The throat is striped.

Fruiting pods begin ripening in August and are up to 5 inches long. They are tapered at both ends, but they are not flattened in the middle like those of crossvine. There are many flat, winged seeds inside.

Twigs are pale and smooth with sizable U-shaped leaf scars. Lateral buds are barely visible, and they are mostly sunken in the twig. Clustered rootlets commonly grow from the nodes as seen here.

Mature bark is usually light-colored and vertically oriented. It can become fairly thick, but over time, it sloughs off. Vines can grow to a large diameter of 4 inches or more.

raccoon grape, heart-leaf ampelopsis

Ampelopsis cordata Michx.

Family: Vitaceae

The genus *Ampelopsis* in the Midwest claims 2 native species. Unfortunately, an introduced species, *A. brevipedunculata*, porcelain-berry, is similar to raccoon grape and has become problematic throughout much of the eastern United States, including our region. *A. arborea*, pepper vine, is a common vine in the Southeast, but reaches into southern Illinois where it is rare. It has finely divided, compound leaves. Raccoon grape is often mistaken for our native grape vines, genus *Vitis*, as its leaves closely resemble them.

Form and Size: Raccoon grape is a large vine that is able to climb into the canopy of large forest trees with its numerous tendrils. It is fast-growing and can reach 60 feet or more (but generally smaller in the Midwest) in order to access more light.

Habitat: This is a vine of bottomland woods and low, woodland borders. It is also found along wooded ponds and streams and is usually growing in light gaps or on top of trees or large shrubs. It prefers well drained, loose to moderately heavy soils and a neutral pH.

Wildlife Uses: Raccoon grape provides cover very similar to that described for *Vitis* species. It tends not to grow as large or climb as high into trees in our region as it does in the South and thus has somewhat lower wildlife value. Its fruits are produced in clusters, but they are much fewer than in *Vitis* clusters (usually less than 15) and relatively dry. These small fruits, however, are eaten by several birds, including woodpeckers (flickers and pileated), thrushes, and bobwhites, and mammals, especially gray squirrels and raccoons.

Landscaping Value: It is hard to find a native vine with more luxuriant, dark green, shiny foliage than that of raccoon grape. It has pale yellow fall color, and its unusual-colored fruit is relished by birds. It is easy to propagate from cuttings, but it can take over an area if not pruned. Japanese beetles relish the foliage and must be controlled in domestic applications.

Similar Species Distinctions:
—**Grape** (*Vitis* spp.) Several species of grape have similar leaves, including *V. riparia* and to some degree *V. palmata*, both of which are found in similar habitats. Both grapes have dark brown, shredding bark. Raccoon grape has chambered pith at the nodes, not diaphragmed, and fruit clusters are less dense than those of grapes.

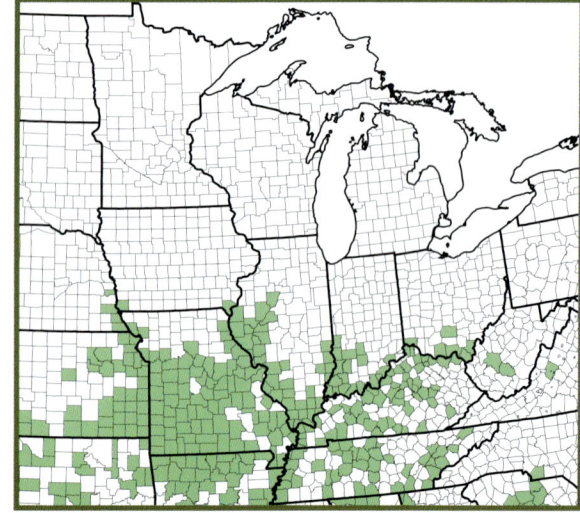

Leaves are up to 4 inches long, ovate to broadly ovate, and they have coarsely toothed margins. The leaves are mostly smooth (some hairs along lower veins are possible) and can produce small lobes near the tip. The upper surface is green; the lower surface is paler. Petiole is hairy.

Flowers appear in June in broad, forked clusters, often emerging from 1 branch of a tendril. They are perfect or unisex, about ¼ inch across, and white.

Fruit ripens in the fall in loose, forked clusters. Berries are about ⅜ inch in diameter, and they transition in color from whitish to gray-blue to fuchsia and eventually dark blue. They are always speckled.

Twigs are tannish-colored and dotted with pale lenticels. Leaf scars are circular and contain a vertical split at the top. Bundle scars form a U-shape in the leaf scar. Lateral buds are small, hairy, and mostly sunken into the twig above the leaf scars. Pith is white and mostly solid throughout the twigs except at the nodes, where it is chambered.

Bark is light brown, tight, and deeply furrowed on old specimens. It does not shred like the bark of grapes.

woolly pipevine

Aristolochia tomentosa Sims

Family: Aristolochiaceae

Woolly pipevine is certainly 1 of the Midwest's more unusual vines with its fuzzy, pale yellow, pipe-shaped flowers. A very similar, more southerly species, *A. macrophylla*, was once commonly planted for use as a screen on sides of porches.

Form and Size: Pipevine is a twining vine, usually found growing on and up large trees and shrubs in bottomland woods and forest edges. It grows quickly and can climb to a height of 75 feet. In the Midwest, a vine diameter of 3 inches is large.

Habitat: This is a bottomland species that prefers moist floodplains of river bottoms or along wooded streams. It grows best in full sun, but it tolerates a great deal of shade.

Wildlife Uses: This unique species often forms dense tangles as it climbs on shrubs and saplings. Its tangled structure is superb cover for nesting songbirds and likely escape cover as well. It has very little food value for wildlife, despite its unique fruit pods, the seeds of which may be used by mammals when they dehisce in the spring (but not recorded). Although the leaves are the sole food of the larvae of the pipevine swallowtail butterfly, they are taken little if any by vertebrate herbivores.

Landscaping Value: Pipevine is hardy to Zone 4, but it is much more common in the southern portions of the Midwest. Although a bottomland species, it grows well on upland sites and adapts well to most soil types. Pipevines grown in full sun are said to create "living curtains" with their large, flat, numerous leaves, and they make great summertime porch screens if grown on a trellis. Their fall color is dull and the unique flowers are small and difficult to see.

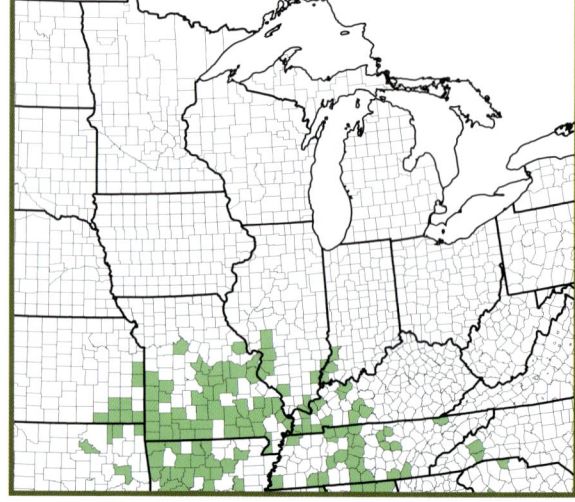

Similar Species Distinctions:
—**Pipevine** (*A. macrophylla*) occurs mainly throughout the Appalachian region. It is essentially hairless, but it can have large, similar-looking leaves.

Leaves are heart-shaped and up to 8 inches long. They have entire margins and a rounded tip. The upper surface is mostly smooth while the lower surface is paler and hairy. Petioles are long, curved, and hairy.

Young twigs are greenish or tannish and round. Lateral buds are circular, densely hairy, and mostly sunken into the swollen nodes. Leaf scars are elongated, U-shaped, and contain 3 bundle scars. Vines have no means of attaching to supporting plants and instead twine around them.

Flowers appear over a several month period in late spring. They are pale yellow, hairy, and curved, very much resembling a curved, smoking pipe. There are 3 short petals surrounding a small, purple-tipped opening and a spotted throat. The solitary flowers are 1 inch in length and appear after the leaves on long peduncles (flower stalk).

Bark is tan-colored and corky with thick, longitudinal, discontinuous ridges that split along the top.

Fruit is a dry, hairy capsule that ripens in the fall. They are 3 inches in length and a bit more than 1 inch in diameter. There are 6 longitudinal ridges, with lesser ridges between, which split upon ripening. The capsule shatters, releasing many flat, triangular, tan-colored seeds that had been stacked in layers the length of the capsule. Also pictured is the more elongated, hairless fruit of *A. macrophylla* (right).

American bittersweet

Celastrus scandens **L.**

Family: Celastraceae

This vine is the source of the unusual orange-colored fruit used in fall flower arrangements. Unfortunately, for many decades it has been over-harvested in areas where bittersweet regeneration is now almost nonexistent. The fall color is a pretty pale yellow, making it easy to spot when growing along a forest edge.

Form and Size: Bittersweet can grow to 40 feet into small treetops along forest edges, but it also takes advantage of fences and telephone poles. It is usually found growing in shady sites, but it tolerates full sun. It is fast growing and probably fairly short-lived.

Habitat: Bittersweet is not particular about where it grows, and soil type and pH seems to be of little concern. It grows best in rich, moist sites, and it is most often seen along forest edges climbing on moderately-sized trees.

Wildlife Uses: While aesthetically satisfying, bittersweet has relatively low wildlife value. When growing on roadside fences, over shrubs, and into low trees, it provides nesting cover for songbirds. In these habitats, fruit production is relatively great because of high light exposure. Fruits are occasionally removed completely in the fall by flocks of cedar waxwings, robins, and starlings, but in most cases they are of low preference and remain into the early winter, when they serve as emergency food for songbirds, game birds, and fox squirrels.

Landscaping Value: Bittersweet is hardy to Zone 3 and is a nice vine for a trellis or fence. It can constrict the living plant on which it is growing with its twisting growth form. It has fall and early winter appeal with its pale yellow leaves and bright orange fruit clusters. At least 1 variety is available in the Midwest; 'Bailiumn' has perfect flowers, so there is no need for male and female plants.

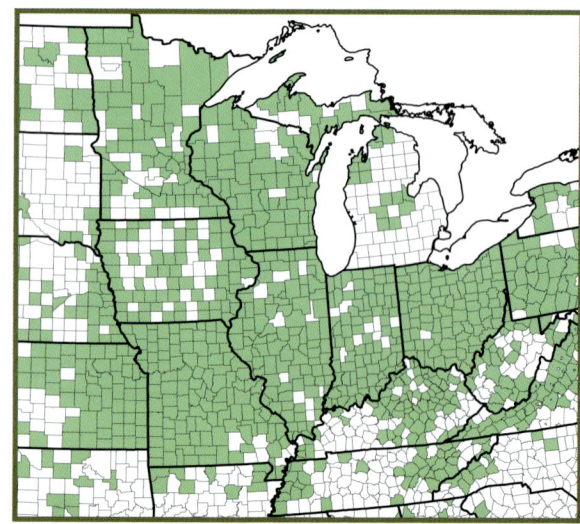

Similar Species Distinctions:
—Oriental bittersweet (*C. orbiculatus*) is an Asian species that has invaded from eastern states. See the "Introduced Species" section for more information.

Fruit ripens in the fall in dangling clusters. Each fleshy, orange-red fruit (aril) is enclosed in a 3-parted orange capsule that splits open when ripe. The fruits are about ⅓ inch in diameter. They are usually persistent through late fall, but they can be found into the winter months.

The alternate leaves are oval, ovate or obovate, and up to 4 inches long. The leaf tip is elongate, and the margin is finely toothed. Leaves are smooth, shiny, and green above, smooth and slightly paler beneath.

Bittersweet begins flowering in early May after leaves are developed. They are in terminal, dangling clusters of small, greenish yellow flowers. Plants are usually separate sex; male is near left, and female is above.

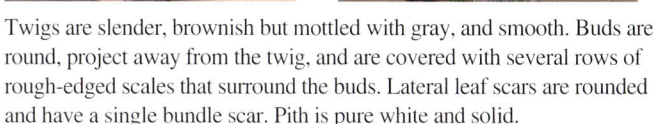

Twigs are slender, brownish but mottled with gray, and smooth. Buds are round, project away from the twig, and are covered with several rows of rough-edged scales that surround the buds. Lateral leaf scars are rounded and have a single bundle scar. Pith is pure white and solid.

Vines are rarely larger than 3 inches in diameter, and they are usually much smaller. The bark is brown and thin. It splits and peels vertically along older stems.

moonseed

Menispermum canadense **L.**
Family: Menispermaceae

This short-lived, twining vine belongs to a family found mainly in the tropics. Many plants in the Menispermaceae family are poisonous to some degree—at least 1 in South America is a source of arrow poison. The fruit of moonseed is reportedly poisonous to humans.

Form and Size: Moonseed is a slender, climbing, twining vine that can reach a height of 12 feet or more. It is considered herbaceous in the northern parts of its range, and it never attains a large size or diameter elsewhere. It is a short-lived, fairly fast-growing vine.

Habitat: This species is commonly found growing over shrubs or on small-diameter trees along forest edges, in road ditches, and on streamsides wherever there is enough light. It tolerates partial shade, but it prefers nearly a full day's sun.

Wildlife Uses: This widely distributed vine is less well known and likely has lower overall wildlife value than many others of our region. It grows well under forest cover, especially in moist sites, but it does best on edge situations where it is sometimes luxuriant, climbing over shrubs and low into trees, creating good nest sites for songbirds. Its fruits are in clusters similar to those of *Vitis*, and, although taken by cedar waxwings and others, are evidently not relished, since some remain on the plant through the winter. Fruits are reportedly toxic to humans, but such toxicities frequently seem not to impact wildlife. Leaves are little-browsed by herbivores.

Landscaping Value: Moonseed is hardy to Zone 2, and it becomes more woody in more southerly parts of its natural range. It trains easily to trellises and grows thick, attractive leaves and shiny green or reddish brown, twisting stems. Flowers are small, but the attractive black fruits hang down in clusters. Individual stems are short-lived, but moonseed produces more annually. It is not sold in nurseries, but it is easily grown from cuttings or seed.

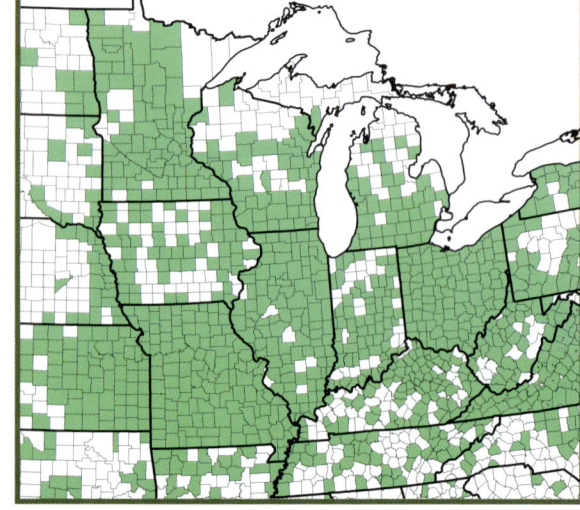

Similar Species Distinctions:
—Yellow passionflower (*Passiflora lutea*) is a woodland perennial that has very similar leaves. Its petiole, however, attaches in a typical manner, rather than like that of moonseed, which is set back from the leave margin. Flowers and fruit are different as well.

The alternate leaves average 3 to 4 inches long and usually have 2 side lobes with entire margins. They are green and smooth above, paler and sometimes hairy beneath. Leaf venation is impressed into the leaf and palmate (like fingers radiating from a palm). The long petiole is attached on the back of the leaf, away from the margin (peltate).

The bluish black, round, grape-like fruit sometimes has a whitish bloom on the surface, and it ripens in the fall. The ridged seeds are flattened and shaped like a crescent moon.

The tiny flowers hang in delicate, dangling clusters after the leaves are fully developed in May and June. They are separate sex flowers on separate plants, greenish white with flower parts of 3. Female is far right, and male is near right.

Twigs are slender, shiny, smooth, and either green or reddish brown. Leaf scars are whitish and nearly circular, but contain a vertical slit on top where the lateral buds are positioned. Buds are small, round, and mostly sunken into the twig. Pith is segmented into sections that look like spokes on a wheel. Moonseed is a small-diameter vine that develops very little bark character. It becomes roughened with raised, tan-colored lenticels that stretch and split. Vines climb by twining; they produce no tendrils.

Greenbriers

Genus: *Smilax*

Smilax is the Midwest's only woody mono-cotyledonous vine, and its parallel leaf venation confirms its identity. The genus was formerly included in the large Liliaceae family, but it recently has been given its own—Smilicaceae. The genus is represented in our region by 5 species of woody vines, plus several that are classified as herbaceous. One southern greenbrier sneaks into eastern Tennessee, *S. laurifolia*, a common, evergreen species with leathery leaves somewhat similar to those of *S. bona-nox*, only without the spiny margin; it is pictured on page 325.

Greenbriers have a bit of a bad-boy reputation with forest managers and anyone who works in the woods. Several species, particularly *S. rotundifolia* and *S. bona-nox*, can dominate open forest understories. They do this by spreading via underground rhizomes, and the long, thorny stems tangle together to create impenetrable thickets. This frequently occurs when any disturbance, such as a tornado or timber harvest, allows additional light to reach the forest floor—greenbrier goes wild.

Determining species of *Smilax* is often tricky, especially based solely on leaves. You must consider your location and the possibilities for that area, then expand your reference base and look not only at leaves, but also at stem and prickle characteristics.

Smilax is a genus that has multiple wildlife benefits, and certain characteristics are shared by all species in our region. Anyone who has worked in the woods, especially where disturbance has opened the forest canopy, will appreciate how formidable a barrier greenbrier vines can present. This thick, vining growth, enhanced by a semi-evergreen nature, presents excellent cover for species from white-tailed deer to nesting catbirds and hooded warblers. The genus also produces excellent food, especially foliage that is eaten by white-tailed deer, not only in spring and summer, but also well into the fall and early winter. Stems are often browsed by cottontails. Fruits are small enough to be taken by songbirds and game birds, but they are also regularly eaten by mammals, including deer and raccoons. Almost every attribute of this genus seems designed to benefit wildlife.

Most greenbriers have delayed leaf drop, and the farther south one travels, the longer leaves are retained. Some are evergreen in the Deep South. Pictured is bristly greenbrier in December in west-central Indiana, dusted with snow.

Leaves of many greenbriers are similar in shape, and determining a species can be difficult. However, the parallel veins should confirm that it is a monocot, and *Smilax* contains the only woody, monocotyledonous vines in the Midwest.

Smilax laurifolia, laurel greenbrier, is a common feature of southern woods with its thick, evergreen leaves, and it can be found in eastern Tennessee.

Fruit of the greenbriers is generally nearly black, abundant, and not highly preferred by wildlife; therefore, it remains to be commonly seen. Bristly greenbrier is seen here growing on a young Ohio buckeye (*Aesculus glabra*).

The base of greenbrier petioles completely encloses the developing buds during the summer, and it is thus "hollowed-out." One tendril is attached to either side. When the leaf falls off, it does not abscise cleanly and leaves a brown, ragged base covering part of the winter bud.

fringed green-brier, catbrier, sawbrier

***Smilax bona-nox* L.**
Family: Smilacaceae

This greenbrier is certainly our most interesting from a leaf standpoint. You really never know what shape you might find next on any given vine, which can also keep you guessing as to which species it is.

Size and Form: Fringed greenbrier climbs high into trees and also grows over understory and forest edge shrubs and small trees. It spreads by underground rhizomes and spiny tubers to create thickets that are almost impossible to traverse.

Habitat: This species is found in dry, open woods, forest edges, thickets, and abandoned fields. It is well adapted to nutrient-poor, heavy soils and usually grows in partial shade.

Wildlife Value: Fringed greenbrier is very similar to *S. rotundifolia* in form and wildlife value. This is a more southern species, although all except S. *tamnoides* are mostly limited to the southern portion of our region, so it serves a slightly different cadre of potential users. While it climbs, it also forms dense, low tangles on forest edges that are favorite nest sites for songbirds and provides concealment for wild turkey nests.

Landscaping Value: Hardy to Zone 6, this southern species is most interesting because of its semi-evergreen to evergreen, fringed leaves that are blotched with white. Its thorns hinder its use as an ornamental, but for a naturalized setting, it might be considered. It is very drought tolerant, and it is undemanding as to soil type. This greenbrier is not offered in nurseries.

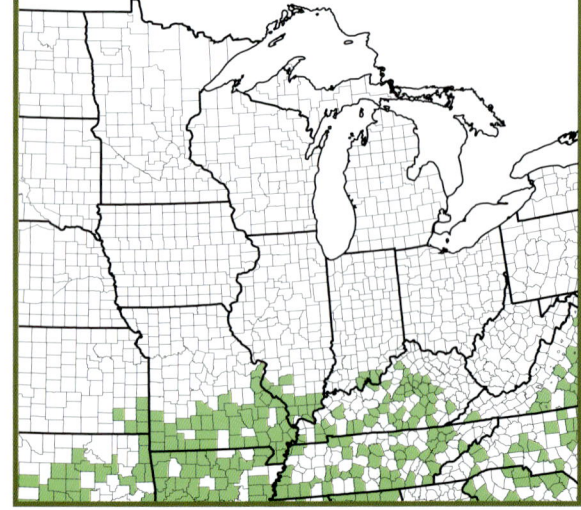

Similar Species Distinctions:
—No other greenbriers in the Midwest have leaves quite like this 1. The fringe along the leaf margin is distinct, as well as are the short prickles along the midrib underneath. *S. glauca* is the only other greenbrier with whitish leaf blotches. Several evergreen greenbrier species can be found in the South, but none have the characteristic blotches.

Leaves are simple, rather thick, and 3 to 4 inches long. They are dark green, smooth, and blotched with white above, paler and smooth beneath. The leaf is somewhat triangular in shape but partially contracted on the sides. Its margins are usually rimmed with tiny spines or finely toothed. The lower midvein is often dotted with fine spines. The base of the petiole is hollow and has a pair of tendrils attached. Leaves are said to be tardily deciduous to semi-evergreen, but the farther South one ventures in its range, the more evergreen they become.

Fruit is a berry that ripens in the fall. Each round, nearly black fruit is about ¼ inch in diameter and often persists through much of the winter. These berries are just beginning to ripen.

Flowers appear in April and May in clusters from leaf axils on the new growth. Male (left) and female (right) flowers are on separate plants. Each greenish yellow flower is about ¼ inch across and has 3 petals and 3 sepals. Stamens are present on the females, but they quickly fall off.

Mature twigs are green and angled with ½-inch-long single or paired, sharp, somewhat flattened prickles. At each node brown, papery remnants of last year's leaf petiole remain attached, along with a pair of tendrils. Buried beneath these remnants is next spring's bud that is green, somewhat 3-sided, and conical with a single scale. Older stems become dotted with clusters of white, star-shaped hairs, especially at the base, which are unique to this greenbrier.

catbrier, sawbrier

Smilax glauca **Walt.**

Family: Smilacaceae

Catbrier is usually 1 of the easier greenbriers to identify, mainly because of its whitened leaf undersides. Its somewhat evergreen nature, as well as its often overturned leaves, provide an easy glimpse of this feature, especially during winter months.

Form and Size: Catbrier is a climber that can reach 20 feet or so into small trees or over shrubs. It spreads by rhizomes from underground tubers that help it form dense tangles that are all but impenetrable. It grows quickly, but individual stems are short-lived.

Habitat: Similar to other greenbriers, catbrier is found along forest edges and openings and in clearings, fallow fields, and pine woods. Sites are dry to slightly wet but well-drained.

Wildlife Uses: Catbrier is very similar to *S. rotundifolia* in its wildlife values. It does tend to be less shade tolerant and exhibits a straggling growth form in the understory. However, when openings occur, it is even more aggressive in growing over and dominating other woody vegetation, giving substantial wildlife cover. Low growing catbrier (and other greenbriers) in single-tree gaps in the forest is the favorite nest site for hooded warblers and other shrub-nesting species. In addition to songbirds, major users of catbrier fruit in winter include wild turkeys and ruffed grouse.

Landscaping Value: Hardy to Zone 5, this vine has several nice features worthy of merit in a naturalized landscape. Its white leaf backs are frequently exposed and are a striking contrast to the green upper surface; the contrasting colors are even more obvious in the fall when leaves turn bright red. Many leaves are retained into the winter months, particularly in the more southern reaches of its range.

Similar Species Distinctions:
—**Roundleaf greenbrier** (*S. rotundifolia*) is probably the species most likely to be mistaken for catbrier, as its leaves can look rather pale green, particularly on the lower surface. Its stems are 4-sided and is has no glaucous bloom on leaves or stems.

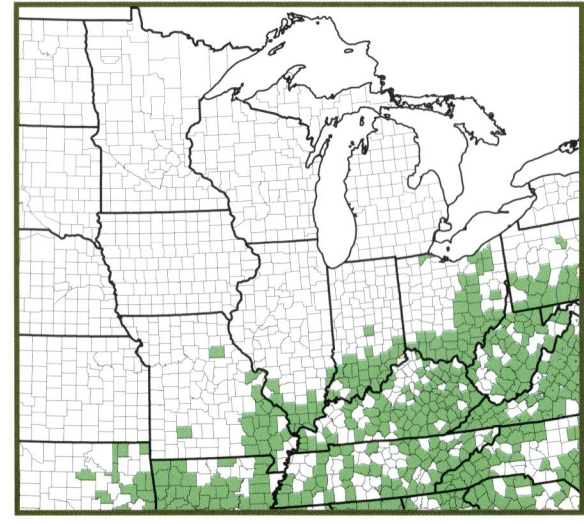

Leaves are simple, nearly round to somewhat triangular or "tongue-shaped," tough, and 3 to 4 inches in length. They are shiny green and smooth above and sometimes marked with whitish blotches; they are white glaucous and smooth beneath. Leaves are semi-evergreen, and they can lose the glaucous coating as winter months pass. Like all greenbriers, the petiole is hollow with a pair of tendrils attached.

Twigs are angular when new, but eventually they become round with a glaucous bloom. Prickles are scattered and straight. Nodes and lateral buds are covered with the base of last year's petiole. Buds are elongated, pointed, green, and smooth, with a single scale. Stem diameter is less than ½ inch and greenish or reddish with glaucous bloom.

The nearly black fruit ripens in the fall in hanging clusters. Each fruit, a berry, is about ⅓ inch in diameter and has a glaucous (white, waxy coating) bloom. Fruits can be persistent into the winter.

Flowers appear in May after the leaves on new growth. They hang down in greenish yellow clusters. Flowers are ¼ inch across and are unisex on separate plants. Since *Smilax* are monocots, flower parts are in 3's. Females are at far left, and males are at near left.

roundleaf greenbrier

Smilax rotundifolia **L.**

Family: Smilacaceae

E. Lucy Braun, in *The Woody Plants of Ohio*, pronounced this "the most viciously spiny of our plants" because of the abundant, thick, sharp prickles along the vine. Dense tangles of this greenbrier are common in open woods, and trying to navigate through it makes one a convert to Braun's way of thinking.

Form and Size: Roundleaf can climb to a height of 20 feet into large shrubs and small trees, but it is more often found growing much closer to the ground, spreading by underground stems, and forming impenetrable thickets in open woods.

Habitat: This is generally a species of open woods and forest edges in dry, well-drained soil, although it can be found in wet sites. It also occurs in fallow fields and along streams. Growth rate can be fast, and it is partial to at least half-day sun.

Wildlife Uses: This species shares many characteristics with its congeners in our region. It seems to be more shade tolerant than others, occurring in forest understories, frequently producing good fruit crops with greater than 50 percent crown closure; it also typically climbs into trees from its understory base. The fruits are eaten by birds and mammals, although they occasionally last into the winter. Leaves are highly preferred by white-tailed deer and are used well into the fall. Stems (and leaves) are regularly browsed by cottontails through the winter. Any light gap creates a response in vegetative growth that adds to its considerable cover value, providing concealment for all wildlife from deer, which will bed down in the thickets and allow hunters to pass, to nesting songbirds, such as indigo buntings and towhees.

Landscaping Value: Hardy to Zone 5, this vine might be appropriate for a naturalized setting along a forest edge to attract wildlife. It is adaptable to most soils types and moisture regimes, other than poorly drained. Leaves are commonly retained well into the winter, and it will tolerate fairly heavy shade to full sun. Greenbriers are not usually available in nurseries, but stem and rhizome cuttings can be propagated.

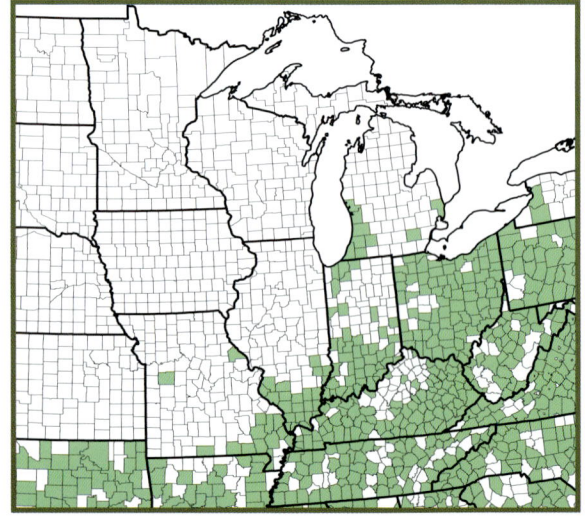

Similar Species Distinctions:
—**Catbrier** (*S. glauca*) can have similar leaf shape, but its leaf backs and stems are glaucous white.

Leaves are usually 3 to 4 inches long, green, glossy, leathery, and smooth on top. The lower surface is pale green and smooth. Leaf margin is minutely toothed. This species has the roundest leaves of our native greenbriers, but shape can vary greatly. Petioles are hollow and have a pair of tendrils attached.

Flowers appear in April and May on the new growth in dangling clusters. They are greenish yellow, about ½ inch across, and unisex on different plants (male and female plants). Female flowers (below, near) have non-functioning stamens.

Fruit ripens in the fall in dangling clusters. The berries are bluish black and about ¼ inch in diameter. They usually have a whitish coating.

Twigs are green, smooth, and 4-sided. Numerous thick, flattened, sharp prickles that are often dark-tipped line the stems. Buds are hidden at the nodes by remnants of the petiole, but they are green, somewhat 3-sided, elongated, and appear as an extension of the stem. They have a single bud scale. Roundleaf stems never grow larger than 1 inch in diameter. They remain green and become less angled near the ground, and the stout prickles seem to grow thicker toward the base.

bristly greenbrier, hispid greenbrier

Smilax tamnoides **L.**
Smilax hispida **Muhl. ex. Torr.**
Family: Smilacaceae

Bristly greenbrier is 1 of the most intimidating *Smilax* species, with its numerous, black, needle-sharp prickles that densely populate the largest stems. They are particularly obvious on the green stems. Like all greenbriers, new spring leaves, tendrils, and vines are quite tasty eaten straight off the vine and in a fresh salad.

Form and Size: The long stems arise from a single location and spread out and over the tops of shrubs and small trees to as high as 40 feet, but they are usually much closer to the ground. The largest stems are less than 1 inch in diameter.

Habitat: This is a very adaptable species that prefers moist, rich soils and is usually found in open woods, forest edges, wooded roadsides, and stream banks. It is shade tolerant but handles full sun. Individual stems are short-lived, but clumps continue to produce new ones.

Wildlife Uses:
Bristly greenbrier is very similar in wildlife value to other greenbriers, especially *S. rotundifolia*. Since it is often the only greenbrier species in the northern part of the Midwest, it supplies its multiple benefits to a slightly different clientele of wildlife users.

Landscaping Value: Hardy to Zone 3, this bristly plant is probably not appropriate for areas where humans cohabitate, but to draw wildlife into an area, it can be beneficial. Its green stems and tardily deciduous leaves are pleasant in the late fall/ early winter. The black fruit usually persists well into the winter, and the black, numerous prickles are actually quite attractive (just don't try to grab the plant anywhere near them). It is a tough plant and generally handles some degree of drought and drier sites. Greenbriers are not normally found in nurseries, but they can be grown from stem and rhizome cuttings.

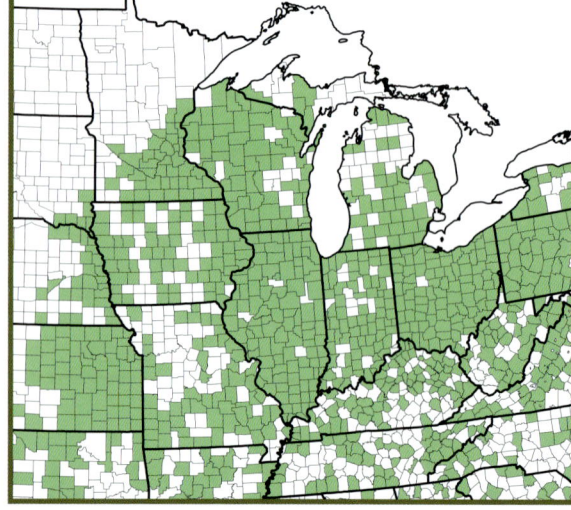

Similar Species Distinctions:
—The numerous, black prickles along the stems distinguishes this greenbrier from all others. No other vine has them.

Leaves are mostly ovate and about 3½ inches in length. They are thin, dark green, and smooth above, and slightly paler and smooth beneath. Leaf margin is entire, minutely toothed, or wavy. The base of the petiole is hollow and has a pair of tendrils attached.

The greenish yellow flowers appear in May on the new growth. They are in clusters of unisex flowers—1 sex per plant. Each flower is about ¼ inch across and has 3 petals and 3 sepals. Males are on the left; females are on the right.

The nearly black berries ripen in the fall in dangling clusters. Each is about ¼ inch in diameter and hangs from a long stalk (peduncle).

Bristly greenbrier never develops bark, but its lower stems are green and densely covered with black prickles.

Twigs are smooth, green, and mostly round in cross section. Sharp, needle-like prickles of 2 lengths line the stems, with progressively more toward the base of the plant. Nodes and lateral buds are partially hidden by the remnant of last year's petiole. Buds are pointed, 3-sided, green, and have a single scale. The curling tendrils are easily seen in the above photograph.

Grapes

Genus: *Vitis*

Grapes are a common feature of midwestern woodlands, and there are 8 species native to the region. They are often cursed by forest managers, though encouraged by wildlife managers, based strictly on differing management objectives. Most grapes are opportunistic, high climbers that make their way into the tops of forest trees. In doing so, they can deform their host tree, and their numerous leaves can substantially reduce the amount of light the tree leaves receive. This is not a problem for the wildlife manager, who knows that the vine is providing cover for songbirds and squirrels as well as food for many species of wildlife. However, if the tree is being grown for timber, then the vine is problematic to the forester.

When a vine climbs into trees, it can form a complex that is very attractive for nesting cover for everything from the aforementioned songbirds and squirrels, to crows and red-tailed hawks. And nothing does this better than the wild grapes. More than any other vine in the Midwest, the grapes also grow large diameter stems, which can seriously deform and compete for light with the trees on which they grow. In addition to their substantial cover value, grapes produce much fruit when they receive sufficient light—fruits that are juicy and big enough to be used by only the largest of songbirds. Wild turkeys and ruffed grouse feed heavily on them, but the most prevalent users are mammals, from chipmunks, to raccoons and foxes, to white-tailed deer. In addition, in the Midwest the leaves form an important part of the summer diet of deer. Finally, the peeling bark of larger stems is often included in the nests of several species, most notably catbirds and tree squirrels.

Grapes continue to be important commercially, and small vineyards are popping up across the Midwest with great regularity. A native, *V. lambrusca*, and its cultivars are still commonly planted for grape production here. Another native species of grape that is planted commercially is *V. rotundifolia*, muscadine or scuppernong; it is a thick-skinned grape, golden brown when ripe, with a unique taste. It is common in the Southeast and found in Kentucky and Tennessee.

Grapes can be highly variable and often difficult to identify. Their flowers, and to some degree the fruit, are not much help, so leaves are the major feature to use. All grapes have brown pith (on older twigs) that is interrupted by slight, solid partitions at the nodes (diaphragmed), which helps to distinguish them from their similar-looking cousins, the *Ampelopsis*, which have chambered pith at their nodes. Young grape twigs (mainly first year's growth) have white or green pith that changes to brown with age.

The variability of leaf shape within a species can be frustrating. Pictured here is *V. aestivalis* with its new leaves of the spring. The mature, summer leaves usually look nothing like this.

Grapes have an unusual pith that is interrupted at the nodes, as shown on this frost grape twig. In this case, it is actually diaphragmed at the node but solid everywhere else.

Stems, as seen in this whittled section of sweet winter grape, have a radiating characteristic common to *Vitis* species that resembles spokes on a wheel.

All grape vines climb using long, slender tendrils that are produced from the nodes. The tendril tips wrap around structures for support as they climb to a light source. Tendril characteristics help identify several species.

Vines that are several years old usually have multiple buds at each node that produce either vegetation alone or vegetation and flowers.

summer grape

Vitis aestivalis **Michx.**

Family: Vitaceae

Like most grape vines, this species grows rapidly and can engulf the plant on which it is growing. Its numerous, large leaves block light that would have otherwise been utilized by its buried "host." Several varieties have been named based on the abundance of hairs on the lower leaf surfaces. As one moves north in summer grape's range, the lower leaf surfaces become practically hairless, and it is referred to as variety *bicolor*.

Form and Size: This is a large, high-climbing vine that can make its way to the canopy of forest trees. It also blankets small trees or shrubs by growing over them. It grows best in full sun but tolerates less.

Habitat: Summer grape is often found on drier sites than some *Vitis* species. It is usually found on wooded ridges and open thickets or forest edges.

Wildlife Uses: This is 1 of the 3 most common grapes in our region and 1 that regularly occurs on upland sites. The other 2, *V. vulpina* and *V. riparia*, are generally bottomland species, growing in alluvial soils. Summer grape frequently climbs into trees in upland woods and, in complex with tree limbs, produces nest sites for large species, such as gray squirrels and red-tailed hawks. Since wildlife values of most *Vitis* species are similar, the reader is referred to the *Vitis* genus page 334.

Landscaping Value: Hardy to Zone 5, this grape could make a nice addition to a naturalized setting if a large area is available. Pruning can be done to control its size, but growth rate is fast, so it might be a frequent job. The 'Norton' variety has become popular with wine growers in the Midwest. Half of this variety's genes are *V. aestivalis*, and it is prized for the wonderful flavor produced and for its high resistance to most fungal pathogens that are problematic with other grape varieties. This variety is available from online nurseries. Otherwise, stem cuttings or seeds from our native grow easily.

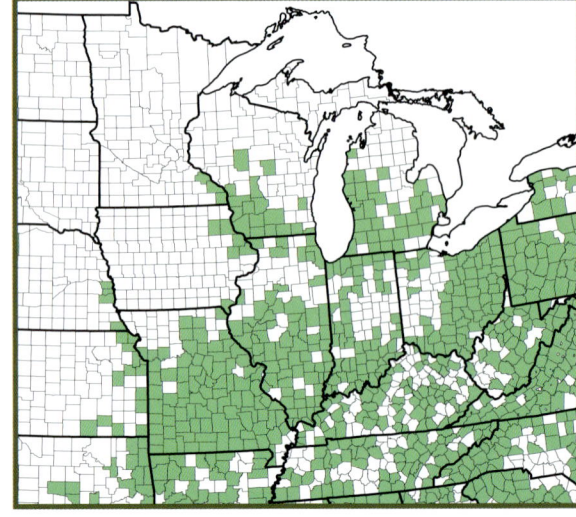

Similar Species Distinctions:

—**Riverbank grape** (*V. riparia*) has similar leaves and fruit, but it has no glaucous leaf back, and the teeth along the margins are usually somewhat elongated and sharply defined.

—**Frost grape** (*V. vulpina*) leaves are similar but usually have hairs along the lower veins only.

Leaves are 3 to 5 lobed and average 6 inches long, with rounded (usually) sinuses. New leaves can be deeply 5-lobed. The margins are toothed. The upper leaf surface is bright green and smooth at maturity. The lower surface is usually glaucous, with a rusty or whitish cobwebby network of hairs. Sometimes these hairs are only along the lower veins, but leaves are almost always glaucous beneath. Petioles are generally hairy.

Fruit is a berry that ripens in the fall in dangling clusters. They are dark blue to black with a slight whitish bloom. Each berry is about ½ inch in diameter and edible after frost.

Flowers appear in dangling clusters in May and June. There are many tiny, yellowish flowers that are "actually or functionally" unisexual, according to Gleason and Cronquist (1991). The petals fall off quickly after opening, leaving just stamens or stigma lobes visible.

The twigs are mostly smooth at maturity (by wintertime), light brown, and round in cross section. Buds are brown and conical-shaped with 2 scales. Pith is brown and interrupted at the nodes. Mature bark is similar to that of most grapes—brown, thin, and shredding into strips.

graybark grape, winter grape

Vitis cinerea (Engelm.) Engelm. ex Millard

Family: Vitaceae

The new leaves of this grape are very attractive. Dense, whitish gray hairs cover the leaves, particularly underneath. Grapes continue to grow throughout much of the summer, so the new, snowy leaves create a nice contrast to the dark green, mature leaves that have lost their upper surface hairs. There are numerous regional varieties recognized across the eastern United States.

Form and Size: This is a large, high-climbing grape that can be found in treetops or sprawling over large shrubs or small trees. It is fast-growing and long-lived.

Habitat: Graybark is common in open woods, forest edges, and along roadside fences, where it can take advantage of nearly full sun. It prefers moist, well-drained sites.

Wildlife Uses: Very similar to *V. aestivalis* and most other grapes in wildlife values; it frequently occurs in bottomlands as well as upland sites. The reader is referred to *Vitis* genus page 334.

Landscaping Value: Hardy to Zone 5, this grape could be used in a naturalized setting in the same manner as it grows in the wild. It is particularly attractive growing along a fence, but it needs plenty of room to grow. Grapes are easily propagated from stem cuttings and seeds.

Similar Species Distinctions:
—**Summer grape** (*V. aestivalis*) has similar leaves and is found growing alongside graybark. Summer grape almost always has a whitish bloom on the underside of the leaves, and usually cobwebby, rusty, or whitish hairs underneath as well.

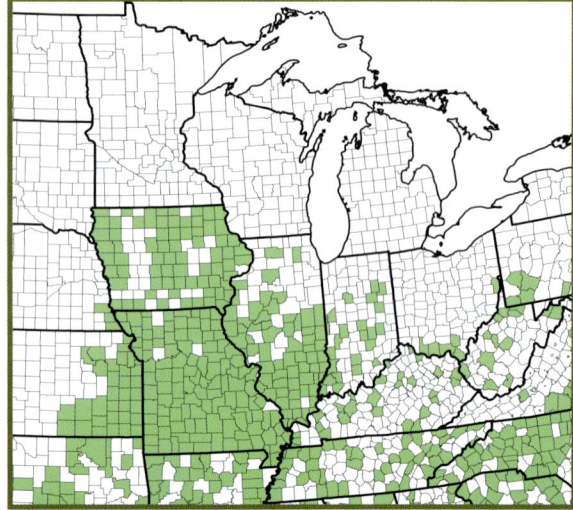

Leaves are usually shallowly, 3-lobed with irregular, small teeth along the margins. They are about 8 inches long with a hairy petiole. The upper leaf surface is dull green and mostly smooth (long, scattered, appressed hairs can be seen, especially along the midvein) at maturity, while the lower surface is densely covered with whitish gray hairs.

Fruits ripen in the fall in clusters. Berries are ¼ inch in diameter and are fleshy and sweet after frost. They are blue-black and usually have no whitish bloom.

Flowers appear in May into June in dangling clusters that are about 8 inches long. Each cluster contains many tiny, yellowish flowers that are a single sex ("actually or functionally unisexual"). The 5 fused petals detach at the base and fall as a single unit.

Bark is thin, brown, and flaky, commonly shredding into strips.

Twigs are angular, especially on new growth, and hairy. Buds are small, with 2 brown, hairy scales. Older vines usually have multiple buds at a node. The brown pith is interrupted at each node. Tendrils (or flowers/fruiting stalks) are produced opposite leaves; every third leaf has neither.

fox grape

Vitis labrusca L.

Family: Vitaceae

Fox grape is the ancestor of the Concord and other commercially important grape varieties. Though many horticultural varieties are available, our wild vines produce large, sweet fruit and can be easily grown in the home garden. Fruits, unlike the wine grapes of Europe, have a slip-skin— when a grape is squeezed between the fingers, the skin splits open and slips easily off the contents.

Form and Size: This is more of a trailing vine, but it does climb, using tendrils for support. It can reach to smaller treetops, but more commonly it grows over shrubs and smaller trees, especially along forest edges, where it can take advantage of sunlight.

Habitat: Fox grape is common in forest openings, edges, and thickets, along roadsides and railroad rights-of-way, and in other disturbed areas. It prefers full sun and tolerates drier conditions than most of our native grapes. Vines are long-lived.

Wildlife Uses: Distribution of this grape is somewhat limited in our region, because it does not prefer a high soil pH. It is very similar in wildlife value to other *Vitis* species; the reader is referred to *Vitis* genus page 334. A major difference from other species is the size of the fruit, quite large (to 1 inch diameter), which limits its use almost exclusively to mammals and wild turkey.

Landscaping Value: Hardy to Zone 4, this grape prefers more acidic soils than other native *Vitis*, which limits, to some degree, planting sites, unless soil is amended. It is easily trained for growth on a trellis or arbor and will grow in a variety of moisture regimes from moist, well-drained to dry sites. There are many cultivated varieties that have more disease resistance than our native, which probably should be planted by the serious grape grower.

> **Similar Species Distinctions:**
> —This is a distinct grape in several ways. It has the largest fruit of any midwestern grape (**Muscadine grape**, *V. rotundifolia*, in the South also has similar-sized fruit), and it has a tendril or inflorescence opposite at least 3 leaves in a row (or more).
> —**Summer grape** (*V. aestivalis*) usually has densely hairy leaf back, but in combination with a glaucous bloom.

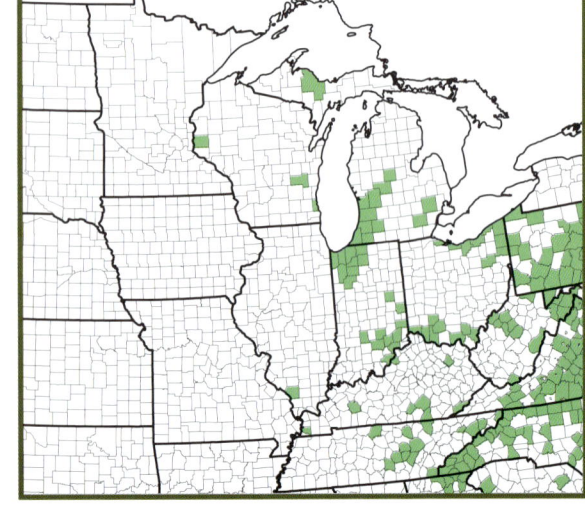

There is a tendril or inflorescence opposite at least 3 leaves in a row, usually all; no other native grape has this tendril pattern. Leaves are usually 3-lobed and up to 6 inches long. They have 2 shallow sinuses and shallow teeth along the margin. The upper surface is dark green, dull, and mostly smooth, while the lower surface is covered with dense, reddish hairs when new that fade to ashy-gray. Petiole is hairy.

Flowers appear beginning in late May and June in an inflorescence shorter than that of other natives. Including the peduncle (fruiting stalk), it is no more than 6 inches long. Individual flowers are tiny with deciduous petals and unisex on separate plants.

Fruit begins ripening in early fall in dangling, densely packed clusters. There are few berries per cluster compared to other native grapes. Each grape is nearly 1 inch in diameter, and it ripens to a wine-red to dark purple color. There is little to no whitish bloom.

Mature bark is similar to that of most native grapes—thin, brown, and scaly.

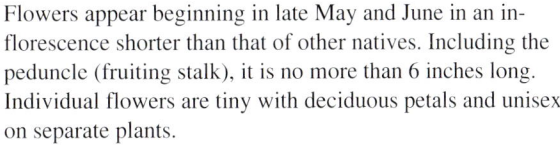

New twig growth is densely covered with rusty or whitish-colored hairs. Buds are teardrop-shaped and covered with 2 brown, smooth scales. Pith is solid and brown except at the nodes; there it is paler and diaphragmed.

catbird grape, red grape

Vitis palmata Vahl.

Vitis rubra Michx.

Family: Vitaceae

Several things jump out at you when you see this grape. The first is its unique, delicate, red vines and tendrils. The second is the elongated, pointed lobes and deep sinuses of the leaves. The combination of these characteristics makes this grape fairly easy to distinguish from others. This species is listed as rare in Indiana.

Form and Size: Catbird grape is usually seen growing over other plants—large shrubs and small trees along waterways. It can climb upwards to 40 feet or more, using its tendrils for support. It grows best in full sun.

Habitat: This is a species of bottomlands in low, wet woods and forest edges, river sloughs, swamps, and pond margins in well- to poorly-drained soils. It has a restricted range, mainly along the bottomlands associated with large waterways such as the Mississippi and Ohio rivers. It grows quickly and is probably long-lived.

Wildlife Uses: It is a relatively uncommon species in our region, but it retains many of the general *Vitis* values (see *Vitis* genus account on page 334). It does, however, differ in a few ways from the general *Vitis* pattern. Its distribution is focused on very wet sites, and its stems are slender, almost delicate. It, thus, supplies relatively unique nesting cover to bottomland songbirds. In addition, its fruits are among the smallest of all grapes, and they are taken regularly by smaller songbirds that can handle this sized fruit.

Landscaping Value: Hardy to Zone 6, this is a grape for wet sites on pond or lake margins or along a wet forest edges. It is an attractive vine with its bright red stems and tendrils and glossy leaves. Cuttings from this species root easily, and seeds can be germinated with little effort.

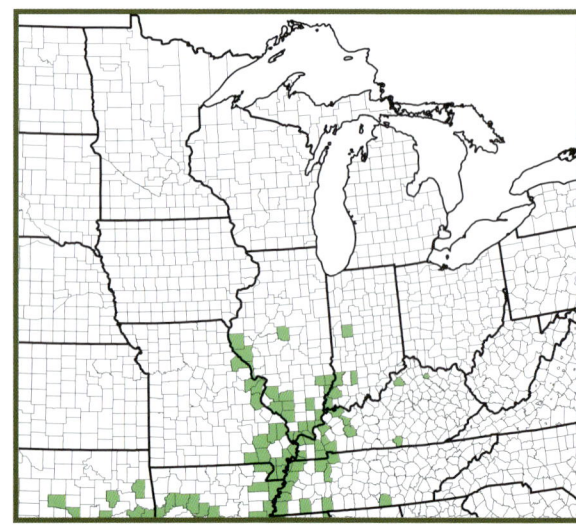

Similar Species Distinctions:
—**Riverbank grape** (*V. riparia*) is a common associate. Its leaves do not develop deep sinuses, and it does not have red stems or tendrils.
—**Raccoon grape** (*Ampelopsis cordata*) is a common associate as well, which can have similar-looking leaves, but its stems and tendrils are not red.

Flower clusters appear beginning in late June in southern Indiana. They are dangling clusters of tiny, yellowish, unisex flowers with deciduous petals that fall upon opening.

Leaves have 5 elongated lobes, the lowest 2 being the smallest. An average leaf size is 4 inches long. The leaf margins are irregularly toothed, and sinuses are deep. The upper leaf surface is bright green and glossy (can be hairy along the veins), while the lower surface is paler and mostly smooth, except perhaps for hairs on the veins or in axils of veins. Petioles are red, grooved, and somewhat hairy.

Fruits ripen in the autumn in clusters with blue-black berries that are rarely coated with a whitish bloom. Individual berries are about ¼ inch in diameter, juicy, and edible, with 1 or 2 sizable seeds each.

Twigs are bright red to purplish red, smooth and shiny, especially when new. Buds are conical and have 2 reddish brown, hairless scales. Pith is brown and interrupted at the nodes. Tendrils are red. Mature bark is reddish brown, thin, and shreddy.

riverbank grape

Vitis riparia **Michx.**

Family: Vitaceae

Riverbank grape is probably the most common eastern species of *Vitis*. Its large diameter vines are found dangling from the tops of bottomland trees as if to accommodate Tarzan's next emergency.

Form and Size: This is a large, high-climbing vine that uses tendrils for support. It can grow into the treetops and drape down the sides of smaller trees along forest edges to the point where the tree is not visible. It grows best in full sun.

Habitat: Riverbank grape grows in areas that its name implies: riverbanks, lake and pond margins, and floodplains of streams and rivers; however, it also occurs in openings such as along roadside fences, in brushy thickets, on sand dunes, and along gravely disturbed areas such as railroad rights-of-way. It is fast-growing and long-lived.

Wildlife Uses: Riverbank grape is similar in wildlife value to most other *Vitis*; the reader is referred to *Vitis* genus page 334. Its distribution is commonly associated with bottomland alluvial soils, a feature also shared with several species (e.g., *V. cinerea*, *V. palmata*, *V. vulpina*). This species' range extends further north in our region than that of any other grape.

Landscaping Value: Hardy to Zone 2, this large vine could be used in a more naturalized setting along a forest edge where the goal is to attract wildlife. It is well adapted to wet sites, but it will tolerate more upland areas. There are several cultivars available, including 'Praecox' and 'Clinton.'

Similar Species Distinctions:
—**Frost grape** (*V. vulpina*) leaves are more heart-shaped in outline with small teeth that are not so sharply pointed. Usually there are no hairs on the leaf margins.
—**Catbird grape** (*V. palmata*) has similar leaves, but its sinuses are rounded. Twigs are red.

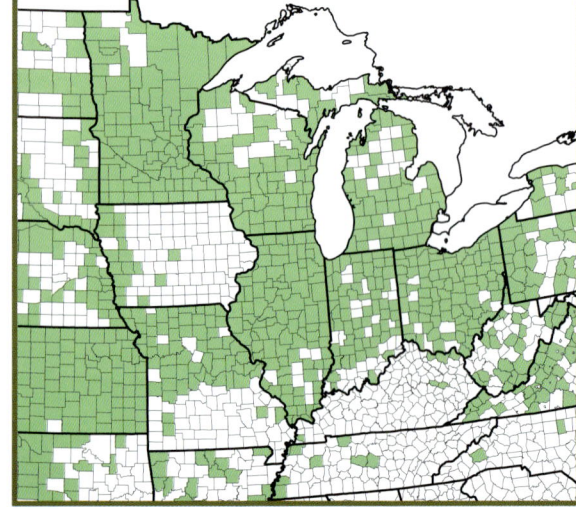

Leaves are distinctly 3-lobed and up to 6 inches long. They have broadly V-shaped sinuses between the 2 outer lobes. Margins have sharp, sizable teeth with a row of short hairs (ciliate) along them. Leaves are green and smooth above, with hairs possible along the main veins. Lower surface is paler and hairy on some or all of the main veins or veinlets. Petiole is hairy.

Flowers appear in May and June in dangling clusters. Individual flowers are tiny, yellowish, and unisex, with petals that fall soon after opening.

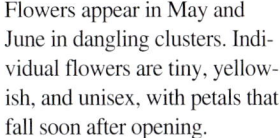

Fruit can ripen earlier than other native grapes, sometimes in mid-summer, but they tend to be sour until frost. They are up to ½ inch in diameter, blue-black, with a whitish bloom.

Twigs are reddish brown, mostly smooth, and round in cross section. Buds are 2-scaled, conical, and hairy. Pith is brown and interrupted by the narrowest diaphragm of any native grape. Bark is brown, shreddy, and shaggy in several layers.

sand grape

Vitis rupestris **Scheele**

Family: Vitaceae

Sand grape could be considered a ground cover and lends itself well to that category. It creeps along the ground growing to a stem length of 15 feet and is easily contained by pruning, which results in a fuller, more beautiful specimen. Sand grape is listed as threatened in Kentucky, proposed endangered in Tennessee, and endangered in Indiana.

Form and Size: This is a low-growing grape that creeps along the ground and occasionally grows over low shrubs, attaching with only a few tendrils. Some stems will grow upward a foot or more.

Habitat: Sand grape is always found sprawling on the ground in gravel bars and along gravelly and rocky streamsides. It is adapted to flooding events in which it is commonly clogged with debris. It thrives in full sun, but it handles partial shade.

Wildlife Uses: Of all the grapes in our region, sand grape is certainly the most unique in form and wildlife values. It is relatively rare and is definitely a trailing vine, rather than a climber, remaining low as it crawls over rocky banks and gravel bars along streams. This low stature and open habitat serves a whole different suite of nesting birds than other *Vitis*, species like song sparrows, yellow-throats, and towhees. Despite its smaller stature, the grapes it produces are relatively large and principally used by mammals, such as opossums, raccoons, coyotes, and deer.

Landscaping Value: Hardy to Zone 5, this interesting vine makes a nice ground cover and is easily pruned to whatever size is desired. It has unique, folded foliage on bright red petioles and stems, with yellow fall color. It is extremely drought tolerant, and it is adapted to well-drained, alkaline sites. Japanese beetles do not care for this grape, as they do other *Vitis* species. Sand grape easily roots from stem cuttings.

Similar Species Distinctions:
—**Muscadine grape** (*V. rotundifolia*) has similar leaves that are not usually folded, and the petiole is indented. Bark is smooth and white-dotted, and fruit is 1 inch in diameter and speckled. It is a grape of southwestern Kentucky and southward.

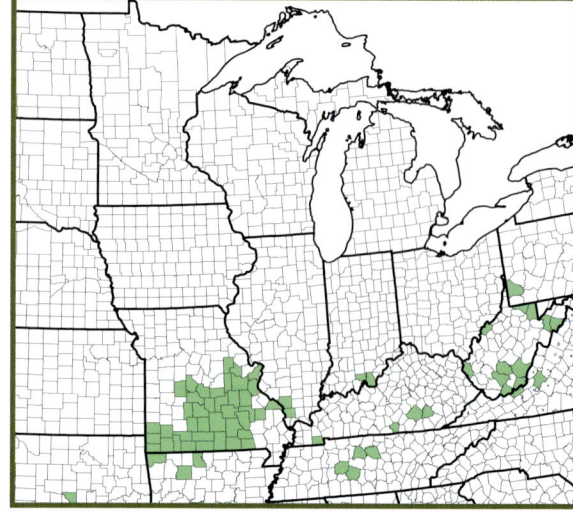

Leaves are 4 to 5 inches long, tough, and commonly partially folded. They have coarsely-toothed margins and an elongated (sometimes) leaf tip. The upper surface is green with a satiny finish and smooth. The lower surface is only slightly paler and mostly smooth, occasionally with hairs along the largest veins. The petiole is grooved and green, or red, or striped with both colors.

Fruits ripen in the fall and are nearly black with no whitish bloom. Each berry is about ⅓ inch in diameter. For whatever reason, the fruits are rarely produced or seen.

Flowers appear in May and June in small clusters that are 4 inches long or less—very short for a native grape. Each cluster has numerous tiny, yellowish, unisex flowers. The 5-lobed petals are fused and deciduous immediately after opening.

Twigs are slender, smooth, and bright red with red, forked tendrils produced only with the uppermost leaves. The brown pith is interrupted at the nodes with diaphragms. The conical-shaped buds are covered with 2 brown, mostly hairless scales. Bark is brownish or gray, thin, and shreddy like that of most grapes. This vine does not develop large diameter "trunks" like those of the climbing grapes.

frost grape

Vitis vulpina **L.**
Vitis cordifolia
Michx.
Family: Vitaceae

Frost grape is another common vine of the eastern United States. The vines, like most grapes, are collected to create woven wreaths and baskets. Like most grapes, the fruit should not be eaten until after a frost sweetens them. This species is listed as threatened in Michigan, endangered in New York.

Form and Size: Frost grape is a large vine that climbs with tendrils. It can be found in the treetops or suffocating smaller trees and large shrubs along forest edges. It grows quickly and performs best in full sun.

Habitat: This is a vine of low, wet ground in forest openings and edges, thickets, roadsides, and fencerows. It prefers full sun and grows quickly.

Wildlife Uses: This is 1 of the common bottomland grape species in the Midwest. It has a small fruit that can be handled by songbirds, as well as the typical users of wild grapes—game birds and mammals. Most wildlife values are very similar to other *Vitis* species; the reader is referred to *Vitis* genus account 334.

Landscaping Value: Hardy to Zone 5, this large grape needs room to grow, and it could be used in a naturalized setting along a forest edge to attract wildlife.

Similar Species Distinctions:
—**Summer grape** (*V. aestivalis*) has similar leaves, but it has whitish bloom beneath and usually a cobwebby network of hair over that entire leaf surface.

Leaves have no lobes or 2 small side-lobes and numerous small teeth along the margins. They are up to 6 inches long and appear heart-shaped. The upper surface is dark green and smooth, while the lower surface is paler with hairs on the veins and in the vein axils. The petiole is smooth.

Fruit ripens in the fall in clusters up to 6 inches long. The individual berries are nearly black, about ¼ inch in diameter, and coated with a whitish bloom. They are juicy and sweet once frosted.

The new growth varies in the degree of hairiness, but it is usually smooth by fall. Twigs are not usually grooved. Lateral buds are conical, with 2 hairy scales. The brown pith is interrupted at the nodes.

Bark is brown, thin, and shreddy as in most grapes.

Flowers appear in May and June in dangling clusters up to 6 inches long. The individual, tiny, yellowish flowers are unisex and have deciduous petals that fall soon after the flowers open.

Virginia Creepers

Genus: *Parthenocissus*

There are 2 species (some taxonomists claim just 1) of creepers in the Midwest that can be difficult to separate. Their ranges overlap throughout most of our region, which adds another complication.

The creepers are common and frequently mistaken for poison-ivy, also a very common vine. However, leaves of the creepers have 5 leaflets while those of poison-ivy have just 3. Both taxa can develop sizable diameter "trunks," as they live for many decades attached to a tree. However, their bark is significantly different.

Creepers are very attractive when grown on a fence or over an arbor, and the more sun they are provided, the more color appears in the fall, and the more fruit they produce . For a colorful, late summer vine, these can hardly be topped. Before most other woody plants begin to don their fall color, the creepers turn from green to light reds and oranges, and finally to deep red. If the vines happen to be loaded with fruit, their orange-red peduncles (fruit stalks) and dark blue berries really make an impact on the landscape.

One commonly planted introduced species seen in many eastern United States towns and cities is *Parthenocissus tricuspidata*, Boston-ivy. It is famously seen on the outfield wall of Wrigley Field in Chicago where the Cubs play baseball.

The 2 species of this genus in the Midwest are almost identical in appearance and in wildlife value. They are woodland and forest edge plants that supply both browse to herbivores and fruits that are relished by a myriad of songbirds and game birds. They do not supply as much cover as do the grapes with which they often share an environment. Both species of *Parthenocissus* will creep along the ground, providing cover for ground-dwelling creatures, and climb into trees. Virginia creeper is more likely to climb to the very tops of trees where it reaches sunlight and fruits prolifically; northern Virginia creeper typically does not climb as high, but it is more likely to cover shrubs, fences, and other obstacles, thereby creating good nesting habitat for songbirds. Both are favorite spring and summer browse species for white-tailed deer.

Unlike Virginia creeper, seen here, with its flat, suction cup-like pads on the end of multiple tendril tips, northern Virginia creeper usually has none, and it wraps its few-forked tendrils around small twigs or itself, similar to the behavior of a grape vine.

Virginia creeper vines are often cut by people who believe they are eradicating poison-ivy. However, the bark of the 2 species is quite different. That of Virginia creeper (pictured left) is thick, brown, and somewhat interlacing, often reminiscent of ash bark. The other 3 vines (pictured right) of the creeper are all poison-ivy. Its bark is thin, gray, and smooth. The 2 should never be confused based on bark.

Parthenocissus have tendrils somewhat similar to those of grape vines, but northern Virginia creeper's are quite long, as seen here wrapping themselves around its own stems.

Creeper leaves have 5 leaflets, compared to 3 found on leaves of poison-ivy. Northern Virginia creeper leaves (right) are commonly shiny, while those of Virginia creeper (left) are dull.

The introduced *Parthenocissus tricuspidata*, Boston-ivy, is commonly planted in the eastern United States. It is easily recognized by its 3-lobed leaves.

Toxicodendron radicans, poison-ivy.

Virginia creeper, 5-leaved ivy

Parthenocissus quinquefolia (L.) Planch.

Family: Vitaceae

This common woody vine is found growing up tree trunks in all midwestern forests, quite often directly beside poison-ivy. Its dark brown, thick, interlacing bark and thick, matted rootlets are nothing like the thin, gray bark and hair-like, sprawling rootlets of poison-ivy.

Form and Size: This is a large, deciduous vine that climbs as high as it can into the treetops. Leaves may cover large sections of the tree's canopy, cutting off potential energy to the tree, but generally are confined to major branches. Large vines have no leaves along their main stem—only at the top. Stem diameter of old vines may be 4 to 5 inches. It is also commonly seen running along the forest floor, its upright leaves resembling those of an herbaceous plant. Once established on a steady support, such as a sound tree, vines are long-lived.

Habitat: Virginia creeper is common throughout most of the eastern United States in woods, clearings, fencerows, and various openings. It is tolerant of shade, but it thrives in full sun.

Wildlife Uses: Virginia creeper is an excellent wildlife plant, providing many benefits. Its leaves are 1 of the major spring/summer foods of white-tailed deer in our region and are taken by cottontails and woodchucks as well. When it receives enough light, as when growing on edges or when the thin vines climb to the tops of canopy trees, they produce good amounts of fruit; these fruits are relished by songbirds and game birds, including thrushes, warblers, vireos, woodpeckers, and wild turkeys. Vines occasionally grow densely enough to provide nesting cover, but not as frequently as does northern Virginia creeper; however, the vines frequently run along the ground sending up leaves periodically, which produces a ground cover of sorts that ground nesters utilize to conceal nests.

Landscaping Value: Hardy to at least Zone 3, this species is tolerant of most soils and moisture regimes. It grows well in almost any light conditions, but it gives better color and more fruit with greater amounts of light. It is the earliest native vine to turn color in the fall, and it is a great choice for an arbor or trellis. However, it is not recommended for planting where it can grow onto a building, for removing the disks is literally impossible. Unfortunately, its leaves are highly preferred by Japanese beetles. Stem cuttings root easily, which is probably the best means of acquiring this species.

Similar Species Distinctions:
—**Northern Virginia creeper** (*V. vitacea*) has similar leaves that are usually shiny, and long, forked tendrils without terminal disks (usually).
—**Ginseng** (*Panax quinquefolium*), an herbaceous plant, has very similar leaves, but its lowest 2 leaflets are much reduced.

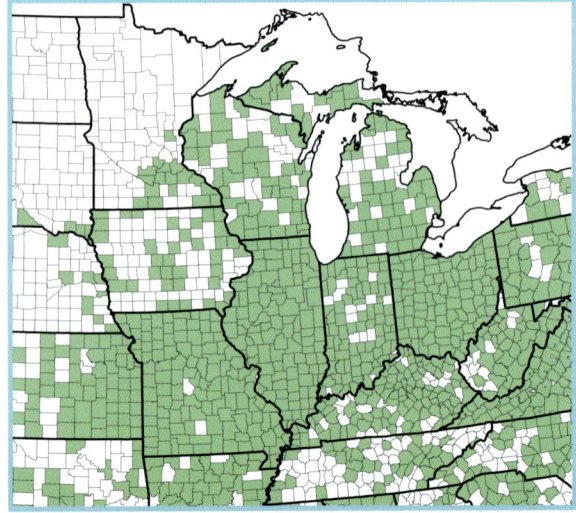

The compound leaves are about 5 inches across and have 5 toothy leaflets. They are dark green and dull above, slightly paler and smooth beneath. The petioles are often very long—up to 12 inches.

Fruits ripen in the fall in dangling clusters. The stalks are red, and the berries are blue, which makes for quite a show against the burgundy-colored leaves at that time of year.

Flowers appear in dangling clusters from May into July, depending on where one lives in the Midwest. They appear after leaves are fully developed on the tips of the new growth. The tiny flowers are greenish yellow and usually more numerous than those of *P. vitacea*.

Twigs are beige or tan-colored and dotted with orange-colored, raised lenticels. New twigs are covered with short, soft hairs. Leaf scars are as wide as the twig, circular, and concave. Bundle scars appear as a "smiley face" along the lower margin of the leaf scar. Buds have several cinnamon-colored scales. Tendrils are short and branched several times at the tips, where flattened disks adhere to whatever is available. Older stems' aerial rootlets create thick, flattened mats on both sides of the vine.

Mature bark is thick, dark brown, with interlacing but discontinuous ridges that resemble the bark of white ash (*Fraxinus americana*).

northern Virginia creeper, woodbine

Parthenocissus vitacea
(Knerr) Hitchc.

Parthenocissus inserta
(Kerner) K. Fritsch

Family: Vitaceae

This *Parthenocissus* looks almost like a hybrid between Virginia creeper and a grape, mainly because it develops lengthy tendrils that wrap around twigs and small diameter branches. Its shiny leaf is usually 1 of its best distinguishing features.

Form and Size: This is not as high a climber as *P. quinquefolia*, but it tends to grow over small trees, shrubs, or fences with ease. Its maximum stem diameter seems to be less as well. It tends to be long-lived once established, and it grows in partial shade to full sun.

Habitat: It is usually found in similar habitats as *P. quinquefolia*, but it tends to grow over rather than up its support. It is a woodland or forest edge species in moist, rich, well-drained soil, but it is also found in more disturbed sites such as openings, roadsides, and fencerows.

Wildlife Uses: Northern Virginia creeper has almost identical wildlife values as Virginia creeper. It does, however, tend not to grow as high into trees, but rather it often creeps (especially in open situations) over bushes, rocks, and fences, producing superior songbird nesting habitat in the process.

Landscaping Value: Hardy to Zone 2, northern Virginia creeper makes an excellent and easily-grown vine for trellises, arbors, or fences. It quickly covers them, and since it has no (or very few) disks, it does not adhere to anything underneath. Instead, its tendrils grow through cracks between boards or around wire. Its fall color mixed with the blue fruit on its red stalks is impressive. Japanese beetles can be a problem. This species is occasionally available through native nurseries.

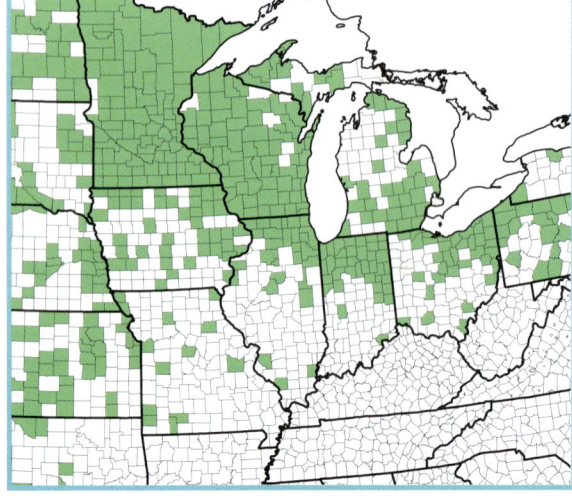

> **Similar Species Distinctions:**
> —**Virginia creeper** (*P. quinquefolia*) usually has leaves with a dull surface and short tendrils that end with disks.

Tendrils are long and designed to wrap around twigs.

The compound leaves are up to 6 inches across with 5 toothy leaflets. The upper leaf surface is dark green, shiny, and smooth, while the lower surface is somewhat paler and mostly smooth. Petioles are up to 8 inches long.

Fruit ripens in the fall in dangling clusters. The blue-black fruit is about ¼ inch in diameter, and it is attached to red stalks. They may have a slight whitish bloom.

Flowers appear in late May into July as one moves north through the Midwest. They are in a much-branched dangling cluster. Flowers are greenish yellow and tiny; they are usually a bit larger than those of *P. quinquefolia*.

Mature bark is dark brown and thick, with discontinuous, interlacing ridges. No aerial rootlets extend from the sides of the large vines.

Twigs are beige or tannish-colored and dotted with raised lenticels. New twig growth is smooth. Leaf scars are as wide as the twigs, circular, and concave. Bundle scars are in a "smiley face" row along the lower portion of the leaf scar. Buds are covered with a few, tannish-colored scales.

poison-ivy

Toxicodendron radicans **(L.) Kuntze**
Rhus radicans **L.**
Family: Anacardiaceae

Poison-ivy, or 3-leaved ivy (which is a misnomer), contains the oily compound toxidendrol in the sap found throughout the plant. It causes an uncomfortable, itchy skin rash that lasts for several weeks. Most people have an allergic reaction or develop 1 later in life. There are several subspecies within its range.

Form and Size: This is a vine that climbs to 60 feet or more into treetops, or it may form a spreading "shrub" that reaches 10 feet high. It also creeps along the forest floor, looking for a tree to support its climbing. Leaves on these creeping vines are often confused with herbaceous plants. "Trunk" diameter can be substantial on old plants—8 inches or more is not uncommon in older plants in bottomlands where disturbance has been minimal for many decades. It is fast-growing and long-lived.

Habitat: It can be found in just about any location from the most disturbed urban area to an old growth forest, in wet to dry sites, full sun to shade.

Wildlife Uses: While it supplies misery to many of the human animal (authors included!), poison-ivy is a wonderful resource for wildlife. Its variable growth forms benefit a myriad of species. The vines that crawl along the ground sending up vertical stems with leaves supply cover for ground nesters in summer and woody twigs that are relished by rabbits in winter. Those stems that grow as small shrubs, especially when supported by a stump or fence post, provide ideal nest sites for songbirds, such as blue grosbeaks and even kingbirds. Vines that climb into trees form a tree/vine complex that provides ideal bird and mammal nest sites. These latter vines, especially if on the forest edge, produce many small fruits that are favorites with songbirds; yellow-rumped warblers are said to winter in our region principally during years of heavy poison-ivy fruit crops. Finally, leaves of poison-ivy form a very important part of the spring/summer diets of white-tailed deer in our region.

Landscaping Value: Hardy to Zone 3, this vine would never be recommended for landscaping. However, leaving it in an area attractive to wildlife is beneficial, if your goal is to supply wildlife food. Its fall color is as spectacular as any of the sumacs.

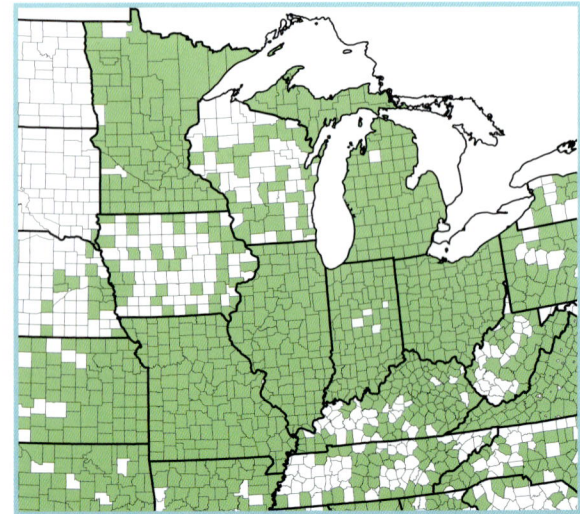

Similar Species Distinctions:
—**Western poison-ivy** (*T. rydbergii*) occurs in the Lake States, northern Ohio, Illinois, and Iowa, and is similar, but it is a knee-high shrub with short, woody stems.

The compound leaves have 3 leaflets (not *leaves* of 3) that are up to 8 inches long. Leaflet shape is variable, but they are usually oval to ovate-shaped. Their margins vary from entire to irregularly toothy. They are dark green and smooth above, paler beneath, with varying amounts of hair. Petioles are long and hairy.

Fruits ripen in the fall. Each berry is about ¼ inch in diameter, tannish or yellowish, appearing translucent, and smooth to hairy. They often persist into the winter months.

Flowers appear in May in loose clusters. There are many small, greenish white, 5-petaled flowers less than ¼ inch across. Bees are very attracted to them.

Mature bark is gray, thin, and slightly scaly or flaky. Aerial rootlets remain slender, hair-like, and matted against the vine and its host tree.

Twigs are tan or beige and softly hairy on the newest growth. Second year twigs are smooth, and they produce fine, brownish, aerial rootlets from the sides. Orange/brown lenticels dot the twigs. Buds are dark tan, hairy, naked, slender, and slightly curved at the tip, appearing finger-like. Lateral buds are smaller, hairy, and appressed. Leaf scars almost U-shaped, containing 6 bundle scars.

Kentucky wisteria, American wisteria

Wisteria frutescens (L.) Poir.

Wisteria macrostachya (Torr. & Gray) Nutt. ex. B. L. Rob. & Fernald

Family: Fabaceae

Kentucky wisteria has been cultivated for several centuries, but exotic species are encountered far more often in urban landscapes. All have beautiful, large clusters of gracefully hanging lavender flowers. Wisteria is listed as rare in Indiana and threatened in Michigan.

Form and Size: This is a twining vine that climbs to 30 feet or more into forest trees. It climbs in a clockwise direction around its support. Growth rate can be fast, and vines are usually long-lived.

Habitat: Wisteria prefers, and is usually found growing in, low wooded areas such as flood plains of large rivers, streams, and river sloughs. It prefers full sun but tolerates partial shade.

Wildlife Uses: This vine is relatively rare with a patchy distribution in the Midwest; its overall importance to the wildlife of our region is quite low. Its leaves and seeds are evidently not preferred and little-used by wildlife. It does climb onto shrubs and into trees, which gives it some cover value for wildlife, especially for supporting nests of birds and squirrels.

Landscaping Value: Hardy to Zone 5, this is a vine well suited for trellises, arbors, or fencing. It is most beautiful when in full bloom. It tolerates flooding and is fairly drought resistant; it grows well in well-drained, upland sites. There are several horticultural varieties with blue and white flowers, but they are adapted to southeastern United States. Japanese beetles can be problematic. Kentucky wisteria is available occasionally through midwestern nurseries.

> **Similar Species Distinctions:**
> —**Chinese wisteria** (*W. sinensis*) has similar leaves, but its fruiting pods are larger and velvety-hairy. All flowers in each cluster open essentially at the same time.
> —**Japanese wisteria** (*W. floribunda*) has similar leaves, and its pod is also velvety-hairy. Its inflorescences are not glandular-hairy.

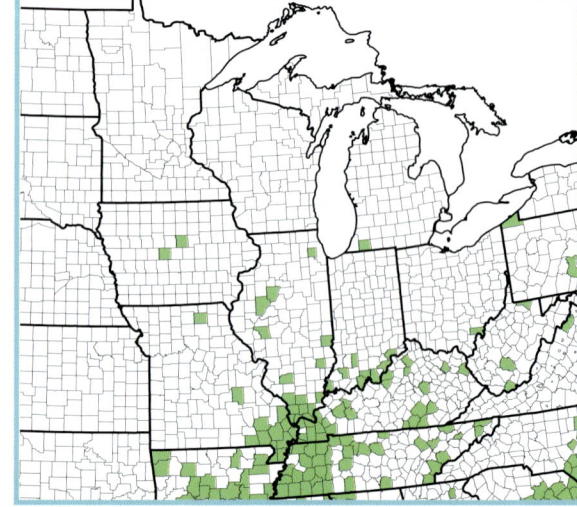

The compound leaves are up to 12 inches in length and have 5 to 11 leaflets. Each leaflet is egg- or lance-shaped with entire margins. They are dark green, dull, and smooth above, paler and hairy beneath.

Fruits begin ripening over the summer and can persist through the winter months. The pods are up to 6 inches long, smooth, reddish brown, and constricted between the seeds. The large seeds are poisonous to humans.

Flowers appear in May in dangling clusters (racemes) up to 12 inches long. There are up to 90 lilac-purple, irregular flowers per cluster. Each flower is about ¾ inch long. The inflorescence is glandular-hairy. Flowers open from the petiole end to the slender inflorescence tip. Flowering occurs over a several week period.

Twigs are brownish, mostly smooth, and rather slender. Lateral buds are rounded, with 2 brown scales that have white, woolly hairs. Leaf scars are somewhat circular and raised, and they contain 5 bundle scars in 2 rows. A stem cross-section reveals concentric rings of pores and numerous radial rays.

Bark is thin, light brown to gray, and dotted with darker lenticels. It peels and curls in small sections.

INTRODUCED SPECIES

Introduced invasive species of all kinds can be found everywhere—zebra mussels in the Great Lakes, white-nose syndrome within our bat populations, and the list goes on and on. For example, in Indiana alone, there are currently 379 exotic, invasive plant species recorded, and probably others not yet discovered. Of those 379, 64 are shrubs and vines. Ten species have the dubious distinction of occurring in all 92 Indiana counties, and 6 of those 10 are covered in this chapter. The only other woody plant on this "Top Ten" list is ailanthus (*Ailanthus altissima),* which is covered in *Native Trees of the Midwest* (Weeks, Weeks, and Parker 2010).

In general, an exotic species is one that has been introduced into an area where it formerly did not naturally occur (for this project, we have essentially written about Asian species brought to the United States). Once this happens, the species usually takes several decades for its numbers to increase to the point where it may become problematic. Some end up being so invasive that they cause problems with biodiversity in native communities; most do not. There are usually no natural controls in place in their new surroundings to keep populations in check, and therefore, there is often nothing to limit their spread. Several major problems arise from this scenario. For example, native insects have not co-evolved with these introduced species, and they often do not utilize the plants, disrupting processes in natural communities. In addition, the exotic may grow so densely that it eliminates many natives that cannot survive in its shade. Thus, the diversity of our natural habitats may eventually be very negatively impacted.

We have included some of the most aggressive species within this volume, but there are other problem exotics across the region, which are listed in our "Other Introduced, Escaped Species in the Midwest" section.

winged wahoo, burning bush

***Euonymus alatus*
(Thung.) Sieb.**

Family: Celastraceae

First introduced into the northeastern United States in the 1860s, this plant has been utilized to death as an ornamental, and it has now become naturalized in some parts of the Midwest. Unfortunately, it is still available at nearly every nursery, greenhouse, and garden center in our region, simply because it is easy to propagate, grows quickly, and sells well.

Form and Size: This is a large shrub that can grow to 20 feet tall and at least that wide. It has numerous twigs that create a dense form.

Habitat: Winged wahoo is very adaptable to almost any site, except those that are poorly-drained. It has escaped into oak-hickory forests, prairies, and urban sites. It prefers full sun but tolerates full shade, and it grows slowly wherever it is.

Wildlife Uses: When grown as an ornamental, this exotic produces a dense crown and is regularly used for nesting by cardinals, catbirds, and other suburban birds. When it escapes into native woods, it still has more substantial structure than native *Euonymus* species (partially because of its winged stems), and it is used more frequently for nesting by songbirds. It is occasionally browsed by deer, and fruits are sparingly used by birds.

Landscaping Value: Hardy to Zone 4, this species, it *seems*, is planted in every lawn in every town in the Midwest. It is commonly used as a specimen plant for its attractive fall color, and it is often used as a hedge or border shrub. Bright fall color occurs regardless of whether plants are in full sun or in heavy shade. There are several cultivars.

Similar Species Distinctions:
—**Eastern burning bush** (*E. atropurpureus*), especially when growing alongside this exotic in a wooded situation, can be difficult to separate. The main reason is that the exotic often does not produce the exaggerated, winged stems in such environments, and leaves can be very similar.

ESCAPED IN THESE STATES

Fruits ripen in the fall. Each red fruit is a 1 to 4 lobed, smooth husk that splits open to reveal a bright reddish orange aril.

Leaves are simple, elliptic to lanceolate-obovate, and up to 3 inches long. Margins are finely toothed or wavy. The upper surface is bright green and smooth; the lower surface is paler and smooth. Petioles are very short and smooth.

Flowers appear in small, 3-flowered clusters (cymes) in the spring once leaves are fully developed. Each flower is about ⅓ inch across with 4 yellowish green parts. Pedicels are about ½ inch long.

Twigs are greenish when new, eventually turning brownish. Vigorous growth begins development of 2-4 tannish-colored wings that continue to grow with age. Buds are ovoid, diverge away from the twig, and are covered with 5 to 6 scales that are the same color as the twig. Bud scars are whitish with 3 to 4 bundle scars.

Mature bark is gray and smooth, with pale inner bark exposed from thin, vertical cracks.

border privet, regel privet

Ligustrum obtusifolium **Sieb. & Zucc.**

Family: Oleaceae

This commonly planted native of Japan has been in the United States since around 1860, used mainly as a formal hedge, because of its ability to grow densely once pruned. In the Midwest, it has not yet become the aggressive invasive that it has in the eastern part of the country, but one can find it with ease.

Form and Size: This can be a large shrub up to 12 feet tall and 15 feet wide. It is usually smaller, and the common variety *L. obtusifolium* var. *regelianum*, regel privet, is usually 4 to 5 feet tall with a more horizontal, spreading form. Branches are fine, numerous, and multi-stemmed at the base.

Habitat: Border privet prefers moist, well-drained sites with full sun. It has escaped into disturbed woods, including city parks and forest fragments, and along streams, wherever birds have deposited seeds. It tolerates partial shade but becomes rather straggly, especially if not pruned.

Wildlife Uses: Border privet is a very dense shrub, frequently planted and pruned to form hedges. In these situations in urban/suburban yards, they are heavily used for nesting by birds, such as chipping sparrows, cardinals, and catbirds. It has escaped into natural areas where it occurs as individuals through most of our region but becomes increasingly a threat to biodiversity at more southerly and easterly locales. Shrubs in natural habitats retain their limby growth form and are used substantially by songbirds as nest sites. Their fruit production can be plentiful in open areas, but they are not heavily used in the fall, an indication of low preference; they are ultimately taken in winter by both songbirds and game birds, such as ring-necked pheasants and wild turkey. The foliage is not preferred browse for herbivores and is little used.

Landscaping Value: Hardy to Zone 4, this common ornamental is very hardy and tolerant of just about any site, from wet to dry, from sunny to shady. Unfortunately, numerous *Ligustrum* species are readily available through midwestern nurseries.

Similar Species Distinctions:
—Numerous privet species and cultivars are found (many escaped) throughout the eastern United States. Those planted in the South, such as *L. japonicum* and *L. sinense*, tend to be evergreen.
—**European privet** (*L. vulgare*) is a commonly planted shrub in the northeastern United States. Its mature twigs are smooth and leaves are narrower.
—**Amur privet** (*L. amurense*) has similar leaves, but new twig color is purple, compared to the green twigs of *L. obtusifolium*. Occasionally planted in the Midwest.

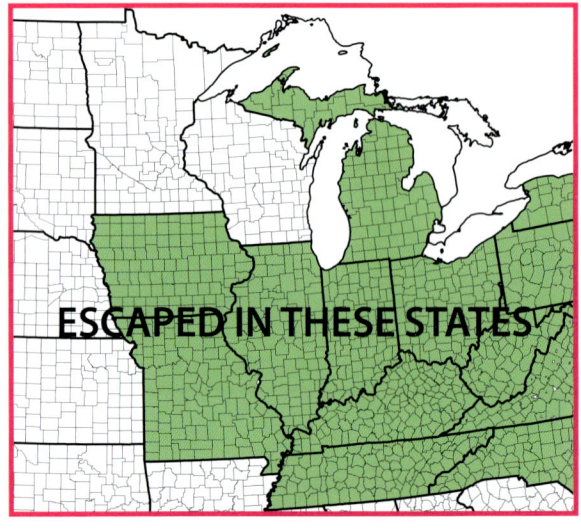

ESCAPED IN THESE STATES

Leaves are simple, about 2 inches long, and elliptic to oblong; the petiole is short. The upper surface is green and smooth; the lower surface is paler and somewhat hairy. Leaf margins are entire, and the tip is almost rounded (not a typical trait on a privet leaf—usually pointed). As one moves south, border privet becomes semi-evergreen.

Flowers appear in terminal clusters in spring after the leaves are fully developed. Each flower is about ⅓ inch long, tubular, and whitish with 4 spreading petals. They have a rather unpleasant odor.

Fruit ripens in the fall in terminal, drooping clusters. Each drupe is about ¼ inch long, blue-black, and can have a bit of a whitish cast. They often persist into the winter.

Twigs are slender, light gray, and hairy with a greenish undertone when young. Buds are small and plump, covered with 6 loose-fitting, tannish-colored scales. Leaf scars are extremely raised, are crescent-moon shaped, and contain a single bundle scar.

Bark is thin, gray, and unpatterned. Some slight fissuring occurs with age that exposes tan-colored inner bark, and some raised or horizontally stretched lenticels are visible.

Bush Honeysuckles

Genus: *Lonicera*

Honeysuckles are commonly planted ornamentals for 1 main reason—their showy, colorful, fragrant flowers. There are many introduced *Lonicera* species in the eastern United States, most of which were brought into the country for landscaping purposes. Several, including Amur, were touted as potentially important to wildlife in terms of food and cover, and they were sold by the USDA Soil and Water Conservation offices and state nurseries for decades. This certainly helped to proliferate the species. Exotic honeysuckles are everywhere, and a few species are so abundant as to be destructive to native plants and their habitats. One disturbing fact from *Plants of the Chicago Region* (Swink and Wilhelm 1994) is that of the 17 species of *Lonicera* documented in that region, 13 are introduced and have escaped cultivation. Another concern is that several other species have escaped cultivation from the large horticultural trade of the East Coast, and have become problematic there— those are presumably heading our way. This is all very disturbing and discouraging to those who are concerned about the welfare of our natural environment.

It is difficult to speak of wildlife benefits of any of these species/hybrids without first disabusing the reader of the notion that we are recommending them in any way. Any benefits accruing to indi- vidual wildlife species in the form of food or cover is miniscule compared to the ecological devastation wrought by these exotic invaders of native habitats. All 3 discussed here were introduced for their hor- ticultural value and quickly escaped; all are without natural controls and have the ability (and tendency) to form substantial thickets under forest canopies, often in dense shade (especially *L. maackii* in the central part of our region). The intense shade pro- duced by this sub-canopy almost totally eliminates understory woody and herbaceous vegetation, and the wildlife diversity dependent on these plant communities. Regardless of the species/hybrids of *Lonicera* involved, the results are more or less the same.

All exotic shrub honeysuckles produce fruits that are small and taken by songbirds. They are not highly preferred, some lasting well into the winter, when they are heavily used then by many species, including wintering purple finches, fox and white- throated sparrows, and hermit thrushes. These plants are little-browsed by herbivores. When these species occur in open areas or in woodland under- stories, they are used by nesting songbirds and as cover for larger mammals (e.g., white-tailed deer), occasionally because no alternative cover exists. For lesser ground dwelling (and nesting) species, however, very little cover exists beneath the shrub matrix, and, thus, wildlife use is essentially nil.

Bush honeysuckles can be very large. This is Amur honeysuckle along a moist, well-drained, wooded stream- side that is 12 feet in height.

Honeysuckles have opposite branching; fruits are paired and almost always red. Shown is Morrow's honeysuckle.

Flowers of honeysuckles are irregular and tubular, with sizable, flared, lobed lips. Most exotic honeysuckles have fragrant flowers—our natives do not. Shown are flowers of Morrow's honeysuckle.

Pictured is Amur honeysuckle on November 15 in west-central Indiana. It has taken over the understory of a degraded bottomland woods.

Amur honeysuckle

Lonicera maackii (Rupr.) Herder
Family: Caprifoliaceae

It seems like every locale has its own infamous bush honeysuckle, and in west-central Indiana where the authors reside, that species is Amur honeysuckle. This is 1 of the easiest honeysuckles to identify because of its elongated leaf tip. It is particularly good at "stealing" light from native species, because it leafs out before almost everything else in the spring, and it retains its green leaves through mid-November.

Form and Size: This is a large bush honeysuckle, easily reaching 15 to 20 feet in height with a large, spreading, much-branched crown. It is multi-stemmed from the base and develops an almost small tree-sized trunk diameter.

Habitat: This species is closely linked to forest cover of some kind, especially open woods, light gaps in otherwise closed-canopy woods, forest edges, and urban parks; however, it also colonizes much disturbed areas such as abandoned fields, utility rights-of-way, and roadsides. It grows quickly in full sun but is amazingly tolerant of full shade. Any soil type, moisture regime, and pH will do for the most part.

Wildlife Uses: Values for all exotic *Lonicera* shrubby species are similar; the reader is referred to the *Lonicera* genus page 366 for details.

Landscaping Value: Hardy to Zone 3, this aggressive honeysuckle is 1 of the worst. It is very large, produces fruit prolifically, and it creates dense thickets that eliminate nearly any native plant with its heavy shade. Currently, it has been planted and/or spread to every state east of the Mississippi, except Alabama and Florida.

Similar Species Distinctions:
—This is the only bush honeysuckle with elongate leaf tips. Check for opposite branching and excavated, brown pith to make sure it is an exotic honeysuckle.

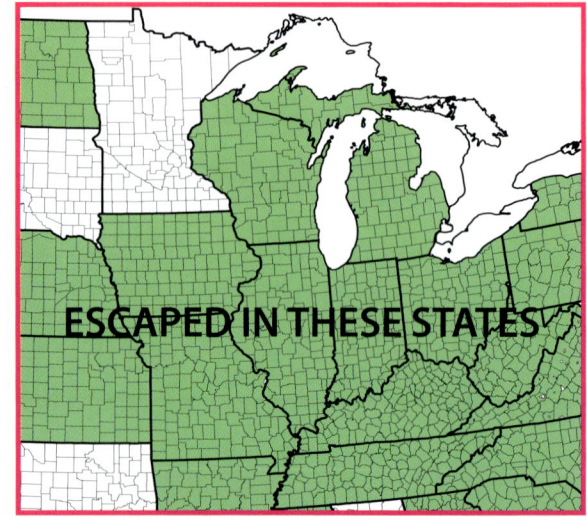

ESCAPED IN THESE STATES

Leaves are simple, ovate to elliptic, and 2 to 3 inches long (can be longer). They are obviously long-tipped with entire, wavy, hairy margins. The upper surface is green and slightly hairy; the lower surface is paler and lightly hairy. Petioles are short and hairy.

Fruit ripens in the fall in paired, axillary clusters that are short-stalked. Each berry is about ¼ inch in diameter, bright red (can be yellowish orange), shiny, and fleshy. Seeds will germinate without cold stratification.

Flowering begins in May after leaves are fully developed, and it continues sporadically—often until frost. Flowers are 1 inch long, tubular, white, aging to yellow. They grow in pairs along the branches from very short stalks (peduncles).

Twigs are pale gray and hairy. Buds are covered with at least 12 hairy scales the same color as the twig. Lateral buds are at nearly right angles from the twig. Pith is brown and excavated through its center. Mature bark (right) is tannish brown or gray-brown with thick, vertical, weaving, somewhat fibrous ridges. Trunk diameter can be substantial, as large as 10 to 12 inches.

Morrow's honeysuckle

Lonicera morrowii A. Gray

Family: Caprifoliaceae

Yet another exotic honeysuckle is Morrow's, which was brought into the United States in the late 1800s for use as a landscape plant. It has naturalized and spread quickly throughout the Northeast into the upper Midwest, invading disturbed and pristine habitats alike. This species is also problematic along the Blue Ridge Mountains of Tennessee and Virginia.

Form and Size: Morrow's is a smaller honeysuckle, reaching heights of 8 feet, and spreading to widths of 10 feet. It has a broad, dense form with many tangled branches, some at ground level.

Habitat: Like our other invasive bush honeysuckles, it has spread to disturbed sites such as grassy areas, open woods, thickets, urban parks, and roadsides. It prefers full sun but tolerates much shade. It seems adaptable to just about any soil type and moisture regime, although it seems not to do well in poorly-drained sites.

Wildlife Uses: Values for all exotic *Lonicera* shrubby species are similar; the reader is referred to the *Lonicera* genus page 366 for details.

Landscaping Value: Hardy from Zones 3–6, this is a shrub of cooler, less humid environments. It has hybridized with another exotic (*L. tatarica*) to produce *L.* x *bella* Zabel, which seems even more aggressive at colonizing our landscapes. It has characteristics that are intermediate between the 2, but it seems to always have some hairs on the lower leaf surface.

Similar Species Distinctions:
—As stated with tartarian honeysuckle, this discussion could be very lengthy. Readers should learn which species are in their region and study their characteristics.

ESCAPED IN THESE STATES

Leaves are simple, ovate to oblong or elliptic, and up to 2½ inches long. They are green to bluish green and have numerous, tiny veins that are slightly impressed into the leaf. The leaf tip is short and pointed, and the margins are entire and hairy. The upper surface is green and hairy; the lower surface is somewhat paler and hairy. Petioles are short and hairy.

Mature bark (not pictured) is brownish or gray with vertical, often weaving, somewhat fibrous ridges. Trunk diameter can be large—6 inches or so. Bark of all these introduced honeysuckles is very similar. Buds and twigs are identical to those of *L. tatarica,* including excavated brown pith.

Fruit ripens in early to late summer in pairs on hairy stalks (peduncles) that are up to ¾ inch long. Each berry is ¼ inch in diameter, yellowish to red, and dull-surfaced.

Flowers appear in the spring after leaves are fully developed. They are white, becoming pinkish with age, ¾ inch in length, and paired on peduncles less than 1 inch long. The bractlets on either side of the ovary are half to the same length as the ovary.

tartarian honeysuckle, tatarian honeysuckle

Lonicera tatarica L.

Family: Caprifoliaceae

This introduced honeysuckle is another escapee from cultivation that has invaded many habitats across the upper Central Plains, Midwest, and northeastern United States. It has been in the United States since the late 1700s and spread from east to west. This species has several distinct characteristics that aid in identification. Its leaves appear bluish green, and it is essentially hairless.

Form and Size: This is a large shrub that can grow to 13 feet or more in height. It has long, arching branches that create a shrub as wide as it is tall. There are numerous, fine branches that give it a full appearance.

Habitat: Tartarian honeysuckle invades many habitats, but especially those already disturbed by humans, such as roadsides, grassy openings, and woodlands, including city parks. It grows best in full sun but tolerates a great deal of shade. There is hardly a site or soil type on which it cannot grow.

Wildlife Uses: Values for all exotic *Lonicera* shrubs are similar; the reader is referred to the *Lonicera* genus page 366 for details.

Landscaping Value: Hardy to Zone 3, this species is spread by birds and continues its movement, especially northward, where it appears to thrive in a cooler environment. It hybridizes with another introduced species (*L. morrowii*) to create *L.* x *bella* Zabel, which has spread rapidly through the northern regions of the Midwest. Fruit is produced prolifically on an annual basis, providing a nonstop seed source year after year.

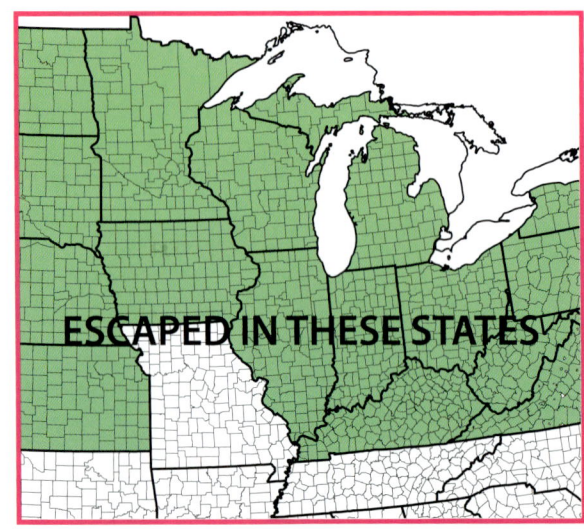

ESCAPED IN THESE STATES

Similar Species Distinctions:
—There could probably be a several page discussion of similar species. Since there are numerous introduced, invasive honeysuckles, it is best for one to know first which species are in your area, and then go to work learning the individual characteristics of each.

Leaves are simple, up to 2½ inches long, and oval to ob-long with entire leaf margins. The leaf tip is short-pointed (sometimes rounded), and the petiole is short and smooth. The upper surface is green and smooth; the lower surface is paler and smooth. The leaf margins may be slightly hairy.

Fruit ripens in pairs from mid- to late sum-mer on stalks (peduncles) up to 1 inch long. Each berry is ¼ inch in diameter, orange or red, and shiny (or translucent).

Flowers appear in May and June in axillary pairs. Each flower is ¾ inch long or less, white, pink or purplish, and tubular, with petals longer than the tube. Ovary has 2 leaf-like scales (bractlets) that are much shorter than it is.

Twigs are slender, light grayish brown, and lightly hairy. Plant down on picture is not typical. Buds are rounded and covered with at least 12 tannish scales. Pith is brown and hollow (escavated).

There seems to be nothing unique about the bark of any of the intro-duced bush honeysuckles. Mature bark is thick, somewhat fibrous, and tannish brown or gray.

European buckthorn, common buckthorn

Rhamnus cathartica L.

Family: Rhamnaceae

Common buckthorn is another weedy species that is classified as a noxious weed in numerous states, including Iowa and Minnesota; it is banned from sale or transport in the latter. Although this species does not spread by rhizomes, it is an extremely prolific fruit producer. Its numerous fruits are not taken by wildlife to any extent, so they drop to the ground and readily germinate. Numbers as high as a half million seedlings per acre have been documented.

Form and Size: This species is considered a large shrub to small tree, often reaching a height of 25 feet. If shrubby, it is few-stemmed with a somewhat open form with few branches.

Habitat: Common buckthorn has escaped and naturalized in well-drained sites, particularly in open woods with a history of disturbance, or even mature oak/hickory woods where it can form thickets too dense to walk through. It tolerates a great deal of shade but thrives in full sun.

Wildlife Uses: This species is highly invasive in the northern parts of our region and locally problematic elsewhere. It is difficult to recite wildlife "values" for a species that is so devastating to natural communities and the wildlife dependent on those communities. Individual common buckthorns make fairly good nest sites for songbirds, and unfortunately, the fruits are taken to a degree (not preferred) by songbirds, especially robins and other thrushes, and small mammals. As this species dominates understories, however, it shades out whole native plant communities and destroys the wildlife communities dependent on those habitats; this dominance is assisted by an almost complete avoidance of the species by browsers, such as white-tailed deer, cottontail, and beaver.

Landscaping Value: Hardy to Zone 3, this aggressive exotic has no landscaping value whatsoever, except that it is able to tolerate some of the worst sites imaginable. There are no cultivars listed.

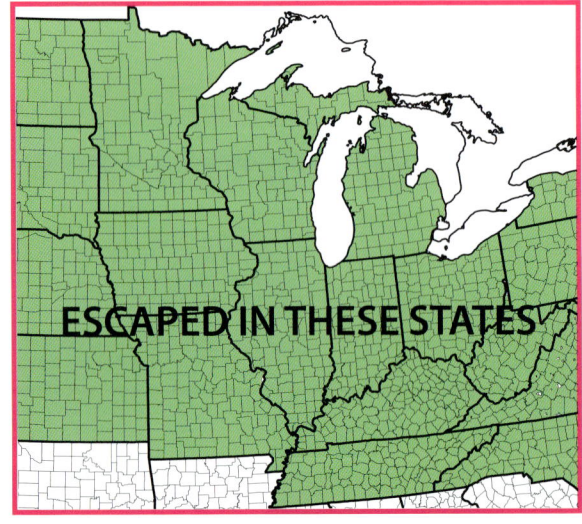

ESCAPED IN THESE STATES

Similar Species Distinctions:
—With a quick look at the twigs, one can rule out all other shrubs by looking for a thorny twig tip and opposite to sub-opposite branching.

Leaves are simple, mostly opposite to sub-opposite, and about 2 inches long. They are ovate to elliptic with 3 to 4 pairs of veins. Margins are finely toothed. The upper surface is dark green with some hairs; the lower surface is paler and hairy. The hairy petiole is over an inch long.

Fruit (below, far left) ripens in the fall. Each fruit (a drupe) is black, glossy, and about ¼ inch in diameter. Fruits remain on the twigs through winter, and seeds remain viable in soil for 2 to 3 years.

Flowers appear in May and June after leaves are fully developed. Each flower is ¼ inch across, yellowish green, and unisex on separate plants. The male flowers (below left) appear to "have all the parts," but they function only as males. Females are below right.

Mature bark (below, right) is much like that of large shrub/small tree-sized *Prunus* species such as American plum (*P. americana*) or Canada plum (*P. nigra*) with brownish, somewhat flaky bark and horizontal lenticels. It also resembles young river birch (*Betula nigra*) bark.

Twigs appear silvery because of an epidermis layer that covers a reddish brown bark beneath. Buds are ovoid and covered with 5 to 6 dark brown scales that are rimmed in a silvery coating. Twigs are usually tipped with a dull, pointed thorn. Lateral buds lay against the twig. Leaf scars are raised, shaped like a stretched triangle, and contain 3 bundle scars. Older twigs (right) develop short shoots with closely-spaced leaf scars.

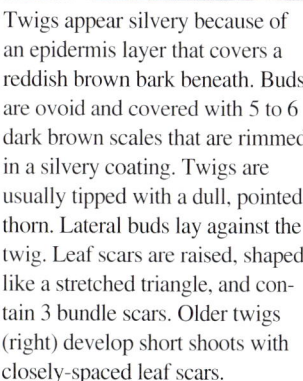

Japanese barberry

Berberis thunbergii **DC.**

Family: Berberidaceae

For a century, several species of barberry have been the targets of a federal eradication program. The native *B. canadensis* and the introduced *B. vulgaris* have been systematically eliminated from the Midwest with amazing efficiency, because they are alternate hosts to black stem rust that kills small grain crops, such as wheat. Japanese barberry, however, is not an alternate host, and it has become 1 of those overly used ornamentals that is escaping from cultivation. It has the ability to alter soil characteristics such that it favors its growth over native plants.

Form and Size: This is a moderately sized shrub that can reach 6 feet in height and just as much in width. It has fine, numerous twigs and is multi-stemmed, giving the shrub a dense appearance. This species produces underground rhizomes, and it is able to spread in that manner.

Habitat: Japanese barberry has escaped cultivation mainly through birds transporting seeds in their feces. It is found in open woods and trails, pastures, upland forests, bottomlands, and old homesites. It is quite adaptable to most sites from wet to dry, shady to sunny. In some areas of the Northeast, it grows in dense, impenetrable thickets in open woods.

Wildlife Uses: Although this exotic may grow to 6 feet, it is typically much smaller when it escapes to native habitats, but retains its dense growth form that supplies good nesting cover for songbirds, such as field and chipping sparrows, catbirds, and veeries. It seems little-browsed by deer and other herbivores, and, in fact, it has been reported as the only remaining woody in areas devastated by heavy deer populations. Fruits are occasionally heavily used by wild turkey and ruffed grouse, but songbirds and small mammals use them only sparingly. They can be classed as not preferred, since they regularly remain on the shrub through the winter.

Landscaping Value: Japanese barberry is hardy to Zone 4, and it is commonly used for borders and as hedges. In the Midwest, it is being found as an escape in woodlands, but not to the degree that it has in the Northeast. There are a shocking number of cultivars that are readily available in nearly every nursery, greenhouse, and garden center in the Midwest. This species and its cultivars are the bread and butter for many nurseries, says Michael A. Dirr and Charles W. Heuser, Jr., in *The Reference Manual of Woody Plant Propagation* (2006).

Similar Species Distinctions:

—**Common barberry** (*B. vulgaris*), if encountered, has gray stems with mostly triple, drooping (but not curved) prickles and toothy leaf margins similar to the native *B. canadensis*.

—**Allegheny barberry** (*B. canadensis*) has prickles (curved) and leaves similar to *B. vulgaris*, but it has reddish brown stems.

ESCAPED IN THESE STATES

Leaves are simple and up to 1 inch long, somewhat oval or spoon-shaped with elongated petioles. They are clustered at the nodes, especially on older twigs. Leaf margins are entire, unlike the native and other exotic barberries. Upper surface is green (or red in some cultivars) and smooth; lower surface is somewhat paler and smooth.

Flowers appear in the spring after leaves are fully developed. Each perfect flower is yellow, about ½ inch across, and in small clusters (usually more than 1) hanging on the underside of the branches.

Fruit ripens in the fall in few-fruited (or single), dangling clusters. Each berry is about ⅓ inch long, egg-shaped, bright red, and dry. They persist through the winter on the shrub, and apparently they have low viability after the first year.

Twigs are ridged and reddish brown, with short, straight, stiff nodal spines that are mostly single (may be triple on oldest stems). Nodes of older plants have clustered buds with numerous, reddish brown scales. The older the twig, the more buds are clustered at the nodes. Mature stems develop a grayish coating that splits vertically to reveal the reddish brown "real" stem color. Inner bark is bright yellow.

autumn olive

Elaeagnus umbellata Thunb.

Family: Elaeagnaceae

This exotic is listed as a serious threat to native communities in many eastern states, and more recently, some in the Midwest. The species was first studied in the 1940s by the Soil Conservation Service, which released the cultivar 'Cardinal' for commercial use. It was touted as a shrub with high potential wildlife food and cover value, and it was sold throughout the eastern United States by state nurseries.

Form and Size: This is a large shrub, reaching 20 feet in height, with a width of equal size. It is multi-stemmed with upright upper branches and spreading, lower limbs. There are usually numerous plants in an area, and they form dense stands that shade out native plants.

Habitat: Autumn olive is found in disturbed habitats or open areas, such as old-fields, open woods, and forest edges; it seems to do especially well in those with soil of low fertility. It thrives in full sun, is unable to tolerate dense shade, and will die out if overtopped by larger woody plants.

Wildlife Uses: Autumn olive has admirably fulfilled the role expected of it; unfortunately, it has proven to be invasive and damaging to biodiversity in native systems, especially early successional habitats. Its very dense growth form is excellent for nesting songbirds, including yellow-billed cuckoos, cedar waxwings, and mourning doves, and patches provide very good protective cover. The fruits are eaten by many songbirds, game birds, and mammals, and they can be considered highly preferred. Leaves are not preferred forage for white-tailed deer, but cottontails consume bark from young stems in the winter.

Landscaping Value: Hardy to Zone 4, this shrub has the ability to increase soil fertility by fixing atmospheric nitrogen, and it possibly displaces natives that are adapted to nutrient-poor soil. It has been planted in mine reclamation sites and as windbreaks where other species cannot survive (mostly western states). Several cultivars exist, including 'Ellagood,' which has yellowish fruits.

Similar Species Distinctions:
—**Russian olive** (*E. angustifolia*) is a small tree and has narrow, willow-like, silvery leaves and silvery fruits.
—**Silverberry** (*E. commutata*), found only in northwestern Minnesota, closely resembles autumn olive but is much shorter (to 6 feet). Its leaves are very similar, but they are more silvery overall, especially more so on the upper surface. Fruits are silvery when ripe.
—**Buffaloberry** (*Shepherdia canadensis*) has similar leaves that are opposite.

ESCAPED IN THESE STATES

Leaves are simple, elliptic to ovate-oblong, and 2 to 3 inches long. Margins are entire and wavy. The upper surface is green with few, silvery speckles along the midrib. The lower surface is densely silvery white from numerous small scales. Petiole is short and silvery.

Flowers appear in May after leaves have fully developed in axillary clusters on the new growth. Each flower is yellowish, funnel-shaped, and very strongly-scented. Flower parts are in 4's.

Fruit ripens in the fall in clusters. Each fruit, a drupe-like achene, is fleshy, orange-red to red when ripe, and dotted with silvery speckles.

Twigs are light-colored or dark cinnamon-brown and covered with rounded, flat, rust- or silvery-colored scales. Buds have 2 fleshy scales the same color as the twig. The lateral buds lay close to the twig; the terminal bud is usually crooked to 1 side. Older stems develop spur twigs that end in thorns. Pith is brown and solid. Mature bark is grayish and flat, with vertical cracks, giving it the appearance of platy bark.

glossy buckthorn

Frangula alnus **P. Mill.**
Rhamnus frangula **L.**
Family: Rhamnaceae

This buckthorn is commonly
referred to as a noxious, prohib-
ited, or restricted weed in many
midwestern and northeastern
states. It is particularly invasive
in wetlands, and once estab-

lished, it is nearly impossible to eradicate. The species grows so densely that is keeps almost all natives
out. Seen here is a wetland in northern Michigan in just such a degraded state.

Form and Size: This large shrub or small tree can reach over 25 feet in height, but it is usually a multi-
stemmed bush. It has few stems per plant, giving it an open form. It does not spread by rhizomes, but it
usually produces a plethora of fruits that drop to the ground and germinate.

Habitat: Glossy buckthorn is partial to wet sites and tolerates temporary root inundation. It is found in
and along marshes, wet meadows, swamps, and fens, mainly in the northern regions of the Midwest. It
prefers full sun but tolerates partial shade.

Wildlife Uses: Glossy buckthorn presents the same problems and has the same marginal benefits for
wildlife that are discussed for *Rhamnus cathartica*. While common buckthorn typically invades uplands,
glossy tends to invade wetland areas, although it will spread to uplands as well.

Landscaping Value: Hardy to Zone 3, this weedy species' cultivar, 'Columnaris,' has been widely
planted for hedges in the upper Midwest. Another, 'Asplenifolia,' has narrow, fern-like leaves and is seen
occasionally in nurseries. This aggressive plant can overtake miles of wetlands and create shade so dense
that sedges cannot grow. Even if planted in a controlled urban area, seeds can be carried by birds.

Similar Species Distinctions:
—**Common buckthorn** (*Rhamnus
cathartica*) really does not have many
similar characteristics other than being a
member of the same family. It has mostly opposite or sub-opposite branching,
leaves with fine teeth along the margins,
and thorns.
—**Carolina buckthorn** (*Rhamnus caroliniana*) is a small tree of the southern
regions of the Midwest (Weeks, Weeks,
and Parker 2010). Its leaves, buds, and
flowers are similar. Its habitat is usually
dry ridges.

ESCAPED IN THESE STATES

Leaves are mostly alternate, ovate-elliptic, and usually 1 to 3 inches long. The upper surface is shiny; the lower surface is dull and smooth (or hairy along the main veins). Leaf margins are entire, and the impressed vein pairs number 6 to 9. Petioles are up to ¾ inch long.

Fruits ripen in the late summer and fall in small axillary clusters. Each black fruit (a drupe) is about ⅓ inch in diameter on a stalk (peduncle) nearly as long. Fruits remain on the shrub throughout the winter, and seeds remain viable in soil for 2 to 3 years.

Twigs are brown and lightly covered with tannish brown hairs, especially near the tip. Terminal buds are naked with 2 fleshy "scales" that are densely hairy. Lateral buds are similar but very small. Leaf scars are somewhat 3-sided and contain 3 bundle scars.

Mature bark is dark gray with numerous, nearly white lenticels, which are of different shapes, including diamonds, circles, horizontal lines, and just plain blobs.

Flowering begins in May and continues sporadically until frost. Each flower is only ¼ inch across, perfect, and yellowish white; it has parts of 5. Flowers are produced in axillary clusters of the new growth.

multiflora rose

Rosa multiflora **Thunb.**

Family: Rosaceae

This is an Asian species that was promoted from 1930 to the 1970s for various beneficial traits, such as controlling soil erosion, providing wildlife food and cover, and creating living fences. It performs these jobs admirably, but it was planted *too* much, and birds have spread seeds from its small fruits just about everywhere. It has become naturalized throughout most of the Midwest, and it is listed as a noxious or nuisance weed in Iowa, Kentucky, Missouri, Pennsylvania, and Wisconsin.

Form and Size: Multiflora has long, arching stems up to 13 feet long that originate from a single root stock, forming broad, loose, spreading shrubs. They continue to sucker from that root stock and form dense, impenetrable thickets.

Habitat: This species is found in many habitats from overgrazed pastures and old-fields to roadsides, open woods, forest edges, and degraded woodlots. The species quickly dominates any habitat where disturbance is not frequent. It prefers full sun but tolerates a great deal of shade. Just about all but the most poorly-drained sites will grow multiflora rose.

Wildlife Uses: This exotic, introduced in the middle part of the last century principally for its wildlife values, has escaped to become 1 of our most invasive woody species; however, it is a superb wildlife species, and its praises were sung for several decades, until its more sinister nature began to express itself. Multiflora's dense growth from a single rootstock and arching branches present an ideal nest site for many species, especially brown thrashers, catbirds, and cardinals. Old nests in shrubs are often secondarily used for nesting by mourning doves or taken over by white-footed mice for nesting or as a feeding platform. The hips are small, smooth, and very attractive to resident and migrant songbirds, which quickly spread the seeds in their droppings.

Landscaping Value: Hardy to Zone 5, this species is naturalized throughout most of the Northeast and Midwest. Fruits are produced on a fairly regular basis each fall, so without removal of the seed source, this species will continue to perpetuate itself. Thorns are so nasty on large stems that they tear away from the plant if contact is made with fabric such as blue jeans. The wearer is left with a painful thorn stuck through the fabric and into skin.

Similar Species Distinctions:
—**Climbing prairie rose** (*R. setigera*) has 3 leaflets on flowering/fruiting stems, and leaf petioles are mostly entire with glands along the margins. Flowers are pink.

ESCAPED IN THESE STATES

The compound leaves have 5 to 11 oval leaflets with finely toothed margins. Leaves are 3 to 4 inches long, dark green and smooth above, paler and softly hairy beneath. Rachises are prickly. Stipules are divided, or feathery, and glandular hairy.

Fruits ripen in the fall in loose, terminal clusters. Individual fruits, known as hips, are smaller than most natives at about ¼ inch in diameter; they are red, hard, and dry. They sometimes persist into the winter. Seeds remain viable in the soil for up to 20 years.

Flowers appear in May after the leaves. They are white (natives are almost always pink) about 1 inch across, and in upright clusters (panicles). Each flower has 5 petals and numerous stamens.

Stems are usually green (often deep red), smooth, and round in cross section. Paired thorns that are the same color as the stems occur at the nodes on young twigs. Terminal buds are ovoid and covered with 3 to 4 bud scales the same color as the stem. Older stems have lateral buds almost completely sunken into the twig with many tan-colored, thick, curved prickles at and between the nodes.

Large stems are rarely more than 1 inch in diameter. The largest ones turn grayish and develop vertical cracks.

wintercreeper

Euonymus fortunei (Turcz.) Hand.-May
Family: Celastraceae

This common plant is technically classified as a liana-shrub, meaning it runs, climbs, and hangs like a vine, but is also shrubby. It has been popular as a ground cover over the last 50 years in the horticulture trade, similar in allure to its shrubby cousin, *E. alatus*. There are at least 50 cultivars according to Michael A. Dirr (1998), and it has been planted across the eastern United States since 1907. Even though it is a menace to our natural environment, it is available in practically every nursery, greenhouse, and garden shop in the Midwest.

Form and Size: This depends to some extent on the cultivar, but it can climb to 70 feet into treetops. It also creeps along the ground, forming dense mats of numerous stems, smothering any plants beneath. It easily roots from stem fragments and grows quickly. Depending on the variety, some are evergreen, others are not.

Habitat: It invades woodlands, disturbed woodlots, and urban parks, and it is not particular about soil type, soil moisture, or light regimes. Whether it produces seeds or not, storms bring stem fragments to the ground, which easily root and help the plant spread even faster.

Wildlife Uses: Wintercreeper little resembles its congeners in form or wildlife values. Where it colonizes native habitats, it creeps along the ground, forming a rather dense matrix that supplies some protective cover (at the expense of the native forest floor species); when it encounters a tree, it often becomes a vigorous climber reaching many feet into the canopy. Older vines can be substantial in size and provide nesting cover typical for tree/vine complexes (see *Vitis*); however, it competes seriously for light with its supporting tree. Leaves and stems have been reported to be heavily browsed by white-tailed deer in the East, but we have observed little use in our region; its fruits are produced only when the plant receives full sun, and as is typical for *Euonymus*, they are not highly preferred, but eaten frequently enough to facilitate its spread.

Landscaping Value: Hardy to Zone 5, wintercreeper is usually used as an evergreen ground cover. Its ability to gradually invade woodlands is rather shocking. Additionally, the sheer size of some arborescent plants is rather impressive. It has been overly planted as an ornamental, and it continues to be offered in nearly all nursery and garden centers in the Midwest.

> **Similar Species Distinctions:**
> —Other *Euonymus* species have similar leaves, but none are vines. **Running wahoo** (*E. obovatus*), a somewhat trailing shrub, has deciduous leaves that are usually broadest above the middle.

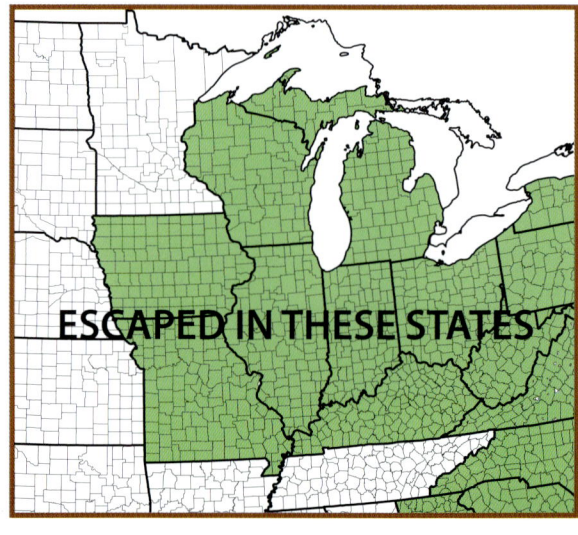

ESCAPED IN THESE STATES

Leaves are semi-evergreen to evergreen, simple, and leathery. They are about 1 inch long with finely toothed margins; leaf shape is ovate-elliptic. Veins are yellowish or silvery. The upper leaf surface is dark green, shiny, and smooth; the lower surface is pale and smooth.

Fruit ripens in the fall in dangling clusters. Each fruit is a 4-lobed capsule of various colors ranging from white to pink to reddish. The capsule splits to reveal a bright orange, fleshy aril in each lobe.

Flowers appear in the summer after the leaves are fully developed. The perfect, greenish flowers have parts of 4.

Twigs are green, smooth, and dotted with numerous, small warts (verrulose texture). Lateral buds are ovate-shaped and covered with numerous, extremely pale green scales. Each scale is rimmed with a brown, frayed margin. Aerial rootlets form at the nodes of vertical stems.

Mature bark is thin, dark brown, and scaly near the base. Over time, outer bark sloughs off, revealing the reddish brown inner bark. Large vines are 4 inches or more in diameter.

periwinkle

Vinca minor **L.**

Family: Apocynaceae

This European vine has been cultivated as an evergreen ground cover since ancient times according to Michael A. Dirr (1998). Closely related to the milkweeds, it exudes sticky, milky sap when broken.

Form and Size: Periwinkle is a short, creeping, slender vine used as an ornamental ground cover. It is up to 8 inches high and forms dense mats with numerous stems. Stems that touch the ground easily root at the nodes.

Habitat: It is usually found near urban areas or at old homesites where it has escaped cultivation at some point. It is common in wooded city parks, degraded woods, and other shady places. Although it grows well in full shade, it handles nearly full sun. Once established, it is long-lived.

Wildlife Uses: A widely available and often planted ground cover, periwinkle can escape and grow under some shade. Whether growing in a suburban yard or native woods, it provides little wildlife cover value, in spite its being evergreen, except to small rodents and shrews and perhaps ground nesting birds like song sparrows. It is regularly browsed in the fall and winter by white-tailed deer and cottontails.

Landscaping Value: Hardy to Zone 4, this is a common ground cover that has been used as an ornamental in urban landscaping and also as a source of erosion control on moderately steep slopes. There are many horticultural varieties that have different flower colors, double flowers, and variegated leaves, among other attributes.

Similar Species Distinctions:
—**Star chickweed** (*Stellaria pubera*) is a woodland herbaceous plant that has opposite, similar leaves that are mostly evergreen. Although it has an upright growth form during summer months, by fall its stems are laying on the ground. It is not woody.

ESCAPED IN THESE STATES

Flowers appear in early spring and continue to be produced sporadically over several months. Each is 1 inch across, lavender-colored, and funnel-shaped with 5 spreading petals that resemble fan blades. The flower's center has a 2-toned, star-like pattern. Fruit is a brown pod (follicle) to 3 inches long; it is rarely produced in cultivation.

Stems are slender, green, shiny, and smooth. Near the base of each stem is a pair of stipules.

Leaves are simple, elliptic-ovate to oblong, and up to 2 inches long. Petioles are extremely short, particularly on the upper stem leaves, and sometimes they appear to be absent. Margins are entire, and the central leaf vein is lighter green than the leaf. The upper surface is green, shiny, and smooth; the lower surface is paler and smooth.

English ivy

Hedera helix **L.**

Family: Araliaceae

English ivy has been cultivated in North America since at least 1800, and there supposedly are 400 cultivars from which to choose. This creeping vine is popular because of its evergreen nature and carefree maintenance in a shady landscape. A problem begins when it oversteps its bounds and begins smothering native plants. The vines can also damage brickwork and building siding if not removed.

Form and Size: There is hardly a limit to the height this vine can grow into treetops and up sides of buildings. The juvenile growth phase is the 1 most encountered creeping along the ground with its 3- to 5-lobed leaves. Once it establishes itself on an upright substrate, it matures into the adult phase. Its leaves change shape, and it is these arborescent, limby vines that flower and fruit.

Habitat: This vine has spread from cultivated plants into a wide range of habitats across the country. It invades urban parks and any habitat adjacent to them. This is an adaptable plant that tolerates a wide range of moisture and light regimes.

Wildlife Uses: English ivy occasionally escapes in our region but rarely grows as luxuriantly as it does in the Southeast. It has very low wildlife value; it provides some evergreen cover as it creeps along forest floors, and when it gets sufficient light and climbs into trees, it adheres closely to the trunk and supplies less cover than most vines, although it is occasionally used for roosting by songbirds. It rarely sets fruits in our region, and when it does, they ripen during the winter and seem to be only occasionally taken by wildlife, although we are sure they are eventually used.

Landscaping Value: Hardy to Zone 4, this vine has the ability to influence succession by blocking our native plant's abilities to regenerate or grow within (or under) the sometimes thick mat of stems. Its often large size on trees generates a great deal of weight that facilitates tree fall, creating light gaps, which provides better conditions for vine growth.

Similar Species Distinctions:
—No vine in the Midwest has similar, evergreen leaves. **Wintercreeper** (*Euonymus fortunei*) has opposite, unlobed, evergreen leaves.

ESCAPED IN THESE STATES

Leaves are thick and up to 4 inches long. Juvenile leaves (left) are 3- to 5-lobed; adult leaves are rather diamond-shaped—both have long petioles. Leaves in general are dark green and shiny with light-colored veins on the upper surfaces. Lower surfaces are pale yellowish green. Both sides are smooth upon maturity.

Flowers appear in late summer in clustered umbels. Each flower is perfect and greenish. A fair amount of sun is required for flower production.

Fruits ripen in late winter and early spring in the Midwest. They are clusters of blue-black berries, each of which appears to be wearing a beanie.

New twigs are green and moderately covered with short, tan-colored hairs. Rootlets develop from the sides of twigs in clusters, and they exude an adhesive substance. Buds are greenish and diverging away from the stem.

Bark is light brown and commonly matted with short rootlets projecting from the vine. Older vines are darker and develop furrowed bark.

Japanese honeysuckle

Lonicera japonica **Thunb.**

Family: Caprifoliaceae

Once planted for its wildlife value and erosion control, Japanese honeysuckle is now listed as a noxious weed in many states across the eastern United States. It climbs across, up, and over anything in its way, similar to kudzu. Fortunately, it is not as large a plant, nor does it grow as quickly. Currently, it is banned from sale in Illinois.

Form and Size: Japanese honeysuckle is a climbing, twining vine that can grow 50 feet or more into tree-tops. Once there, it covers large portions of the tree's canopy, effectively blocking its sunlight. It creeps along the ground until it encounters something to twine around and on which to grow upward. It often nearly carpets the ground with its numerous stems.

Habitat: This vine is not particular about habitat, and it will grow anywhere but in poorly-drained and coarse, sandy sites. It prefers full sun but tolerates some shade. It is found in forest openings, disturbed sites, thickets, and abandoned properties where it has escaped cultivation.

Wildlife Uses: Of all the *Lonicera* species, the foliage of Japanese honeysuckle is the most preferred by white-tailed deer and other herbivores. It is high in protein and even in our region is partially evergreen; it is used into the early part of the winter. Its small, black fruits are readily taken by game birds and songbirds. The vine climbs onto shrubs, fences, and other obstacles, and its dense tangles supply nesting for many songbirds and, in the southern parts of the Midwest, the golden mouse. Its growth presents only minor problems in the northern part of the Midwest, but in the southern reaches, its aggressiveness often leads to its covering and killing supporting shrubs and small trees, somewhat damping its generally positive wildlife values.

Landscaping Value: Hardy to Zone 5, this vine seems to be moving north as winter conditions become less severe in our region. It is aggressive and hard to kill without both mechanical and chemical treatments. Fire does seem to keep it in check. It begins flowering and fruiting as early as 3 years of age. There are several cultivars available, but this vine should not be planted.

Similar Species Distinctions:
—This is the only vining honeysuckle in the Midwest with hairy, brown stems and black fruit.

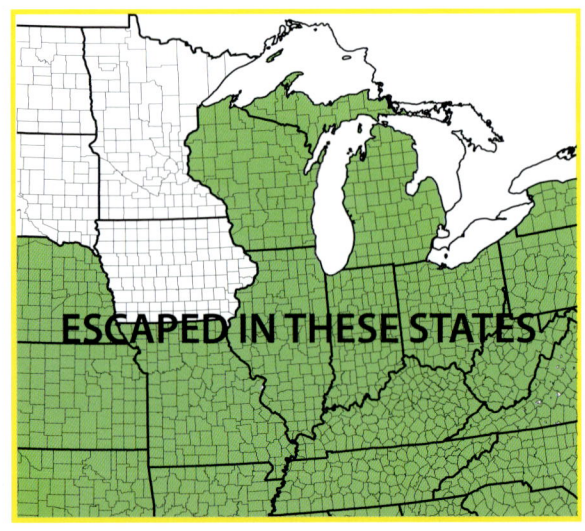

ESCAPED IN THESE STATES

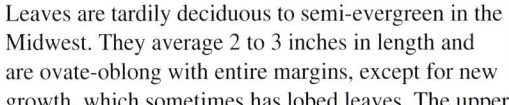

Leaves are tardily deciduous to semi-evergreen in the Midwest. They average 2 to 3 inches in length and are ovate-oblong with entire margins, except for new growth, which sometimes has lobed leaves. The upper surface is dark green with scattered, flattened hairs; the lower surface is slightly paler with hairs. Petioles are short and hairy.

Fruit ripens in the fall in axillary clusters. Each berry is about ¼ inch in diameter, black, rounded, and shiny.

Flowering begins in May and continues sporadically for several months. Paired, tubular, irregular, stalkless flowers about 2 inches long develop in leaf axils. The upper lip has 4 lobes that flair upward, and there are 5 long stamens that turn upward like an elephant's trunk. Flowers are off-white, turning yellowish with age.

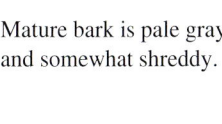

Mature bark is pale gray and somewhat shreddy.

Winter twigs are light brown and hairy, becoming mostly smooth as they age. Second year twigs develop longitudinal splits in their thin, outer bark to reveal reddish brown inner bark. Leaf scars are slender and shallowly V-shaped.

oriental bittersweet, Chinese bittersweet

Celastrus orbiculatus Thunb.

Family: Celastraceae

Native to East Asia, this invasive vine is problematic on 2 major levels. The first is its rapid growth rate—"kudzu of the north" as it is dubbed, because of its ability to suffocate anything in its path very quickly. The second is its ability to hybridize with American bittersweet, *C. scandens*, which threatens our native's genetic integrity and has created more aggressive offspring in the process.

Form and Size: This climbing, twining vine can reach 65 feet into treetops, but it also sprawls and spreads over just about anything sitting still. It wraps tightly around almost anything, and eventually it cuts through tree bark to cambium and kills its support. It freely root suckers and forms dense canopies over other living plants, cutting off sunlight and reducing photosynthesis.

Habitat: Oriental bittersweet is found in a variety of habitats, most of which are disturbed. It is found in open woods, along roadsides, in thickets, on dunes, and so on. It thrives on all soil except poorly-drained. Although it is found in heavy shade, it thrives in full sun.

Wildlife Uses: Oriental bittersweet has essentially the same wildlife values as American bittersweet. It is, however, much more aggressive and tends to run along the ground, rooting at nodes, in addition to climbing. This creates more ground-level protective cover, but at the expense of native flora that it tends to out-compete and smother. Control is difficult to effect because of its dispersed and well-developed root system.

Landscaping Value: Hardy to Zone 4, this bittersweet has been used as "a more attractive" substitute for our native. Many nurseries today accidentally sell this aggressive species, as it is commonly misidentified by many in the horticulture trade.

Similar Species Distinctions:
—**American bittersweet** (*C. scandens*) has more ovate or oval leaves and an elongated leaf tip. Flowers and fruit develop from the terminal end of stems, not leaf axils.

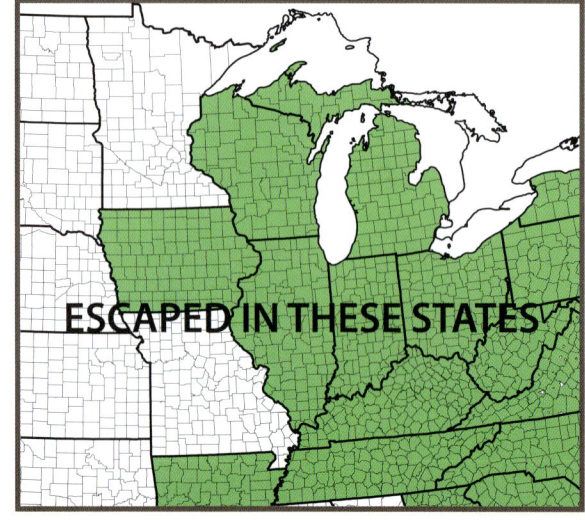

ESCAPED IN THESE STATES

The simple leaves are obovate to orbicular and up to 5 inches long. They are nearly as wide as they are long with finely toothed margins. The upper surface is glossy and smooth, while the lower surface is somewhat paler and smooth.

Flowers appear in summer once leaves are fully developed. They hang in axillary clusters. Individual flowers are greenish yellow with 5 flower parts. Plants, as well as flowers, are single-sex; males pictured at far right, females at near right.

Twigs are mottled brown and gray and dotted with numerous lenticels. Pith is solid and white. Buds are small, rounded (with at least 4 brown scales), and at right angles to the twig. Leaf scar is slightly raised and shaped like a rounded triangle; it contains a single row of bundle scars that are hooked on each side.

Mature bark is thin, grayish, and dotted with nearly black expanded lenticels that appear almost diamond-shaped. The thin bark splits irregularly between and around the diamonds.

Fruit ripens in the fall in dangling, axillary clusters. Each orange fruit has a 3-parted, thin capsule that splits upon maturity to reveal a single, deep orange, fleshy aril.

kudzu

Pueraria montana (Lour.) Merr. var. *lobata*
(Willd.) Maesen & S. Almeida
Family: Fabaceae

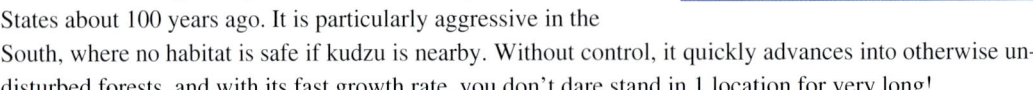

The term "noxious weed" comes up a lot when kudzu is the topic of discussion. Originating in China, it was brought to the United States about 100 years ago. It is particularly aggressive in the South, where no habitat is safe if kudzu is nearby. Without control, it quickly advances into otherwise undisturbed forests, and with its fast growth rate, you don't dare stand in 1 location for very long!

Form and Size: This creeping, climbing vine literally grows over anything that does not move, eventually growing up and over treetops, often using other vines for support. It can climb to 50 feet or more. Vine growth rates have been documented at a foot a day and up to 99 feet in a growing season. They can form a continuous blanket over all vegetation in their path. Creeping ground vines are herbaceous, and they die back to the root stock after frost. Vertical, climbing vines are semi-woody and overwinter to leaf out again the following spring.

Habitat: Kudzu grows best in well-drained, moist soil, but it is very adaptable. It invades open, disturbed areas, clearings, abandoned homesteads, forest edges, and roadsides. Full sun is preferred, but it tolerates shade, although its growth is slowed.

Wildlife Uses: In the southern parts of the Midwest and even more so southward, this is the consummate invasive vine—something out of a Stephen King novel. While temperature limits its northern expansion, it does occupy considerable acres in our region, dominating landscapes and destroying other woody vegetation by covering and smothering. These dense tangles supply nesting and escape cover, especially on edges of concentrations, for everything from songbirds to white-tailed deer fawns. The leaves are a favorite food of deer and woodchucks, the latter frequently choosing these patches for their den sites. The seeds are almost certainly high in protein and likely used by a range of small mammals and game birds; the winter diet of bobwhite in an area in Georgia was over 50 percent kudzu seeds.

Landscaping Value: Hardy to Zone 5, this unbelievably aggressive vine seems to be moving north as warmer winters allow. It has been planted in Illinois and Indiana with success, and control measures have been taken to remove it. Fortunately for midwesterners, the summer growing season does not last as long as in the South.

Similar Species Distinctions:
—Nothing else resembles kudzu from a leaf standpoint. Its sizable, twining winter vines might be confused with wisteria, but they are brown with less obvious brown lenticels.

ESCAPED IN THESE STATES

Leaves are compound and have 3, 3-lobed leaflets each. The leaves are up to 8 inches long and are dark green with white, appressed hairs on the surface and along the margins. The lower leaf surface is pale and hairy. Each leaflet has a single, hairy bract that is curved upward like half of a handlebar mustache. Leaves exhibit what is known as "day-time leaf movement," which allows them to position themselves relative to the sun. In this manner, they can adjust the intensity of the sun's radiation in order to capture more light, or move away from it in order to conserve water.

Fruit is a 2-inch-long brown pod with long, brown hairs. Pods have several brown, pea-like seeds each. There is usually low fruit production in naturalized plants.

Flowers appear in late summer in axillary clusters (racemes) and are 4 to 8 inches long. Each flower is about 1 inch long, irregular, and pinkish purple.

Kudzu vines can grow to a diameter of around 3 inches. There is no substantial bark, but vines become dark brown and rough from raised lenticels.

Twigs are brown and densely covered with rather long, brown hairs that seem to point in many directions. Leaf scars are nearly round and contain 3 groups of bundle scars. Pith is white and spongy, and it is surrounded by a wide ring of pores.

Japanese wisteria

Wisteria floribunda **(Willd.) DC.**

Family: Fabaceae

A native of Japan, this wisteria was introduced into the United States around 1830. It has been extensively used as an ornamental on porches, gazebos, and garden walls in the South, and it has escaped cultivation. It is truly spectacular to see in flower, and truly shocking to see it infesting a large area of natural habitat.

Form and Size: This is a high-climbing vine that easily reaches 60 feet into treetops. It is thicket-forming, and over time, it creates tangles of impenetrable vines across the ground and over all plants. Its stems send out runners (stolons) that creep along the ground, and it suckers from the roots as well. Vines twist around supports in a counterclockwise manner, and it can literally constrict the life out of its victim. Vines are long-lived once established.

Habitat: Japanese wisteria is more of a problem in the South, but it is becoming a headache in the lower Midwest. It can be found in forest edges, open or degraded woods, along roadsides, and elsewhere. It prefers rich, well-drained soils, but it is tolerant of almost any site.

Wildlife Uses: Wildlife values of Japanese wisteria are somewhat similar to those of our native *W. frutescens*, but they differ by an order of magnitude in morphology and aggressiveness. This exotic is long-lived, and in open areas it runs, and in forested areas it climbs. Any openings in infested areas yield a great proliferation of the vines. These wisteria-dominated openings are used by shrubland birds, such as towhees, but its dominance quickly lowers plant diversity. When vines climb into trees, the tree/vine complex provides nest sites for gray squirrels, red-shouldered hawks, and others. However, these top-heavy trees are prone to wind throw, which opens more area for wisteria dominance. The species supplies little wildlife food; leaves are browsed lightly by white-tailed deer (and likely other herbivores), and seeds have been recorded being used by bobwhite. Other game birds and rodents likely use them as well.

Landscaping Value: Hardy to Zone 5, this vine is very tolerant of almost any site. It tolerates full sun to full shade, from dry sites to wet sites. There are several dozen cultivars available. Another exotic species, *W. sinensis*, Chinese wisteria, has also been extensively planted in the Southeast. It freely hybridizes with Japanese wisteria, and it is said that the majority of naturalized plants are hybrids between the 2.

Similar Species Distinctions:
—**Chinese wisteria** (*W. sinensis*) has gray, hairy stems (new growth). Flower clusters are short, only 8 inches long, and all flowers tend to open at about the same time.
—**Kentucky wisteria** (*W. frutescens*) has smooth, brown stems, produces flowers after leaves are fully developed, and has smooth fruiting pods.

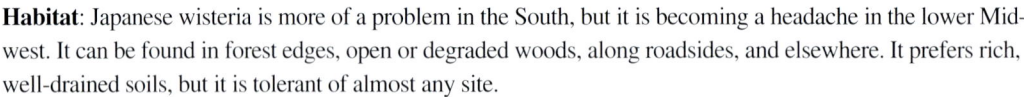

ESCAPED IN THESE STATES

Fruit is a green pod (turning brown at ripening) up to 7½ inches long that ripens in the fall. The persistent pods are covered with short, velvety hairs. Constrictions along the pod often separate seeds.

Leaves are compound and up to 1 foot long. There are 11 to 19 ovate leaflets that are long-tipped. Leaflet margins are wavy. The upper leaf surface is bright green and smooth; the lower surface is paler and hairy.

Twigs are brown, hairless, and dotted by raised lenticels. Lateral buds are pointed and appressed against the twig. Buds are covered with 3 brown scales; the largest outer scale is impressed with lines that give it the appearance of a seashell. Leaf scars are shaped like a shallow U and contain a single, central row of bundle scars. Mature bark is thin, light-colored, and dotted with numerous, somewhat elongated lenticels that have the appearance of braille.

Flowers appear in early spring with the leaves in dangling clusters (racemes) that are up to 3 feet in length. Each flower is irregular, perfect, and various shades of blue-lavender, violet, yellow, or white. They open from the top (the end closest to the vine) first and then move down the inflorescence.

Other Introduced, Escaped Species in the Midwest

Unfortunately, the list of exotic species that have proven problematic through escaping into native habitats of the Midwest seems to increase each year. Below, we list several woody species that have demonstrated such potential somewhere in the Midwest and that you might encounter. If interested in a more complete listing of invasive species of all taxa, visit the website for Early Detection and Distribution Mapping System at www.eddmaps.org/. This database was developed by the University of Georgia's Center for Invasive Species and Ecosystem Health.

Ampelopsis brevipedunculata (Maxim.) Trautv.	porcelainberry
Buddleja davidii Franch.	butterflybush
Cytisus scoparius (L.) Link	Scotch broom
Elaeagnus multiflora Thunb.	cherry silverberry
Elaeagnus pungens Thunb.	thorny olive
Euonymus europaeus L.	European spindletree
Hibiscus syriacus L.	rose of Sharon
Lonicera standishii Jacques.	Standish's honeysuckle
Lonicera xylosteum L.	dwarf honeysuckle
Mahonia bealei (Fortune) Carr.	leatherleaf mahonia
Rhamnus utilis Decne.	Chinese buckthorn
Rhodotypos scandens (Thunb.) Makino	jetbead
Robinia hispida L.	bristly locust
Rubus phoenicolasius Maxim.	wineberry
Salix purpurea L.	purpleosier willow
Spiraea japonica L. f.	Japanese spirea
Syringa vulgaris L.	common lilac
Viburnum lantana L.	wayfaringtree
Vinca major L.	greater periwinkle
Wisteria sinensis (Sims) DC.	Chinese wisteria

Native-Plant Nurseries in the Midwest

This is a partial listing of nurseries throughout the Midwest that sell native shrubs and trees, all of which have Internet sites. There are many other smaller nurseries, some of which can be found on The Lady Bird Johnson Wildflower Center website at www.wildflower.org under "Explore Plants" and "Suppliers." Many nurseries that specialize in native, herbaceous plants also carry some native shrubs.

Possibility Place
7548 W. Monee-Manhatten Rd.
Monee, IL 60449
www.possibilityplace.com
Phone: 708-534-3988
By appointment only

Cardno JF New
708 Roosevelt Rd.
PO Box 243
Walkerton, IN 46574
www.cardnojfnew.com
Phone: 574-586-2142

Shooting Star Nursery
160 Soards Rd.
Georgetown, KY 40324
www.shootingstarnursery.com
Phone: 502-867-7979

Simpson Nursery Co.
1504 Old Wheatland Rd.
Vincennes, IN 47591
Phone: 812-882-2441

Out Back Nursery and Landscaping
Wholesale and Retail
15280 110th St. South
Hastings, MN 55033
www.outbacknursery.com
Phone: 651-438-2771

Klyn Nurseries, Inc.
Wholesale only
3322 South Ridge Rd. (Rt. 84)
Perry, OH 44081
www.klynnurseries.com
Phone: 1-800-860-8104

Bluestone Perennials
7211 Middle Ridge Rd.
Madison, OH 44057
www.bluestoneperennials.com
Phone: 1-800-852-5243

Mary's Plant Farm
2410 Lanes Mill Rd.
Hamilton, OH 45013
www.marysplantfarm.com
Phone: 513-894-0022

Ernst Conservation Seeds
9006 Mercer Pike
Meadville, PA 16335
www.ernstseed.com
Phone: 1-800-873-3321

Sunlight Gardens
174 Golden Ln.
Andersonville, TN 37705
www.sunlightgardens.com
Phone: 1-800-272-7396

TN Wholesale Nursery
Wholesale only
12845 St. Rt. 108
Altamont, TN 37301
www.tnnursery.net
Phone: 931-692-4252

Johnson's Nursery
W. 180 N. 6275 Marcy Rd.
Menomonee Fall, WI 53051
www.johnsonsnursery.com
Phone: 262-252-4988

Natural Landscapes Nursery
354 N. Jennersville Road
West Grove, PA 19390
www.naturallandscapesnursery.com
Phone: 610-869-3788

Spence Restoration Nursery
Wholesale only
2220 E. Fuson Road
Muncie, IN 47302
www.spencenursery.com
Phone: 765-286-7154

How to Use Keys

Keys are integral tools in plant identification, especially when the plant in question is totally unfamiliar to the observer. Yet few lay persons use keys, largely because of the highly technical terms that are frequently employed. We encourage everyone to try to "key-out" unknown shrubs and vines; to facilitate that process we have striven to make these keys as user-friendly and devoid of technical terms as possible. Nevertheless, some terminology that may be foreign to the general user is necessary and in those cases, the reader should refer to our glossary, which depicts the features in question. Additionally, close observation is occasionally required, and everyone is encouraged to obtain and use a magnifying lens, ideally a compact 10-power hand lens, available at most college bookstores.

This book includes two major, relatively lengthy keys—one for use in summer and one for winter, when deciduous shrubs and vines are without their leaves. These keys, after their successful negotiation, will deliver the user to the correct species or, in cases where two or more species in a single genus are included, to the correct genus. For the latter cases, additional, shorter species keys are presented following the larger keys and placed in alphabetical order by genus. Once the correct genus is established by using the general keys, one can determine the correct species with these species keys.

All keys are composed of similarly numbered couplets that require a selection between contrasting features. Each decision results in the identification of the correct genus or species *or* in the direction to another couplet, where the process is repeated. A number in parentheses following the first member of a couplet indicates the previous couplet, which directed the user to the current couplet. Following each genus or species designation is a page number that directs the user to the text page where that taxon is found.

As an additional aid in identification, parts of the key are color-coded to match the color-coded sections of the book. This allows the user who is not sure of his/her choice to go to the section that contains other species with generally similar features as their target species, allowing visual comparisons among similar species.

Many keys exist to assist in the identification of shrubs and vines, and because the major features by which species are separated are static, most keys are very similar in many respects. In addition to our own expertise, we have assimilated features from other keys into ours. For further information and additional keys, readers are referred to Billington (1949), Soper and Heimburger (1982), Petrides (1986), Braun (1989), and Kurz (1997).

One final word of instruction and caution. These keys contain shrub and vine species native to our midwestern region and a variety of introduced species that have frequently escaped into the wild and are often viewed as problematic. Many other species are planted horticulturally, and may occasionally be encountered in seemingly natural situations. These species will not be found in our keys, but the keys may deliver you to the correct "group," which should assist in ultimate identification. Additionally, since only native and widely escaped introduced species are included, the general keys occasionally terminate at a particular species, based solely on features of genera, since only one species of that genus is included in the book. In such cases, descriptions in the text will not match, and reference to a more general shrub and vine identification book will be necessary. A similar problem might be encountered if the plant in question is a young tree, rather than a shrub. If that is suspected (e.g., if the plant has a single, straight trunk), the user should resort to the keys in *Native Trees of the Midwest* (Weeks, Weeks, and Parker 2010) for proper identification. For your convenience, we include titles of other reference books in our Bibliography.

KEYS

Summary Key to Genera, and Occasionally Species

		go to couplet	page
1.	Leaves ≤ ⅛ inch wide	2	
1.	Leaves > ⅛ inch wide	6	
2 (1).	Leaves needle-like, scale-like, and < ⅛ inch wide	3	
2.	Leaves ≤ ⅛ inch wide; plants low, in bogs	5	
3 (2).	Leaves scale-like, very hairy; plant is small (usually < 8 inches), many-branched shrub only in sterile, deep-sand habitat	*Hudsonia tomentosa*	148
3.	Leaves not hairy; larger shrub	4	
4 (3).	Leaves linear, alternate, with sharp tip	*Taxus canadensis*	14
4.	Leaves small, needle-like, in whorls of 3	*Juniperus communis*	12
5 (2).	Leaves of normal proportions (length about 2.5x width)	*Vaccinium* (key page 442)	234
5.	Leaves long, narrow (length about 8.5x width); edges strongly curled	*Andromeda polifolia*	22
6 (1).	Plant a vine; long, linear limbs; trailing or climbing	7	
6.	Plant a shrub; more or less upright and self-supporting	28	
7 (6).	Leaves tiny, < ½ inch long; trailing plant of bogs	*Vaccinium* (key page 442)	234
7.	Leaves larger	8	
8 (7).	Leaves compound	9	
8.	Leaves simple	17	
9 (8).	Leaves opposite	10	
9.	Leaves alternate	11	
10 (9).	Leaves pinnately compound with multiple (7-13) leaflets	*Campsis radicans*	314
10.	Leaves with only 2 leaflets per leaf	*Bignonia capreolata*	298
11 (9).	Leaves pinnately compound	12	
11.	Leaves palmately compound	14	
12 (11).	Leaves with serrated leaflets, stipules at base; plant usually shrub-like, with prickles	*Rosa* (key page 436)	268
12.	Leaves with leaflets with entire margins; no thorns	13	

13 (12).	Leaves 5-12 inches long with 5-15 leaflets; twines clockwise; pods smooth	*Wisteria frutescens*	358
13.	Leaves 8-12 inches long with 13-19 leaflets; twines counterclockwise; pods densely pubescent	*Wisteria floribunda*	396
14 (11).	Leaf petioles with prickles	*Rubus* (key page 437)	278
14.	Leaf petioles without prickles	15	
15 (14).	Leaves with 5 leaflets; climbs with aerial rootlets and/or tendrils	*Parthenocissus* (key page 432)	350
15.	Leaves with 3 leaflets	16	
16 (15).	Leaves densely hairy, golden above, and silver beneath; leaves with swollen bases and 2 stipules ; usually growing in dense, exclusive patches	*Pueraria montana*	394
16.	Leaves with light but variable pubescence, no stipules; variable form, sometimes shrubby; climbs with aerial rootlets	*Toxicodendron radicans*	356
17 (8).	Leaves opposite	18	
17.	Leaves alternate	21	
18 (17).	Leaves with serrated margins	*Euonymus fortunei*	384
18.	Leaves not serrated	19	
19 (18).	Vines prostrate, running along ground; upper surface of leaves shiny	20	
19.	Vines typically climbing, occasionally bush-like; leaves not shiny, medium green (occasionally yellow-green), often irregularly shaped toward end of stem	*Lonicera* (key page 430)	306
20 (19).	Leaves dark green, linear, and widest toward middle; broken petioles exude milky sap	*Vinca minor*	386
20.	Leaves circular to egg-shaped with blunt tip; no milky sap	*Mitchella repens*	18
21 (17).	Major veins parallel; stems armed with prominent prickles	*Smilax* (key page 441)	324
21.	Major veins branched; stems without prickles	22	
22 (21).	Stems climbing by twining; no tendrils	23	
22.	Stems with tendrils or aerial roots by which it climbs	26	
23 (22).	Leaves with serrations and/or lance-shaped	24	
23.	Leaf margins entire; leaves heart-shaped or lobed	25	
24 (23).	Leaves oval to lanceolate, with long, pointed tip and margins entire or with minute serrations; flowers and fruits terminal	*Celastrus scandens*	320
24.	Leaves circular to egg-shaped, blunt tipped with small, rounded serrations; flowers and fruit from axes of leaves	*Celastrus orbiculatus*	392

25 (23).	Leaves heart-shaped, without lobes, and densely pubescent; unique pipe-shaped flowers and cylindrical fruits (capsule)	*Aristolochia tomentosa*	318
25.	Leaves broad with 3-7 lobes, smooth to lightly pubescent; petiole attachment away from leaf margin	*Menispermum canadense*	322
26 (22).	Leaf margins entire; leaves 3-5 lobed (or diamond-shaped), thick, dark green, and shiny on upper surface with prominent, light-colored veins; climbs with aerial roots	*Hedera helix*	388
26.	Leaf margins toothed; climbs with tendrils from leaf bases	27	
27 (26).	Leaves coarsely or sharply toothed, not lobed; pith white; bark smooth with lenticels	*Ampelopsis cordata*	316
27.	Leaves coarsely toothed, often with lobes; pith brown; bark shredding on all but newest stems	*Vitis* (key page 446)	334
28 (6).	Shrubs dull green; semi-parasitic on deciduous trees	*Phoradendron tomentosum*	20
28.	Shrubs not growing on other woody plants	29	
29 (28).	Leaves opposite or whorled	30	
29.	Leaves alternate	45	
30 (29).	Leaves whorled (occasionally opposite); shrubs often growing in water; ball-like fruits often present from summer to following spring	*Cephalanthus occidentalis*	34
30.	Leaves not whorled	31	
31 (30).	Prostrate shrub, creeping vine-like along ground; leaves round or egg-shaped	*Mitchella repens*	18
31.	Form not creeping	32	
32 (31).	Leaves compound	33	
32.	Leaves simple	34	
33 (32).	Leaves palmately compound with 3 leaflets	*Staphylea trifolia*	104
33.	Leaves pinnately compound with 5-11 leaflets	*Sambucus* (key page 441)	98
34 (32).	Leaf margins entire; leaves not lobed	35	
34.	Leaf margins serrate; leaves lobed or not lobed	40	
35 (34).	Leaves with small, translucent glands (view upper surface while backlighted)	*Hypericum* (key page 429)	64
35.	Leaves without glands	36	
36 (35).	Undersurface of leaves and twigs with rusty scales	*Shepherdia canadensis*	74
36.	Leaves without scales	37	
37 (36).	Leaves with major veins bending toward tip of leaf and running parallel to margin	*Cornus* (key page 426)	36

37.	Leaves with major veins not as above	38
38 (37).	Leaves small (< 2 inches in length); lanceolate to ovate	39
38.	Leaves variable in size, shape, and pubescence; usually > 2 inches in length	*Lonicera* (key page 430) 306
39 (38).	Leaves smooth beneath, thick, leathery, and usually at least 2x longer than wide	***Ligustrum obtusifolium*** 364
39.	Leaves pubescent beneath, usually less than 2x as long as wide; bark on twigs typically peeling	*Symphoricarpos orbiculatus* 76
40 (34).	Leaves with 3-5 lobes	*Viburnum* (key page 444) 78
40.	Leaves not lobed	41
41 (40).	Leaves with 3-4 veins on each side of midrib, tending to emerge from basal half; some leaves alternate; twigs often ending in thorns	***Rhamnus*** (key page 433) 174
41.	Leaf veins more numerous and uniform in placement	42
42 (41).	Leaves finely serrated, greater than 2x longer than wide; very short (<⅛ inch) petiole	43
42.	Leaves with substantial teeth on margins, less than 2x as long as wide; petioles longer than above	44
43 (42).	Older branches remain green; unique, regular flowers and/or hanging, lobed capsule present from early summer	*Euonymus* (key page 428) 54
43.	Older branches light brown, although current year's are green with red where exposed to sun; tubular flowers and/or fruit (slender, pointed capsule) present from early summer	*Diervilla lonicera* 52
44 (42).	Shrub many-stemmed and 3-4 feet tall; petiole long (1-4 inches); terminal flower head has large, sterile flowers as well as small, fertile ones; seed heads remain through winter and into next year	*Hydrangea arborescens* 62
44.	Shrubs tend toward a single stem (or at least single root collar) and often greater than 3-4 feet; terminal flower cyme, yields fruit (drupe) that is present from early summer; leaves variable with species, but with petioles generally ≤ 1 inch	*Viburnum* (key page 444) 78
45 (29).	Leaves compound	46
45.	Leaves simple	55
46 (45).	Leaves pinnately compound (see also couplet 54)	47
46.	Leaves palmately compound	52
47 (46).	Stems (and usually petioles) with prickles	48
47.	Stems without prickles	49

48 (47). Leaves with 5-11 leaflets, entire or with fine, rounded teeth, aromatic when crushed, with glands dotting upper surface; stem prickles in pairs only at nodes ... *Zanthoxylum americanum* 294

48. Leaves with 5-9 toothed leaflets and stipules at base of petiole; stem prickles at nodes and frequently elsewhere as well (only at base on 1 species) ... *Rosa* (key page 436) 268

49 (47). Leaflets small (< 1½ inches long) and numerous (9-49) with entire margins ... *Amorpha* (key page 425) 248

49. Leaflets > 1½ inches long, entire or toothed ... 50

50 (49). Rachis winged; 7-21 leaflets with entire or few-toothed margins ... *Rhus* (key page 434) 258

50. Rachis not winged ... 51

51 (50). Tall, upright shrub in swamps and marshes; 7-13 leaflets, entire margins; produces contact dermatitis ... *Toxicodendron vernix* 292

51. Upright shrubs with thick stems in uplands; large leaves (to 18 inches) with 11-31 serrated leaflets ... *Rhus* (key page 434) 258

52 (46). Stems with prickles; leaflets may be 3 or 5; petioles also with prickles ... *Rubus* (key page 437) 278

52. Stems without prickles ... 53

53 (52). Leaves relatively large (≥ 3 inches) with 3 leaflets ... 54

53. Leaves very small (< 1¼ inches) with 3 or 5 leaflets; terminal (central) leaflet deeply lobed, giving appearance of 3 leaflets; 5-petaled, yellow flowers present all summer ... *Dasiphora fruticosa* 254

54 (53). Upright, often single-stemmed shrub; leaves large with leaflets 4-6 inches long, smelling unpleasant when crushed; margins entire or finely serrated; wafer-like fruits present by early summer and persist through winter ... *Ptelea trifoliata* 256

54. Multi-stemmed shrub, typically with some limbs on or near ground; leaves with leaflets to 2½ inches long, fragrant when crushed; margins coarsely toothed ... *Rhus* (key page 434) 258

55 (45). Shrubs prostrate, creeping, or low growing (≤ 1½ feet) ... 56

55. Shrubs upright ... 61

56 (55). Plants exclusively found in bogs ... 57

56. Plants in upland sites, often sandy substrates ... 58

57 (56). Creeping, vine-like, with diminutive leaves (< ½ inch long) ... *Vaccinium* (key page 442) 234

57. Multi-stemmed with long (1½ inches), narrow (⅛ inch), thick leaves; undersurface with dense, white hairs; edges strongly rolled ... *Andromeda polifolia* 22

58 (56).	Plants of deciduous woods; covering ground in herbaceous-like clumps connected by rhizomes	*Pachysandra procumbens*	162
58.	Plants of open, sandy areas; creeping woody plants	59	
59 (58).	Plants have underground stems from which short, erect stems with a few leaves protrude; leaves thick, obscurely serrate, and smell of wintergreen when crushed	*Gaultheria procumbens*	30
59.	Plants with creeping, above-ground stems	60	
60 (59).	Leaves oval, 1-3 inches long, pubescent	*Epigaea repens*	28
60.	Leaves obovate, ≤ 1 inch long, glabrous; forming dense patches	*Arctostaphylos uva-ursi*	24
61 (55).	Leaves lobed	62	
61.	Leaves not lobed	64	
62 (61).	Leaves shallowly or obscurely lobed	63	
62.	Leaves usually deeply lobed, 1-3 inches long; stems usually with prickles, especially on new growth	*Ribes* (key page 435)	180
63 (62).	Leaves large (5-10 inches long), shallowly 3- or 5-lobed; new stems with gland-tipped hairs; old stems with peeling bark	*Rubus* (key page 437)	278
63.	Leaves 2-4 inches long, some with no lobes and most with obscure lobes; all stems (except the newest) have peeling bark; fruits are groups of drooping, inflated capsules present most of summer until following spring	*Physocarpus opulifolius*	164
64 (61).	Stems with spines	65	
64.	Stems without spines	67	
65 (64).	Larger shrubs (10-20 feet) with leaves that are irregularly toothed and serrated, some approaching being lobed; spines long (to 1 inch), hard, and smooth (no buds), scattered along stem with single spine from a leaf node	*Crataegus* **spp.**	140
65.	Smaller shrub to 5 feet with single branched or unbranched spines that are ≤ ½ inch long at each node	66	
66 (65).	Multi-stemmed shrub to 5 feet tall; 3-branched, somewhat curved, spines at nodes, with side branches as long as central; bark brown; leaves with bristle-tipped, widely spaced (5-10 per side) serrations	*Berberis canadensis*	118
66.	Multi-limbed (often with single stem) shrub; unbranched spines at nodes (occasionally with 2 small side branches); leaves entire	*Berberis thunbergii*	376
67 (64).	Leaves entire	68	
67.	Leaves serrated or toothed	80	

68 (67).   Leaves thick, leathery, and evergreen (judged in summer by presence of leaves on both new growth and previous year's growth)   69

68.   Leaves not as above   70

69 (68).   Leaves entire or with hint of rounded serrations and upper surface covered with small dots (scales); a 1-4 foot high shrub of bogs where it forms dense colonies   *Chamaedaphne calyculata*   26

69.   Leaves entire and glabrous on both surfaces; fruits, when present, glabrous capsules; scattered or densely growing shrubs of upland understory   *Kalmia latifolia*   32

70 (68).   Bases of leaves lopsided, with 1 side larger or longer than the other   *Celtis tenuifolia*   128

70.   Leaf bases more or less balanced   71

71 (70).   Leaves with numerous, sticky resinous dots on undersurface   *Gaylussacia baccata*   144

71.   Leaves without resin dots beneath   72

72 (71).   Branches enlarged at nodes, each year's growth appears to come from flared, funnel-like ending of previous year's growth; branches very tough and flexible (cannot be broken)   *Dirca palustris*   142

72.   Stems not as above   73

73 (72).   Branches when scratched, or leaves when crushed, yield spicy fragrance   *Lindera benzoin*   160

73.   Scratched branches not aromatic, although possibly strong smelling   74

74 (73).   Leaves and stems scurfy with silvery scales, especially on underside of leaves; occasionally with thorns   *Elaeagnus umbellata*   378

74.   Leaves not silvery–scurfy   75

75 (74).   Leaves with major veins curving and running toward tip parallel to margin; leaves crowded near ends of stem   *Cornus* (key page 426)   36

75.   Leaves not as above   76

76 (75).   Leaves narrow, more than 2.5x longer than wide; buds with single, cap-like scale (lateral buds forming by early summer)   *Salix* (key page 438)   192

76.   Leaves broader (< 2.5x longer than wide)   77

77 (76).   Petiole short (< ¼ inch) or leaf sessile   78

77.   Petiole longer (≥ ⅜ inch)   79

78 (77).   Multi-branched shrubs; petiole very short; stem and leaves may be smooth or pubescent   *Vaccinium* (key page 442)   234

78. Widely branched shrub; petioles about ¼ inch long with a few gland-tipped (brown) hairs along margin; young stems and undersurface of leaves (especially mid-vein) have white, stellate hairs; at least some leaves on shrub with scattered serrations — *Styrax* (key page 442) 228

79 (77). Lateral veins prominent and unbranched until near margin where they curve sharply toward tip to join next vein — *Frangula alnus* 380

79. Veins not as above; fruits, when present, berry-like and on long pedicels from leaf axes; northern shrub of bogs and wetlands — *Ilex* (key page 430) 150

80 (67). Leaves asymmetrical at base; large teeth — *Hamamelis virginiana* 146

80. Leaves symmetrical at base — 81

81 (80). Some branches ending in thorns — *Rhamnus* (key page 433) 174

81. Shrubs without thorns — 82

82 (81). Leaves with very large teeth, occasionally called lobes — 83

82. Leaves with serrations (small teeth) — 84

83 (82). Leaves with many teeth (15-25 per side) lining parallel sides of leaf; crushed leaves aromatic; small northern shrub of sandy areas — *Comptonia peregrina* 130

83. Large shrub; leaves with 4-8 teeth (occasionally more) on each margin; twigs have clustered terminal buds (form by early summer); fruit is acorn — *Quercus prinoides* 172

84 (82). Lateral veins go to leaf margin, major vein to larger teeth and smaller veins, forked from main vein, go to smaller teeth — 85

84. Lateral veins curve toward tip before reaching margin, occasionally combining with next vein — 88

85 (84). Shrubs of wet areas; mature plants with cone-like fruits — 86

85. Shrubs of more upland areas, although soil may be moist at times — 87

86 (85). Leaves finely toothed; conelets are persistent into following year and very distinctive — *Alnus* (key page 425) 106

86. Leaves coarsely toothed; major lateral veins few (3-4); small shrub with cones with deciduous bracts — *Betula pumila* 120

87 (85). Many-stemmed shrub; leaves large (3-6 inches) and often broad (<2x as long as broad), doubly serrate; young twigs with gland-tipped hairs, perhaps only at nodes; fruit a husk-covered nut — *Corylus* (key page 428) 134

87. Small shrubs with linear leaves (1-2½ inches), the lengths of which are > 2x width, singly serrate; often retain terminal fruiting heads (panicles) until new ones form the following summer — *Spiraea* (key page 442) 222

88 (84). Lateral veins curving strongly toward tip and unbranched (except for small branchlets) so that they are the only prominent veins   89

88. Lateral veins branch, often losing prominence toward margin, or branches unite with branches of other veins to enclose sections of leaf blade   91

89 (88). Lowest pair of veins very prominent so that leaf has only 3 major veins; previous year's fruiting panicles present in early summer   *Ceanothus* (key page 426)   122

89. Lateral veins about equal   90

90 (89). Upper surface of leaf dark green, lower lighter; major veins prominently raised beneath; teeth generally rounded; fruit in berry   *Rhamnus* (key page 433)   174

90. Upper and lower surface of leaf pale green; teeth sharp; terminal spikes of white flowers, produce spikes of 2-celled, dry capsules that remain into the following summer; occurs in wet areas   *Itea virginica*   158

91 (88). Leaf undersurface (especially along veins) and petiole with white stellate hairs (substantial magnification needed); leaves may be entire   *Styrax* (key page 442)   228

91. Leaves without stellate hairs   92

92 (91). Petioles with obvious glands; leaves may have serrations only on half nearest tip   *Prunus* (key page 432)   166

92. Petioles without glands   93

93 (92). Leaves variable but usually linear ($\geq$ 3x longer than wide); lateral buds with single, cap-like scale (buds form by early summer)   *Salix* (key page 438)   192

93. Leaves broader (length usually $\leq$ 2x width); buds, once they form, with 2 or more scales   94

94 (93). Leaves somewhat leathery; base of petiole with small stipules; midrib of upper leaf surface with numerous linear, brown glands   *Aronia prunifolia*   116

94. Leaves without stipules or midrib glands   95

95 (94). Leaf margin obscurely toothed; many-branched shrubs; fruit, often present by early summer, blue with calyx-lobes present on tip   *Vaccinium* (key page 442)   234

95. Leaf margin obviously toothed   96

96 (95). Leaf margin usually somewhat doubly serrate but quite variable; petiole round; leaf widest in middle or toward base; usually a loose clump of stems   *Amelanchier humilis*   114

96. Leaf margin obviously serrate; leaf widest in middle or toward tip; petioles flattened or grooved on top; midrib of leaf often sunken below general leaf surface   *Ilex* (key page 430)   150

Winter Key to Genera, and Occasionally Species

		go to couplet	page
1.	Evergreen or semi-evergreen; green, or off-color supple leaves present	2	
1.	Deciduous (dead, dry leaves may adhere to some plants)	23	
2 (1).	Leaves ≤ ⅛ inch wide	3	
2.	Leaves > ⅛ inch wide	7	
3 (2).	Leaves needle-like or scale-like and < ⅛ inch wide	4	
3.	Leaves ≤ ⅛ inch wide; plants low, in bogs	6	
4 (3).	Leaves scale-like, gray-green, pubescent, persistent; plant a small (usually < 8 inches), many-branched shrub only in sterile, deep-sand habitats	*Hudsonia tomentosa*	148
4.	Leaves not hairy; larger shrubs	5	
5 (4).	Leaves linear, alternate, with sharp tip	*Taxus canadensis*	14
5.	Leaves small, needle-like, in whorls of 3	*Juniperus communis*	12
6 (3).	Leaves normal in proportion (length about 2.5x width); low creeping vine	*Vaccinium* (key page 442)	234
6.	Leaves very narrow (length about 8.5x width); edges strongly curled (flattened width may be slightly greater than ⅛ inch)	*Andromeda polifolia*	22
7 (2).	Leaves opposite	8	
7.	Leaves alternate	13	
8 (7).	Vines, trailing along ground or climbing	9	
8.	Semi-evergreen sub-shrub; underground rhizomes send up stems (to 8 inches high) with clusters of typical dogwood leaves (major veins turning toward tip before reaching margin); occasionally sufficiently abundant to form ground cover	*Cornus canadensis*	38
9 (8).	Small vine running over ground or over rocky substrate; leaves shiny and circular to egg-shaped.	*Mitchella repens*	18
9.	Larger vine, often climbing or covering ground	10	
10 (9).	Leaves compound with only 2 leaflets; winter leaves darken with purplish cast	*Bignonia capreolata*	298
10.	Leaves simple	11	
11 (10).	Semi-evergreen with leaves slowly getting darker and freeze-damaged as winter progresses; leaves and twigs pubescent; vine climbs over fences, shrubs, and other obstacles	*Lonicera* (key page 430)	306
11.	Leaves green all winter, smooth	12	

12 (11).	Vine forms dense ground cover, but also climbs vigorously into trees; leaves irregularly toothed	*Euonymus* (key page 428)	54
12.	Vine forms dense ground cover but does not climb; margin of leaves entire; petiole exudes milky sap when broken	*Vinca minor*	386
13 (7).	Vines that may run along ground or climb	14	
13.	Shrubs, although may be strongly prostrate	16	
14 (13).	Creeping plant of bogs that does not climb; leaves very small (< ½ inch long) with entire margins	*Vaccinium* (key page 442)	234
14.	Robust vines that climb	15	
15 (14).	Leaves with 3-5 lobes (or diamond-shaped), thick, dark green, and shiny on upper surface, with prominent, light-colored veins; runs along ground and climbs with aerial roots	*Hedera helix*	388
15.	Leaves without lobes and with parallel veins originating from petiole attachment point; vines have substantial prickles; winter leaves vary from green to bronze-colored; climbs with tendrils	*Smilax* (key page 441)	324
16 (13).	Plants usually covering ground with herbaceous-like clumps connected by rhizomes; a plant of deciduous woods	*Pachysandra procumbens*	162
16.	Plants with definite woody stems	17	
17 (16).	Shrub semi-parasitic and appearing as green clumps in deciduous trees; leaves and stems dull green	*Phoradendron tomentosum*	20
17.	Shrubs not parasitic on trees	18	
18 (17).	Shrubs exclusively found in bogs	19	
18.	Shrubs of other habitats, often dry, sandy areas	20	
19 (18).	Leaves long and narrow (length about 8x width), with margins rolled, gray-green, with dense, white pubescence below	*Andromeda polifolia*	22
19.	Leaves entire or with hint of rounded serrations, upper surface covered with small dots (scales); a 1-4 foot tall shrub of bogs where it forms dense colonies; leaves turn bronze-colored in winter	*Chamaedaphne calyculata*	26
20 (18).	Upright shrubs, leaves entire and glabrous on both surfaces; fruits, when present, glabrous capsules; scattered or densely growing shrubs in understory of upland woods	*Kalmia latifolia*	32
20.	Prostrate or very low growing (generally < 6 inches high) in open, sandy habitats or other acidic soils	21	
21 (20).	Plants have underground stems from which short, erect stems with few leaves emerge; leaves thick, obscurely serrate, and smell of wintergreen when crushed; dry, red fruits also smell spicy and often remain over winter	*Gaultheria procumbens*	30

21.	Plants with creeping above-ground stems		22
22 (21).	Leaves oval, 1-3 inches long, pubescent; fruit a dry capsule	*Epigaea repens*	28
22.	Leaves broadest toward tip; ≤ 1 inch long, glabrous; forming dense patches; red fruits often remain all winter	*Arctostaphylos uva-ursi*	24
23 (1).	Plant a vine with running or climbing stems		24
23.	Plant a shrub, although it may have a prostrate growth form		38
24 (23).	Leaf scars opposite; dark cream-colored stems run along ground or climb with aerial roots; bark peeling on older stems; usually retain woody fruiting capsules (4-6 inches long) through winter, although they typically have already split in half to release seeds	*Campsis radicans*	314
24.	Leaf scars alternate		25
25 (24).	Vines have prickles		26
25.	Vines without prickles		28
26 (25).	Plants running along ground, stems may have gland-tipped bristles or curved prickles with broad bases scattered along stems; bases of petioles remain on plant	*Rubus* (key page 437)	278
26.	Plant not running along ground; climbing onto shrubs and other objects		27
27 (26).	Plant climbing onto shrubs and fences without tendrils or twining; stout, curved prickles on stem, most frequently in pairs at nodes; fruit heads often present, even after fruits (hips) have been removed by wildlife.	*Rosa* (key page 436)	268
27.	Plant climbing with tendrils; stems, even older ones, green, armed with prickles that are generally straight, may be broad-based or black and bristle-like; may be partially evergreen	*Smilax* (key page 441)	324
28 (25).	Vines climb by twining		29
28.	Vines climb by tendrils or aerial roots		35
29 (28).	Trailing and climbing in very dense patches that quickly kill vegetation that it covers; yellow-green stem without swollen nodes (often rooting where nodes touch ground), young with dense, golden pubescence, later with silver, matted hairiness	*Pueraria montana*	394
29.	Plant not as above		30
30 (29).	Clustered, wooly buds above leaf scars that have 3 bundle scars; older stems somewhat grooved, often with clustered side twigs; new stems are hairy and usually die back a few feet in winter; unusual, cylinder-shaped capsules may remain through winter	*Aristolochia tomentosa*	318
30.	Stems without clustered, hairy lateral buds		31

31 (30). Leaf scars with single bundle scars   32

31. Leaf scars with 3 bundle scars; stems are green, and 2 lateral buds are stacked above each leaf scar; fruits may remain into winter, blue with glaucous coat   *Menispermum canadense*   322

32 (31). Leaf scars with warty knob (horns) on either side; lateral buds smooth or hairy; fruit pods may remain   33

32. Leaf scars not flanked by warty knobs; buds at almost right angles to stem; fruit capsules remain, sometimes with fruit still attached   34

33 (32). Vines twine clockwise, climb, and only rarely run along ground; stems to 1-2 inches in diameter; pods smooth; lateral buds short and pubescent   *Wisteria frutescens*   358

33. Vines twine counterclockwise; stems to 4 inches in diameter; stems often run along ground in openings as well as climb; pods densely pubescent; lateral buds elongated and smooth   *Wisteria floribunda*   396

34 (32). Residual fruit heads terminal with orange capsule and orange-red fruit within   *Celastrus scandens*   320

34. Residual fruit heads axillary (along stem from just above leaf scars) with yellow-orange capsule and red-orange fruit within   *Celastrus orbiculatus*   392

35 (28). Buds without scales and hairy; leaf scars with up to 6 bundle scars; when it climbs, it develops dense, aerial roots   *Toxicodendron radicans*   356

35. Buds not naked   36

36 (35). Bud scars very prominent, round to oval, and raised with 6 or more bundle scars arranged in a circle; climb with tendrils and in 1 species with aerial roots as well; bark on older stems brown and roughened, deeply fissured on largest   *Parthenocissus* (key page 432)   350

36. Vine different from above   37

37 (36). Vine climbing with numerous divided tendrils; bark on older stems peeling in strips; leaf scars offset from prominent, cone-shaped bud with 2 bud-scales; pith brown   *Vitis* (key page 446)   334

37. Vines climbing with tendrils that are few and unforked; bark tight and deeply fissured in older stems; leaf scars divided (vertical partition) with partially sunken bud above; pith white   *Ampelopsis cordata*   316

38 (23). Lead scars opposite or whorled   39

38. Leaf scars alternate   55

39 (38). Leaf scars whorled, occasionally opposite; shrub in very wet conditions, often in standing water; round fruit heads often present, or if seeds have dispersed, receptacles with reddish hairs remain   *Cephalanthus occidentalis*   34

39.	Leaf scars opposite	40	
40 (39).	Some twigs ending in thorns; most leaf scars opposite, but some may be alternate	*Rhamnus* (key page 433)	174
40.	Shrubs without thorns	41	
41 (40).	Shrub somewhat vine-like with upright growth with long stems or a declining growth form; tips of stems die back annually; stems hollow	*Lonicera* (key page 430)	306
41.	Shrub upright	42	
42 (41).	True terminal bud absent, 2 buds at tip; apparent terminal buds always associated with leaf scars; bundle scars 4 to many	43	
42.	Terminal bud present	44	
43 (42).	Opposite leaf scars meeting or connected by short ridge; stems few and thick with prominent raised lenticels; pith thick	*Sambucus* (key page 441)	98
43.	Opposite leaf scars not connected by ridge; stems relatively thin, lenticels few or absent; pith thin; unique papery, fruit pods remain all winter	*Staphylea trifolia*	104
44 (42).	Bundle scar 1	45	
44.	Bundle scar 3	49	
45 (44).	Stem with rusty-colored scales and scurfy texture	*Shepherdia canadensis*	74
45.	Stem without scales	46	
46 (45).	Twigs green, occasionally tinged with red, even on older stems	*Euonymus* (key page 428)	54
46.	Twigs brownish or grayish	47	
47 (46).	Scales (or peeling bark) present at bases of current-year twigs	48	
47.	Scales not present at base of twigs; buds somewhat elongate with 4 or more papery scales; leaf scars much raised; stems numerous and thin; black fruits occasionally last through winter	*Ligustrum obtusifolium*	364
48 (47).	Densely growing shrub with numerous, small branches that usually droop; pinkish-red fruits abundant (clusters adhere to stem) and remain through winter	*Symphoricarpos orbiculatus*	76
48.	Individual shrubs either with very upright limbs that typically retain upright capsules or with sprawling growth form; young stems angled or ridged; buds naked, opening partially in winter to yield very small, leafy shoots	*Hypericum* (key page 429)	64
49 (44).	Buds naked, without well-developed scales; tips of limbs with pair of miniature leaves that enclose the flower bud	*Viburnum* (key page 444)	78
49.	Buds with scales	50	

50 (49). Paired leaf scars not connected by lines; bud scales 2 or more (usually 4 or less); fleshy fruits may remain on shrub into winter *Viburnum* (key page 444)   78

50.   Paired leaf scars connected by lines   51

51 (50). Papery scales present at base of current year's twigs   52

51.   No paper scales at base of twigs   54

52 (51). Thin ridges (generally 2, better felt than seen) extend along twig; bark not peeling and papery *Diervilla lonicera*   52

52.   Twigs without ridges; bark papery and peeling   53

53 (52). Twigs stout; low shrub to 3-4 feet; broad, flat-topped fruiting head remains all winter *Hydrangea arborescens*   62

53.   Twigs slender; many stems from a single rootstock; shrubs to 12-15 feet; red fruits may be present into winter *Lonicera* (key page 430)   306

54 (51). Bud scales 3-6; fruits (or fruitless heads) present through winter *Viburnum* (key page 444)   78

54.   Bud scales 2; leaf scars raised well above stems *Cornus* (key page 426)   36

55 (38). Stems with thorns, spines, or prickles   56

55.   Stems lack thorns, spines, or prickles   64

56 (55). Thorns or spines 1-2 inches long   57

56.   Prickles much shorter, generally < ½ inch   58

57 (56). Irregularly armed with stout, sharp thorns to 2 inches long; twigs covered in silvery scales *Elaeagnus umbellata*   378

57.   Irregularly armed with slender, hard spines that emerge near leaf scar, most 1-2 inches long and straight to slightly curved; twigs without scales *Crataegus* spp.   140

58 (56). Prickles only at nodes   59

58.   Prickles scattered over stem; may have prickles at nodes as well   61

59 (58). Prickles on either side of semi-circular leaf scar; lateral buds red and hairy *Zanthoxylum americanum* 294

59.   Single prickle, straight or branched, below spurs that supported leaf clusters   60

60 (59). Single, straight spine below spur shoot; branches somewhat ridged; red fruit often persists through winter *Berberis thunbergii*   376

60.   Spines branched with 2 side branches about as long as central, which is often curved toward stem; bark brown and warty *Berberis canadensis*   118

61 (58). Prickles of various size scattered over the surface of the stem, none obviously focused on nodes   62

61.	Prickles at nodes and elsewhere on stem as well	63	
62 (61).	Leaf scars shriveled and on old petiole bases, along with stipule remains; bundle scars obscure; stems round or ridged, generally with numerous prickles (or glandular hairs); buds on first-year canes elongated	*Rubus* (key page 437)	278
62.	Leaf scars normally placed on stem, crescent-shaped with 3 bundle scars; buds small, often red	*Rosa* (key page 436)	268
63 (61).	Small shrubs with brown splitting, sometimes peeling bark; strong prickles at nodes that may be simple or branched into 2 or 3 parts; lesser prickles frequently on rest of stem, especially new basal stems	**Ribes** (key page 435)	180
63.	Shrubs typically with greenish or reddish smooth stems (older stems are brown with fissured bark); stout prickles occur on either side of the crescent-shaped leaf scar and similar or smaller prickles are scattered on remainder of stem	*Rosa* (key page 436)	268
64 (55).	Leaf scars lacking or leaf scars present with 1 bundle scar	65	
64.	Leaf scars with more than 1 bundle scar	76	
65 (64).	Leaf scars not evident because petioles break off leaving bases attached; shrubs with long branches with glandular hairs on new growth and thin, flaking bark on older	*Rubus* (key page 437)	278
65.	Leaf scars with 1 bundle scar	66	
66 (65).	Buds lack scales, scurfy; 2 buds stacked above leaf scar; current-year stems may retain scurfiness with stellate hairs	**Styrax** (key page 442)	228
66.	Buds with scales	67	
67 (66).	No true terminal buds (apparent terminal bud subtended by leaf scar)	68	
67.	True terminal bud	72	
68 (67).	Twigs typically green, occasionally with reddish blush, smooth or hairy; newer twigs covered with numerous warty (or blister-like) speckles (use hand lens)	**Vaccinium** (key page 442)	234
68.	Twigs not covered with speckles	69	
69 (68).	Buds with 2 scales (some may show 3)	70	
69.	Buds with 3 or more scales	71	
70 (69).	Twigs predominantly black (in winter) with buds of 2 sizes (1 flower, other leaf, latter may have 3 scales); small shrub to 3 feet	**Gaylussacia baccata**	144
70.	Twigs greenish to brown with buds all 1 size; large shrub to small tree (6-12 feet high)	**Vaccinium** (key page 442)	234

71 (69).	Leaf scars raised, occasionally with ridges (lines) running down stem from center or corners of scar; upright plants typically have residual pointed seed heads at twig ends	*Spiraea* (key page 442)	222
71.	Leaf scars not raised; twigs blackish with 2 bud sizes; small (to 3 feet) multi-limbed shrub, often in colonies	*Gaylussacia baccata*	144
72 (67).	Short side-branches often present; small, black, linear stipule present on either side of leaf scar (use hand lens)	*Ilex* (key page 430)	150
72.	Side branches and small, black stipules absent	73	
73 (72).	Buds with 2-3 hairless scales; shrub frequently found in very wet or streamside/lakeside habitats in the northern part of the Midwest	*Ilex* (key page 430)	150
73.	Buds with 4 or more scales	74	
74 (73).	Bud surrounded at base with unique, greatly raised leaf scar and papery stipules that are appressed to the stem	*Dasiphora fruticosa*	254
74.	Buds not as above	75	
75 (74).	Plants upright with slender twigs; leaf scars raised with ridges (lines) running down stem from center or corner of scar; dense, pointed seed heads often residual at twig ends	*Spiraea* (key page 442)	222
75.	Plants low and spreading; leaf scars not significantly raised; fruit heads, usually from leaf axils, remain in winter; round fruits usually gone but the saucer-like receptacles remain	*Ceanothus* (key page 426)	122
76 (64).	Leaf scars with 3 bundle scars	77	
76.	Leaf scars with more than 3 bundle scars, occasionally the numerous scars occur in 3 groups, yielding some confusion	98	
77 (76).	Pith chambered	78	
77.	Pith continuous	79	
78 (77).	Twigs green; narrow, linear seed heads remain through winter; generally a shrub of wet areas	*Itea virginica*	158
78.	Twigs slender and reddish brown; large shrub or small tree; bark of limbs often with corky ridges; trunk bark gray with warty growths	*Celtis tenuifolia*	128
79 (77).	Buds naked (without scales) and hairy	80	
79.	Buds with scales	81	
80 (79).	Buds stalked; yellowish brown flowers in early winter; woody fruit capsules often present in clusters; several crooked trunks grow from single rootstock	*Hamamelis virginiana*	146
80.	Buds not stalked, brown and pubescent; large shrub/small tree to 20 feet tall; older twigs and branches with numerous, large, light-colored lenticels	*Frangula alnus*	380

81 (79).	Buds long and slender; 3 ridges running from leaf scars down the stem (1 from each end and 1 from middle); newest stems have small, yellow resin-dots (use hand lens)	*Ribes* (key page 435) 180
81.	Stems without yellow resin-dots	82
82 (81).	Terminal buds (true) present	83
82.	True terminal buds absent	91
83 (82).	Buds with 1-3 scales	84
83.	Buds with 4 or more scales	87
84 (83).	Buds with single cap-like scales	*Salix* (key page 438) 192
84.	Buds with 2 or more scales	85
85 (84).	Buds reddish, with 2-3 scales, and obviously stalked; large shrub that often forms dense colonies in wet areas	*Alnus* (key page 425) 106
85.	Plants without stalked buds	86
86 (85).	Spur-shoots with many leaf scars on oldest stems (toward base of shrub); bud scales 2-3, not paired; youngest stems with long hairs and vertical raised lenticels; older stems with horizontal lenticels	*Betula pumila* 120
86.	Shrubs without spur-shoots, lenticels present but none horizontal; 2 bud scales; leaf scars crowded toward tip of glabrous stem	*Cornus* (key page 426) 36
87 (83).	Twigs with 3 lines descending from each leaf scar; multiple stemmed shrub with bark that peels prolifically; often with fruiting clusters of inflated, papery capsules present all winter	*Physocarpus opulifolius* 164
87.	Bark not peeling	88
88 (87).	Buds long and slender; usually with reddish cast	89
88.	Buds stout, usually not reddish	90
89 (88).	Bud scales have black tips and often appear twisted; colonies of gray-barked stems is the typical growth form; shrubs reaching 6 feet in height; fruits not present in winter	*Amelanchier humilis* 114
89.	Bud scales not twisted but have notched tips; dense, multi-limbed shrubs with gray-brown bark; terminal fruit clusters often present into winter	*Aronia prunifolia* 116
90 (88).	Older bark with horizontal lenticels but otherwise smooth and occasionally shiny; scratched twigs yield unpleasant smell; buds vary by species; pith circular	*Prunus* (key page 432) 166
90.	Older bark has warty, light-colored lenticels, but they do not contrast strongly with the gray-brown bark; buds very dark above triangular leaf scars; pith triangular in cross-section; upright "conelet" fruiting bodies last through winter (may droop with time)	*Alnus* (key page 425) 106

91 (82).   Buds with 2-3 scales   92

91.   Buds with 4 or more scales   94

92 (91).   Flower buds roundish and numerous at various locations on
         stem; leaf scars crescent-shaped; greenish stems have spicy
         smell when scratched   *Lindera benzoin*   160

92.   Scratched twigs do not yield spicy smell   93

93 (92).   Leaf scars U-shaped and brownish; buds with silky hairs and
         flattened against stem; clusters of wafer-like fruits often
         present all winter; plant often single-stemmed   *Ptelea trifoliata*   256

93.   Leaf scars shield shaped; buds (3) often stacked above leaf
         scar; linear seed heads often retained through winter; plants
         of either prairie or water's edge   *Amorpha* (key page 425)   248

94 (91).   Scratched twig has spicy smell; small shrub of dry, sandy
         sites; twigs usually have immature male catkins appressed
         to terminal stems   *Comptonia peregrina*   130

94.   Scratched twigs without spicy smell   95

95 (94).   Low shrubs (to 3 feet) with hairy twigs and buds; fruit-heads
         from leaf axils or terminal, remain in winter; round fruits
         usually gone, but the saucer-like receptacles remain   *Ceanothus* (key page 426)   122

95.   Larger shrubs without fruit heads remaining   96

96 (95).   Twigs zigzag and often have corky protrusions; buds small,
         triangular, and tightly appressed against twig; smooth gray
         bark of trunk is warty; pith occasionally chambered; usually
         single-trunk   *Celtis tenuifolia*   128

96.   Buds not tightly appressed to stems; pith continuous   97

97 (96).   Leaf scars in 2 rows along stems; many stems arising from 1
         rootstock; twigs pubescent with straight, brown hairs or
         smooth (depends on species); male catkins present in winter   *Corylus* (key page 428)   134

97.   Leaf scars not regularly placed, raised; lateral buds
         (5-6 scales) cradled in "pocket" created by raised leaf scar   *Rhamnus* (key page 433)   174

98 (76).   Buds stalked (narrow at base) with 2-3 smooth bud scales;
         shrubs often occurring in wet areas   *Alnus* (key page 425)   106

98.   Buds not stalked   99

99 (98).   Buds clustered at twig tips, which are enlarged to receive
         them; acorns produced on small "trees"   *Quercus prinoides*   172

99.   Buds not clustered at twig tip   100

100 (99). Lateral buds hidden in fold at top of nearly circular
         leaf-scar; clusters of red, hairy fruits usually still present   *Rhus* (key page 434)   258

100.   Lateral buds not hidden   101

101 (100). Buds completely (or nearly) circled by leaf scars   102

101.   Buds above leaf scars   103

102 (101). Buds pyramidal; leaf scar pale and completely circling bud;
           dark bundle scars; twigs brown and flexible (will not break)   *Dirca palustris*   142

102.   Buds nearly enclosed by leaf scar; twigs thick with large
           pith; sap milky   *Rhus* (key page 434)   258

103 (101). Bud scales 2-3; leaf scars crescent-shaped with many bundle
           scars in 3 groups; twigs stout; shrub (small tree) of wet areas *Toxicodendron vernix*   292

103.   Buds with 4 or more visible scales; twigs not stout   104

104 (103). Terminal buds present; twigs with peeling bark and lines
           running down stem from bud scars; fruit clusters with papery
           capsules persisting through winter   *Physocarpus opulifolius* 164

104.   Terminal buds false; twigs not lined below leaf scars; twigs
           often with immature male catkins; multi-stemmed shrub   *Corylus* (key page 428)   134

SPECIES KEYS

Keys to the genus *Alnus*

Summer

1.	Leaves finely serrate; female catkins very upright in rather compact clusters; conelets also upright on stalks; often on upland sites	*Alnus viridis*	112
1.	Leaves doubly serrate; female catkins and conelets not on upright stalks; moist locations	2	
2 (1).	Leaves with bases that are broad; stems dark with many transverse, white lenticels; female catkins and resulting conelets dangle; more northerly distribution in our region	*Alnus incana*	108
2.	Leaves with wedge-shaped bases, edges may not be double-serrate but are wavy; female catkins and resulting conelets not dangling, but splayed with some upright; dark stems lack transverse lenticels or have few short ones	*Alnus serrulata*	110

Winter

1.	Buds not stalked (or short-stalked), dark and pointed; male catkins sessile; female catkins upright on long stalks; conelets from previous summer fruiting persistent and move from upright to hanging over time; often on upland sites	*Alnus viridis*	112
1.	Buds long-stalked and not sharp-pointed; male catkins stalked; female catkins essentially sessile; wet areas	2	
2 (1).	New female catkins and old "conelets" dangle (point downward); species in northern part of Midwest; dark stems with many transverse, white lenticels	*Alnus incana*	108
2.	New female catkins and old conelets are splayed (point in all directions, some upward); species with a more southerly distribution in our region; dark stems lack transverse lenticels or have a few short ones	*Alnus serrulata*	110

Keys to the genus *Amorpha*

Summer

1.	Large shrub to 12 feet tall, in moist areas along streams, ponds, etc.; leaves 2½-10 inches long, with 9-27 leaflets, each ¾-1½ inch long	*Amorpha fruticosa*	250
1.	Small shrub in prairies and savannahs; leaves 1½-5 inches long with 13-49 small leaflets, each ½-¾ inch long	*Amorpha canescens*	252

Winter

1. Large shrub to 12 feet tall, in moist areas along streams, ponds, etc.; linear fruiting heads remain through winter; individual pods > ¼ inch long; smooth with resinous dots *Amorpha fruticosa* 250

1. Small shrub of prairies and savannahs; dense fruiting heads remain through winter; individual pods < ¼ inch long and densely hairy *Amorpha canescens* 252

Keys to the genus *Ceanothus*

Summer

1. Leaves broad (< 2x longer than wide); flower and seed heads elongated on leafless stalk *Ceanothus americanus* 124

1. Leaves narrow (> 2x longer than wide); flower and seed heads short and broad, usually on short stalk with leaves *Ceanothus herbaceus* 126

Winter

1. Plant of open areas such as prairies, glades, and savannahs; commonly occurs along railroad rights-of-way in our region; fruit heads on stalks 2-10 inches long that were leafless; twigs flexible, dark gray-green, and densely hairy *Ceanothus americanus* 124

1. Uncommon plant of open sandy or rocky areas; fruit heads on stalks that are generally < 1 inch long (stalks have small leaves in summer and thus will have leaf scars); twigs brittle, greenish to straw-colored, and lightly hairy, becoming smooth *Ceanothus herbaceus* 126

Keys to the genus *Cornus*

Summer

1. Leaves alternate; petioles very long (to 2 inches); leaves appear clustered at tip of stem *Cornus alternifolia* 132

1. Leaves opposite 2

2 (1). Very small prostrate plant, with erect stems (to 6-8 inches) arising periodically from rhizomes, sometimes forming colonies; each stem has 4-6 typical dogwood leaves *Cornus canadensis* 38

2. Plant on upright shrub 3

3 (2). Lateral veins often just 3 pairs (sometimes 4); flower head rounded, globe shaped; shrub with definite ascending branches, often in clones with tallest in center; fruit white, stalks conspicuously red *Cornus racemosa* 46

3. Leaves with generally 4 or more lateral veins; flower heads tend to be flat topped or slightly rounded 4

4 (3). Leaves with 4-6 pairs of lateral veins; pith on older stems (~2 years) brown or cream-colored 5

4.	Leaves generally with larger number (5-9) of pairs of lateral veins (only 3-4 in *C. foemina*); white pith on older stems	6	
5 (4).	Leaves rough on upper surface, at least partially because of many short, stiff hairs; pubescent below, especially on midrib and veins; fruits white, stems reddish	*Cornus drummondii*	42
5.	Leaves not rough, some pubescence beneath tending to be brownish on midrib; current-year twig with brown pubescence; fruit blue	*Cornus obliqua*	44
6 (4).	Leaves very broad (< 2x longer than wide), with many pairs (typically 6-8) of lateral veins; current year twigs green with large red spots; fruits blue	*Cornus rugosa*	48
6.	Leaves linear (> 2x longer than wide)	7	
7 (6).	Leaves whitish, pubescent beneath; 5-6 pairs of lateral veins; branches red; flower for extended period, often ripe fruit (white) and flowers simultaneously	*Cornus sericea*	50
7.	Undersurface of leaves green and glabrous with fewer veins (3-4 pairs); shrub of wet areas; fruit blue	*Cornus foemina*	40

Winter

1.	Leaf scars alternate and crowded toward tip of stems; large shrub or small tree	*Cornus alternifolia*	132
1.	Leaf scars opposite	2	
2 (1).	Very small prostrate plant; erect stems to 6-8 inches arising from rhizomes, sometimes forming colonies; plants semi-evergreen	*Cornus canadensis*	38
2.	Upright growth form	3	
3 (2).	Pith white on young and older twigs	4	
3.	Pith tan or brown on older twigs, may be white in first-year twigs	6	
4 (3).	Twigs bright red, including second and third year growth; shrub of moist to wet habitats in the northern part of our region	*Cornus sericea*	50
4.	Twigs may be reddish but not bright red	5	
5 (4).	Twigs red on top and green on bottom; shrub of wet, usually wooded bottomlands; southern part of our region	*Cornus foemina*	40
5.	Twigs green with patches of pink or light purple; new stems with numerous thin, linear, dark markings	*Cornus rugosa*	48
6 (3).	Twigs essentially glabrous; new twigs reddish but older twigs gray	*Cornus racemosa*	46
6.	Twigs hairy	7	

| 7 (6). | Twigs reddish or wine-colored, older twigs retain this color but begin to get streaks of brown; newest twigs with obvious white pubescence, hairs soft and give only a little feel of roughness | *Cornus obliqua* | 44 |
| 7. | Twigs reddish brown with very short, tannish hairs that are oriented toward tip to the degree that if one rubs stems away from the tip, there is considerable resistance (roughness) | *Cornus drummondii* | 42 |

Keys to the genus *Corylus*

Summer

| 1. | New branches and petioles with brown hairs with glands; fruits (nuts) in group, each nut enveloped by several individual bracts | *Corylus americana* | 138 |
| 1. | New branches and petioles without hairs (or a few, not gland-tipped); fruits (nuts) individual or in small clusters, each enveloped in fused, bristly bracts (flask-shaped) with long beak; northern part of our region | *Corylus cornuta* | 136 |

Winter

| 1. | Newest branches have short, brown hairs with glands; next growing season's male catkins relatively long (about 1½ inches) before spring stretch | *Corylus americana* | 138 |
| 1. | Newest branches without gland-tipped hairs; next growing season's male catkins short (< 1 inch) before spring stretch | *Corylus cornuta* | 136 |

Keys to the genus *Euonymus*

Summer

1.	Plant a trailing, climbing vine with thick, evergreen leaves	*Euonymus fortunei*	384
1.	Plant a shrub	2	
2 (1).	Prostrate shrub, less than 1½ foot high	*Euonymus obovatus*	60
2.	Upright shrub	3	
3 (2).	Limbs with 2-4 light brown corky wings that are ¼-½ inch wide	*Euonymus alatus*	362
3.	Limbs with ridges but without wings	4	
4 (3).	Leaves pale beneath with pubescence; petioles ½-1 inch long; flowers purple; capsules 4-lobed, smooth; exposed fruits red	*Euonymus atropurpureus*	58
4.	Leaves with smooth undersurface (may be slightly hairy on veins); petioles short (≤ ⅛ inch); flowers greenish; capsules round, slightly flattened and warty; exposed fruits orange	*Euonymus americanus*	56

Winter

1.	Plant an evergreen trailing and climbing vine with thick leaves; upper surface green with lighter-colored veins	*Euonymus fortunei*	384
1.	Plant not a vine	2	
2 (1).	Prostrate creeping plant to 18 inches tall with stems to 3 feet; may root at nodes along stem	*Euonymus obovatus*	60
2.	Shrub upright	3	
3 (2).	Twigs with substantial corky wings (to ½ inch wide) that are tan in color and contrast with green stem	*Euonymus alatus*	362
3.	Twigs may be angled but not winged	4	
4 (3).	Buds red with many (about 8) scales that are not tightly appressed; any remaining fruiting capsules are warty	*Euonymus americanus*	56
4.	Buds green to reddish with fewer scales (5-6) that are tightly appressed; any remaining fruiting capsules are smooth	*Euonymus atropurpureus*	58

Key to the genus *Hypericum*

1.	Prostrate, creeping shrub, typically no more than 6 inches high; 4-petaled flowers with petals grouped 2 per side; dry sites	*Hypericum hypericoides*	66
1.	Upright shrubs	2	
2 (1).	Shrub usually to 3 feet; leaves to 2 inches long; twigs 2-edged but older branches 4-edged; flowers with 5 styles and fruits 5-parted capsules, all terminal; restricted to northern part of our region, largely in dune habitats	*Hypericum kalmianum*	68
2.	Shrub to 6 feet; leaves to 3 inches long; twigs and older branches 2-edged; flowers with 3 styles and fruits 3-parted capsules; fairly general distribution	*Hypericum prolificum*	70

Winter

1.	Prostrate, many-branched creeping shrub to 6 inches high; stems brownish and with a ridge on 2 sides, giving the impression of a flattened twig; in older twigs bark shreds along the ridges	*Hypericum hypericoides*	66
1.	Upright shrub, 3-6 feet tall	2	
2 (1).	Shrub with twigs and branchlets (second year growth) 2-edged; height to 6 feet; fruiting capsules remain through winter, 3-parted and about ½ inch long	*Hypericum prolificum*	70
2.	Shrub with twigs 2-edged and branchlets 4-edged; height to 3 feet; fruiting capsules remain through winter, 5-parted and < ½ inch long	*Hypericum kalmianum*	68

Keys to the genus *Ilex*

Summer

1.	Spur branches absent; no small, black stipules on either side of petiole attachment; long-peduncled fruit; wet areas in northern part of our region	*Ilex mucronata*	154
1.	Spur branches with leaves may be present on lower stems; small, black stipules (use hand lens) on either side of petiole attachment;short peduncled fruit	2	
2 (1).	Leaves with tip rounded or blunt; margin with mostly round or blunt teeth; main vein indented from surface; petioles up to 1 inch long; principally in southern part of our region	*Ilex decidua*	152
2.	Leaves with tip pointed and marginal teeth sharp; petioles generally about ¼ inch long; principally in central and northern parts of our region	*Ilex verticillata*	156

Winter

1.	Twigs grayish; buds rather long and pointed; leaf scars without small stipules; fruits on long peduncles and generally do not last into winter; a shrub of very wet areas in the northern part of our region	*Ilex mucronata*	154
1.	Twigs grayish or brownish; leaf scars have very small (use hand lens), black stipules on either side; red fruits persist into, and often through, the winter (on females)	2	
2 (1).	Twigs stiff, gray, often with numerous, short side-twigs (like spur-shoots); buds somewhat pointed; shrub of wet areas in the southern part of our region	*Ilex decidua*	152
2.	Twigs brown, without short side-twigs; buds blunt; shrubs of wet areas throughout our region	*Ilex verticillata*	156

Key to the genus *Lonicera*

Summer

1.	Plant an upright shrub, possibly to 20 feet	2	
1.	Plant a vine or vine-like with trailing, somewhat climbing stems	5	
2 (1).	Leaves oval with tip often blunt; pith is solid and white; fruits oval (elongated rather than round) in pairs on peduncles to 1 inch long	*Lonicera canadensis*	72
2.	Leaf tips usually pointed; pith brown and hollow in center; fruits round in pairs (peduncles various lengths)	3	
3 (2).	Leaves oval and usually bluntly tipped; upper and lower surface of leaves hairy, as are petioles; fruit peduncle usually about ⅓-½ inch long	*Lonicera morrowii*	370

3.   Leaves tend to be more pointed at tip and mostly without
            hairs except along margin   4

4 (3).   Leaves tend to be oval with moderately pointed tip, glabrous
            both top and bottom (hairs along margins); peduncle of
            paired fruits long (to 1 inch)   *Lonicera tatarica*   372

4.   Leaves broadest in middle, tapering to both ends with tip
            abruptly pointed; petiole hairy but leaf glabrous except for
            hairs on veins and leaf margins; peduncle very short
            (almost ⅛ inch) so that fruits appear sessile   *Lonicera maackii*   368

5 (1).   Plant a climbing vine, moving into and over fences, shrubs,
            and trees; entire leaf, petiole, and stem hairy; tends to be
            partially evergreen, especially in southern part of our
            region; no leaves joined at bases (connate)   *Lonicera japonica*   390

5.   Plants usually more a sprawling vine, but some individuals/
            species climb; at least last pair of leaves below flowers connate   6

6 (5).   Top set of leaves connate, with each leaf triangle in shape,
            no other leaves united with short or no petioles; tubular
            flowers red and present continuously until late summer;
            stems twining   *Lonicera sempervirens*   312

6.   Stems straggling or slightly twining or climbing; flowers
            yellow (occasionally with reddish or purplish blush);
            connate leaves not as above   7

7 (6).   Connate disk almost circular (or oval); glaucous above and
            below; petioles short or absent on remaining leaves that
            often have clasping bases; fruits in 2-6 separated whorls
            on stem   *Lonicera reticulata*   310

7.   Connate disk oval, each end with blunt point, green but with
            slight glaucous coat above; lower leaves not united and
            longer than broad (about 2x) and with white, waxy coating
            below; no (or very short) petiole and leaf bases give hint of
            clasping; fruits in 1-3 crowded whorls on stem   *Lonicera dioica*   308

Winter

1.   Plant an upright shrub, possibly to 20 feet   2
1.   Plant a vine, or vine-like with trailing, somewhat climbing stems   5

2 (1).   Pith solid and white; fruits rarely remain into winter;
            typically understory shrub of the northern part of our region *Lonicera canadensis* 72
2.   Pith brown with hollow center; fruits round and in pairs   3

3 (2).   Newest twigs glabrous; fruits do not remain into winter   *Lonicera tatarica*   372
3.   Newest twigs somewhat pubescent   4

4 (3).   Newest twigs soft pubescent; fruits do not remain into
            winter   *Lonicera morrowii*   370

4.	Newest twigs lightly pubescent, becoming smooth; fruits in pairs from each leaf axil (4 per node), have very short peduncles, and appear sessile, remain on shrub well into winter	*Lonicera maackii*	368
5 (1).	Plant a true climbing, twining vine with hairy, brown twigs and semi-evergreen, pubescent leaves; pith brown with hollow center	*Lonicera japonica*	390
5.	Plant a sprawling or climbing vine; pith white with hollow center	6	
6 (5).	Definitely a climbing vine, twining around self and objects	*Lonicera sempervirens*	312
6.	Sprawling and twining vines, scarcely climbing	7	
7 (6).	Newest stems somewhat greenish, with shades of reddish brown; older stems gray, shredding	*Lonicera reticulata*	310
7.	Newest stems light to dark brown; older stems yellowish brown, shredding	*Lonicera dioica*	308

Keys to the genus *Parthenocissus*

Summer

1.	Vine with short (to 1½ inches), multi-branched tendrils that often end in adhesive disks; aerial roots along stem when climbing common; flower stalk has strong central axis; upper surface of leaf dull	*Parthenocissus quinquefolia*	352
1.	Vine with long tendrils (to 6 inches) that branch only once or twice and do not end in adhesive disks (may occasionally have few disks); no aerial roots; flower stalk without strong central axis (branches dichotomously); upper surface of leaf shiny	*Parthenocissus vitacea*	354

Winter

1.	Newest growth pubescent (very short hairs, use magnification); tendrils < 1½ inches long and multiple-branched (2-10 branches) and ending in adhesive disks; when main stem climbs, aerial roots develop	*Parthenocissus quinquefolia*	352
1.	Newest growth glabrous; tendrils 1½-6 inches long with few (1-3) branches, not generally ending in adhesive disks (some may occasionally develop); no aerial roots	*Parthenocissus vitacea*	354

Keys to the genus *Prunus*

Summer

1.	Low growing shrub, generally 3-4 feet maximum height; leaves with serrations missing or reduced on basal ⅓ of leaf; flowers in umbel; fruits relatively large (⅜ inch diameter); in northern part of our region	*Prunus pumila*	168

1. Larger shrub (to 25 feet); leaves with fine, sharp serrations on entire margin; flowers in raceme (central stalk); fruits usually about ¼ inch in diameter; generally distributed in central and northern portions of our region   *Prunus virginiana*   170

Winter

1. Low-growing shrub, generally 3-4 feet maximum height; twigs are reddish (later gray) and glabrous; buds reddish with numerous, tightly adherent scales   *Prunus pumila*   168

1. Large shrub or small tree, 6-25 feet; twigs reddish brown but covered with whitish, removable epidermis; attractive buds dull brown with scale margins straw-colored; scratched twigs with very strong, unpleasant odor   *Prunus virginiana*   170

Keys to the genus *Rhamnus*

Summer

1. Leaves relatively broad (2x as long as wide) with only 3-4 veins on each side of midrib; most leaves opposite but some appear alternate; many twigs ending in thorns   *Rhamnus cathartica*   374

1. Leaves alternate with 6-9 veins on each side of midrib; no thorns   2

2 (1). Leaves rather broad (about 2x as long as broad); multi-stem low shrub (≤ 3 feet in height) in wet places in northern part of our region   *Rhamnus alnifolia*   176

2. Leaves rather narrow (> 2x as long as broad); generally a single, erect stem to 10 feet fall; usually in drier sites throughout central portion of our region   *Rhamnus lanceolata*   178

Winter

1. Large shrub or small tree; twigs often ending in thorns; black fruits may persist into winter; often in dense concentrations   *Rhamnus cathartica*   374

1. Limbs without twigs ending in thorns   2

2 (1). Large shrub to 12 feet; twigs grayish; buds ovoid, less than 2x as long as wide; upland sites but often on rocky, gravelly banks of streams   *Rhamnus lanceolata*   178

2. Small shrub to about 3 feet in height; occurs in very wet sites; buds linear, ≥ 2x longer than wide   *Rhamnus alnifolia*   176

Keys to the genus *Rhus*

Summer

1.	Small shrub (to 8 feet but generally 3-4 feet) with limbs that may be ascending or lying on ground; leaves with 3 palmate, stalkless leaflets, aromatic when crushed	*Rhus aromatica*	260
1.	Upright shrub; leaves pinnately compound with more than 3 leaflets	2	
2 (1).	Shrub with relatively slender branches (~ ¼ inch diameter near tip) with prominent raised lenticels; midrib of leaf (7-17 leaflets) with obvious wings	*Rhus copallinum*	262
2.	Shrubs with thick branches (~ ½ inch diameter near tip); midrib of leaf (15-27 leaflets) not winged	3	
3 (2).	Twigs smooth with white coating; petioles and undersurface of leaf glabrous	*Rhus glabra*	264
3.	Twigs covered with dense, tan pubescence (like deer antlers in velvet) as are petioles of leaves; stems exude milky, light-colored sap when cut	*Rhus hirta*	266

Winter

1.	Small shrub (to 8 feet tall but generally 3-4 feet) with limbs that are ascending and some usually lying on the ground as well; scratched stem aromatic; lateral buds hidden by bark fold above leaf scar; terminal male catkins usually present	*Rhus aromatica*	260
1.	Upright shrub or small tree; lateral buds not hidden; no spicy smell for scratched stems; previous season's fruiting heads usually present	2	
2 (1).	Shrub of relatively slender branches (about ¼ inch diameter near tip) with prominent, raised lenticels; leaf scars surround bud by about half; fruiting heads, while still upright, have tips drooping	*Rhus copallinum*	262
2.	Shrubs with comparatively stout branches (about ½ inch diameter near tip); fruiting heads red, conical, and upright; leaf scar almost completely surrounding bud	3	
3 (2).	Twigs smooth with glaucous coat; small lenticels	*Rhus glabra*	264
3.	Twigs covered with dense, tan pubescence, similar to the look of deer antlers in velvet; cut stem oozes milky, light yellow sap	*Rhus hirta*	266

Keys to the genus *Ribes*

Summer

1.	Stems lacking bristles or spines; flowers 2-12 per linear cluster; leaves with 3-5 lobes with pointed tips (reminiscent of red maple); undersurface of leaf with minute, orange glands	*Ribes americanum*	182
1.	Stems with bristles and/or spines, at least on new growth; flowers 1-3 per cluster; lobes of leaves with rounded tips	2	
2 (1).	Leaves with pubescence on both surfaces, light on upper surface; ovary with stiff, gland-tipped hairs that become prickles in fruit; nodes have spines and there are frequently internodal prickles, especially on new growth and lower stem of plant	*Ribes cynosbati*	186
2.	Leaves with pubescence below, especially on vein, but not above; ovary and fruits without bristles	3	
3 (2).	Leaves deeply cleft into 3-5 lobes; nodal spines absent or small; internodal prickles present, especially on new growth, but sparse; branches glabrous	*Ribes hirtellum*	188
3.	Leaves shallowly cleft into 3-5 lobes; nodal spines very large (to ¾ inch) and prominent, even on new growth; lower portions of plant and new growth have long, slender internodal prickles; flower unique with long, swept back sepals and elongated stamens and style that looks like a beak	*Ribes missouriense*	184

Winter

1.	Plant without prickles or nodal spines; newest twigs and scales of buds with yellow resin-dots	*Ribes americanum*	182
1.	Plant usually with prickles; no resin-dots	2	
2 (1).	Plant without large (> ¼ inches) nodal spines; may have single, short (< ⅛ inch) spine at node; prickles absent on upper stems, although they may occur at bases of older stems and on new basal sprouts; stiff, upright stems 188	*Ribes hirtellum*	
2.	Plant with nodal spines (1-3) and substantial prickles on stems	3	
3 (2).	Nodal spines formidable, stout, and to ¾ inch long; buds slender and diverge 45 degrees or more from stem, straw-colored; margins of bud scales with few hairs	*Ribes missouriense*	184
3.	Nodal spines sharp, but < ½ inch and slim; buds conical and diverge < 45 degrees from stem, light brown or tan; bud scale margins have hairs	*Ribes cynosbati*	186

Keys to the genus *Rosa*

Summer

1.	Stems long and arching, whip-like; vines climb into shrubs and trees if given opportunity; styles united and project from center of numerous stamens		2
1.	Shrubs; stems do not climb; styles not united and are short		3
2 (1).	Leaflets 7-9, stipules very toothed; flowers white; fruits small, about ¼ inch in diameter	*Rosa multiflora*	382
2.	Leaflets typically 3 (may be 5 on new stems); flowers large, pink; fruits ⅜ inch in diameter	*Rosa setigera*	276
3 (1).	Stems upright to 3 feet tall, almost totally without prickles except near base of plant and on new shoots from rhizomes (no stout prickles)	*Rosa blanda*	270
3.	Stems upright and armed with substantial prickles		4
4 (3).	Leaves with coarsely toothed margin; petioles with flat stipules; short (½-3 feet) stem covered with short, straight prickles; dry, open, upland sites	*Rosa carolina*	272
4.	Leaves with finely serrated margins; stipules about 1 inch long, narrow and rolled with very little petiole above it; stems erect to 8 feet, usually with paired stout prickles at nodes and elsewhere on stem; wet areas	*Rosa palustris*	274

Winter

1.	Stems long and arching, often whip-like; vines climb into shrubs and trees if given the opportunity; stems with stout, recurved prickles, often in pairs at nodes and scattered elsewhere		2
1.	Shrubs with stems that do not climb and are upright		3
2 (1).	Flower heads with multiple (10-20) small (about ¼ inch long) fruits that are smooth; no residual sepals	*Rosa multiflora*	382
2.	Flower heads with multiple (but smaller numbers than above—about 5-10) fruits that are about ⅔ inch long, usually covered with gland-tipped hairs; no residual sepals	*Rosa setigera*	276
3 (1).	Stems upright to 3 feet tall; often totally without prickles except near base of plant and on newer shoots from base (but some individuals may have substantial prickles elsewhere on stem); fruits about ½ inch long, smooth, and with residual sepals	*Rosa blanda*	270
3.	Stems substantially armed		4
4 (3).	Tall shrubs to 8 feet; paired prickles at leaf scar decurved and flattened at base but not particularly stout; prickles elsewhere are sparse, except at base; fruits about ½ inch long, covered with gland-tipped hairs (occasionally smooth), and generally retain long sepals; plant of wet areas	*Rosa palustris*	274

4.   Short shrubs to 3 feet; prickles at and between nodes circular
      in cross section, straight and thin; prickles may be abundant
      or sparse; fruit to ½ inch long, with scattered, gland-tipped
      hairs and no residual sepals; in dry upland areas   *Rosa carolina*   272

Key to the genus *Rubus*

Summer

1.   Shrub with large, 3- to 5-lobed, simple leaves; stems
      with gland-tipped hairs and no prickles   *Rubus odoratus*   190
1.   Shrubs with palmately compound leaves; stems with prickles   2

2 (1). Stems trailing along ground, climbing over low objects;
        flowering stems short, upright   3
2.   Stems not trailing; plants upright   4

3 (2). Stems mostly round (can be angled) with scattered, stout,
        curved prickles; leaves with 3 or 5 sharp-toothed leaflets *Rubus flagellaris*   282
3.   Stems mostly round with numerous slim prickles; leaves
      with 3, occasionally 5, blunt-toothed leaflets; growing in
      very wet areas in the northern part of our region   *Rubus hispidus*   284
4 (2). Stems round   5
4.   Stems ridged   6

5 (4). Stems glaucous; prickles substantial (hooked) with broad
        base; leaves with 3 (occasionally 5) palmately compound
        leaflets, very white beneath   *Rubus occidentalis*   288
5.   Stems not glaucous (occasionally so on very young stems);
      leaves with 3 leaflets, occasionally 5 (in such cases
      pinnately compound); terminal leaflet may be 3-lobed;
      stem covered with dense bristles (no broad-based prickles) *Rubus idaeus*   286

6 (4). Leaves with 3 or 5 leaflets; margin irregularly toothed with
        some large teeth; leaflets densely hairy below; large bract
        at base of each short flower stalk; flower stalks densely soft
        hairy   *Rubus pensilvanicus*   290
6.   Leaves with 3 or 5 leaflets; margin sharply toothed; lower
      surface with hairs and a few gland-tipped hairs on veins;
      flower cluster linear (2-4x longer than wide); flower stalks
      with gland-tipped hairs   *Rubus alleghertiensis*   280

Winter

Note: Rubus is a genus that has species that may be difficult to identify in summer, much less in winter
when leaves are not available to assist. Identification of highbush blackberries is especially problematic
and is not attempted here. Reader should mark questionable individual plants and return several times
during the year to identify the species with certainty.

1.   Stems creeping or prostrate   2
1.   Stems erect or arching, not creeping along ground   3

2 (1).	Stems with numerous prickles (bristly) and upright stems to a foot high; some leaves may persist into winter; a plant of wet areas	*Rubus hispidus*	284
2.	Stems not bristly, but with few, stout, curved prickles; stems often angled but may be round	*Rubus flagellaris*	282
3 (1).	Stems with gland-tipped hairs and no prickles; bark on older stems becoming loose and peeling	*Rubus odoratus*	190
3.	Stems with prickles	4	
4 (3).	Stems ridged; plants upright, not generally arching	Blackberry group	
4.	Stems round	5	
5 (4).	Stems strongly glaucous; slightly hooked prickles substantial with broad bases; stems long and arching	*Rubus occidentalis*	288
5.	Stems not glaucous, bristly, with a few larger prickles that may be curved but round in cross section; stems ascending, not generally arching	*Rubus idaeus*	286

Key to the genus *Salix*

Note: This key was kindly supplied by George Argus, without doubt the "willow expert of North America." Because of the complexity of willow identification, terminology may be a little more complicated than in our typical key; the reader is referred to our glossary for unfamiliar terms. Even an expert like George Argus does not frequently identify willows in winter; indeed, he suggests that individual specimens be marked and returned to repeatedly throughout the year to examine a myriad of features to confirm identification. We suggest that approach to users and do not attempt to include a winter key.

Summer

1.	Bud scale margins distinct and overlapping	*Salix caroliniana*	198
1.	Bud scale margins united	2	
2 (1).	Petioles glandular dotted or lobed at distal end	3	
2.	Petioles not glandular at distal end	4	
3 (2).	Stipules on first leaves absent or minute rudiments, on later and vigorous shoots minute rudiments or leaf-like; juvenile leaves glabrous; largest medial leaf blades generally broader, 2-6x as long as wide; branchlets glabrous; bud scale inner membranaceous layer free but not separating from outer layer; petioles with paired, spherical glands at distal end	*Salix serissima*	220
3.	Stipules on all leaves leaf-like; juvenile leaves with white and rust-colored hair; largest medial leaf blades generally narrower, 3.1-9.5x as long as wide; branchlets generally hairy, sometimes becoming glabrous; bud scale inner membranaceous layer free and separating from outer layer; petioles glandular at distal end with clusters of spherical, sometimes stalked, glands or leaf-like glands	*Salix lucida*	210
4 (2).	Largest medial leaf blades less than 4x as long as wide	5	
4.	Largest medial leaf blades greater than 4x as long as wide	12	

5 (4).	Ovaries glabrous	6	
5.	Ovaries hairy	8	
6 (5).	Leaves not glaucous on lower surface	*Salix cordata*	200
6.	Leaves glaucous on lower surface	7	
7 (6).	Largest medial leaf blades with white hairs only, margins toothed; stigmas 2 plump lobes; female catkins densely or moderately densely flowered	*Salix eriocephala*	202
7.	Largest medial leaf blades often with rust-colored hairs, margins entire or toothed; stigmas flat or slender-cylindrical lobes; female catkins loosely flowered	*Salix myricoides*	212
8 (5).	Branchlets white wooly; juvenile leaves very densely tomentose; petioles tomentose or wooly; largest medial leaf blades tomentose on top surface, and tend to be narrower, 3.3-12x as long as wide; ovaries very densely tomentose or wooly	*Salix candida*	196
8.	Branchlets glabrous or variously hairy, but not wooly; juvenile leaves usually sparsely hairy but rarely tomentose; petioles glabrous or variously hairy but not tomentose or wooly; largest medial leaf blades pubescent to almost glabrous on upper surface, tend to be broader, 1.7-9x as long as wide; ovaries sparsely to very densely covered with short, silky hairs	9	
9 (8).	Floral bracts tawny; stipes to ¼ inch long	*Salix bebbiana*	194
9.	Floral bracts brown, black, or 2-colored; stipes < ⅛ inch long	10	
10 (9).	Leaves densely hairy on the lower surface with short, appressed, straight hairs; branches highly brittle at base	*Salix sericea*	218
10.	Leaves range from glabrous to densely hairy with spreading, wavy hairs; branches flexible at base	11	
11 (10).	Leaves glabrous or with sparse long, soft hairs on lower surface; top surface of petiole sparsely tomentose	*Salix discolor*	204
11.	Leaves moderately to very densely tomentose or wooly on lower surface; upper surface of petiole with straight, soft hairs	*Salix humilis*	206
12 (4).	Ovaries glabrous	13	
12.	Ovaries hairy	16	
13 (12).	Leaf margins entire; plants of bogs or fens	*Salix pedicellaris*	214
13.	Leaf margins toothed; plants of other habitats	14	
14 (13).	Leaves linear with essentially parallel sides; margins with very fine, forward-pointing teeth with sharp tips	*Salix interior*	208
14.	Leaves narrow but of shapes without parallel sides; margins with pointed or rounded serrations	15	

15 (14). Largest medial leaf blades with only white hairs, margins toothed; stigmas 2 plump lobes; female catkins densely to moderately densely flowered   *Salix eriocephala*   202

15.   Largest medial leaf blades often with rust-colored hairs; margins entire or toothed; stigmas flat or slender-cylindrical lobes; female catkins loosely flowered   *Salix myricoides*   212

16 (12). Branchlets white wooly; juvenile leaves very densely tomentose; petioles tomentose or wooly; largest medial leaf blades tomentose on upper surface; ovaries very densely white tomentose or wooly   *Salix candida*   196

16.   Branchlets glabrous or variously hairy but not wooly; juvenile leaves usually sparsely hairy but rarely tomentose; petioles glabrous or variously hairy but not tomentose or wooly; largest medial leaf blades range from pubescent to almost glabrous on upper surface; ovaries sparsely to very densely short-silky   17

17 (16). Leaves usually linear with very small serrations that are sharp-tipped and point toward leaf tip; forms clones by root-shoots   *Salix interior*   208

17.   Plants not as above; not clonal or forming clones by stem fragmentation or layering   18

18 (17). Largest medial leaf blades with short-silky hairs on lower surface, juveniles have similar hairs on both surfaces; branches highly brittle at base; ovaries ovoid; ripe fruiting capsules to a little more than ⅛ inch long   *Salix sericea*   218

18.   Largest medial leaf blade may range from almost glabrous, to long-silky, to tomentose on lower surface; branches not brittle at base; juvenile leaves almost glabrous, tomentose, or long-silky on both surfaces; ovaries pear-shaped; ripe fruiting capsules to nearly ½ inch long   19

19 (18). Largest medial leaf blades very narrow, with parallel sides or slightly wider in middle, serrations gland-tipped, almost glabrous or long-silky on lower surface; juvenile leaves long-silky; tall shrubs, not colonial; flowering as leaves emerge   *Salix petiolaris*   216

19.   Largest medial leaf blades various narrow shapes but always widest at or beyond middle of leaf; small glands not quite on margins; undersurface of leaf tomentose or wooly; juvenile leaves may be almost glabrous to tomentose; low shrubs (to 8 feet) forming colonies by layering; flowering before leaves emerge   *Salix humilis*   206

Key to the genus *Sambucus*

Summer

1.	New twigs and leaves smooth; pith white; flower and fruiting heads broad and flat-topped to convex; fruits dark purple	*Sambucus nigra*	100
1.	New twigs and leaves pubescent; pith light brown; flower and fruiting head oblong; fruits red	*Sambucus racemosa*	102

Winter

1.	Raised lenticels brown; pith white	*Sambucus nigra*	100
1.	Raised lenticels orange; pith light brown	*Sambucus racemosa*	102

Key to the genus *Smilax*

Summer

1.	Stems round in cross section; variably armed with slender prickles	2	
1.	Stems angled in cross section (but occasionally round); armed with stout prickles that are flattened at base	3	
2 (1).	Leaves dark green above (sometimes with lighter blotches) and white glaucous beneath, with 3 main veins from base; prickles sparse or numerous and slender, ¼ inch long	*Smilax glauca*	328
2.	Leaves dark green above and lighter green below, with 5-7 main veins that are sunken from upper surface; stem more or less covered with black, round, bristly prickles of various sizes	*Smilax tamnoides*	332
3 (1).	Leaves of variable shape but occasionally about as wide as long; stem 4-angled or round (not with grooves); leaf margin entire or with occasional teeth, margin not thickened; pale prickles often with black tip, to ½ inch long	*Smilax rotundifolia*	330
3.	Leaves variable, but often triangular or fiddle-shaped; margins entire, usually noticeably thickened with stiff prickles; upper leaf surface often with whitish patches; stems often with 1 strong angle, rough with stellate hairs at base; dark prickles to ¼ inch long	*Smilax bona-nox*	326

Winter

1.	Stems round in cross section; variably armed with slender prickles	2	
1.	Stems usually angled in cross section, armed with stout prickles that are flattened at the base	3	
2 (1).	Prickles few (usually) and scattered, round in cross section and straight or slightly curved, usually light-colored (greenish) tipped in dark pigment	*Smilax glauca*	328

2.   Prickles generally few or lacking on newer growth but
extremely dense, variable-sized (bristly stem), and black
at bases of stems   *Smilax tamnoides*   332

3 (1).   Stems strongly angled, bases (near ground) rough with
tufts of stellate hairs; prickles stout, variably colored, and
short (to ¼ inch long)   *Smilax bona-nox*   326

3.   Stems not strongly angled; prickles scattered but very stout,
light-colored (usually with dark tip), and to ½ inch long   *Smilax rotundifolia*   330

Key to the genus *Spiraea*

Summer

1.   Leaves irregularly toothed, green on top with wooly,
white/rusty pubescence below, widest toward base; shrub
rather broad with pink flowers   *Spiraea tomentosa*   226

1.   Leaves serrated, green on top and lighter green beneath,
widest toward tip; shrubs rather narrow in form with white
flowers   *Spiraea alba*   224

Winter

1.   Stems densely tomentose (matted hairs), as are the outsides
of the individual seed capsules on the residual fruiting heads;
heads 6-8 inches tall   *Spiraea tomentosa*   226

1.   Stems smooth as are the individual fruiting capsules; heads
12 inches or more tall   *Spiraea alba*   224

Key to the genus *Styrax*

Summer

1.   Shrubs to 9 feet; small branches nearly glabrous; leaves to
4 inches long, smooth on upper surface and smooth to
slightly hairy on lower; flower clusters have 1-4 flowers   *Styrax americana*   230

1.   Shrubs to 25 feet; small branches densely pubescent with
stellate hairs; leaves to 6 inches in length with dense
pubescence beneath; flower clusters have numerous flowers   *Styrax grandifolius*   232

Winter

1.   Lateral buds rather pointed, sometimes stacked in groups
of 3; young twigs densely hairy (becoming smooth later)   *Styrax grandifolius*   232

1.   Lateral buds rather blunt; stacked in groups of 2; young
twigs glabrous or with a few stellate hairs   *Styrax americana*   230

Key to the genus *Vaccinium*

Summer

1.   Plant creeping and vine-like; habitat restricted to bogs or
associated wetlands   2

1.	Plant upright shrubs; typically on drier sites	3	
2 (1).	Leaves pale beneath, but not glaucous, edges slightly rolled; flower pedicel from axils of leaves along branch (not at tip) and has 2 green bracts above middle of stalk	*Vaccinium macrocarpon*	302
2.	Leaves whitish beneath and edges strongly rolled; flower pedicel terminal with 2 reddish bracts at or below the middle of the stalk	*Vaccinium oxycoccos*	304
3 (1).	Generally large shrubs (8-12 feet tall); flowers bell-shaped rather than urn-shaped, and diverging solitarily from leaf axils on pedicels ½-1 inch long; twigs without very numerous small "warts"	4	
3.	Generally small shrubs (< 3 feet tall, except *V. corymbosum*) with urn-shaped flowers in clusters, either terminal or from leaf axils; twigs with very numerous small "warts" (verrulose), best seen with magnification	5	
4 (3).	Leaves mostly rounded or blunt at tip, thick and leathery, glossy above, entire or slightly toothed; flower (fruit) stalks without leafy bract at base; often tall, lanky, small tree	*Vaccinium arboreum*	238
4.	Leaves with pointed tips, thin and light green, dull above, entire with marginal hairs; flower/fruit stalks with leafy bract at base; usually with more limbs than *V. arboreum*	*Vaccinium stamineum*	246
5 (3).	Tall shrub (3-12 feet tall); leaves 2-3 inches long, usually without teeth; multiple habitats	*Vaccinium corymbosum*	240
5.	Small shrubs (≤ 3 feet tall); leaves usually < 2 inches long; generally in dry, upland habitats	6	
6 (5).	Leaves serrate, small (≤ 1¼ inch long), shining, and smooth on upper surface; twigs smooth	*Vaccinium angustifolium*	236
6.	Leaves entire (occasionally with few teeth near tip)	7	
7 (6).	Leaves small (≤ 1¼ inches long) with underside densely pubescent, as are twigs	*Vaccinium myrtilloides*	242
7.	Leaves larger (1¼-3 inches long), undersurface smooth to somewhat hairy (but not densely pubescent); twigs smooth or with some hairs	*Vaccinium pallidum*	244

Winter

Note: The small blueberries are inherently difficult to identify, and results of keying should always be checked in spring and summer if possible to absolutely confirm species identity.

1.	Plant creeping, vine-like, and evergreen; habitat restricted to bogs or associated wetlands	2	
1.	Plant upright, not evergreen; usually on drier sites	3	
2 (1).	Leaves pale beneath, but not glaucous; edges of leaves slightly rolled; fruits persistent and pedicels originate from leaf axils (not at tip)	*Vaccinium macrocarpon*	302

2.	Leaves whitish (glaucous) beneath and edges strongly rolled; fruit persistent and pedicel originates terminally	*Vaccinium oxycoccos*	304
3 (1).	Generally large shrubs (3-12 feet tall); twigs without very numerous small "warts" or "blisters"; buds diverge from twigs at 45 degrees or more	4	
3.	Generally small shrubs (< 3 feet tall, except for *V. corymbosum*); twigs with very numerous small "warts" (verrulose), best seen with magnification; buds generally rather appressed to stems, but if diverge, at less than 45 degrees	5	
4 (3).	A multi-branched shrub, usually to 6-9 feet tall with spreading branches; newest twigs green or gray-green; trunks gray-brown, fissures forming with some peeling with age	*Vaccinium stamineum*	246
4.	Typically a single trunk with crooked branches; twigs brownish gray becoming shiny brown with light lenticels; bark brown, sometimes with purplish cast, sloughing in irregular plates to reveal smooth, reddish brown, inner bark; somewhat evergreen in the more southerly latitudes of our region	*Vaccinium arboreum*	238
5 (3).	Large shrub, to 12 feet tall; twigs green with reddish features; leaf scar very thin transverse slit; leaf buds with scales with very thin tips that do not adhere to bud	*Vaccinium corymbosum*	240
5.	Small shrubs (< 3 feet tall)	6	
6 (5).	Twigs densely pubescent; leaf buds smooth and pointed	*Vaccinium myrtilloides*	242
6.	Twigs smooth or with a few hairs	7	
7 (6).	Twigs smooth; plants to about 2 feet or a little taller; often in extensive colonies	*Vaccinium pallidum*	244
7.	Twigs occasionally sparsely hairy in patches or lines, usually smooth; plants very small (to about 8-12 inches tall); form rhizomatous colonies	*Vaccinium angustifolium*	236

Key to the genus *Viburnum*

Summer

1.	Leaves 3-lobed	2	
1.	Leaves may have large teeth but are unlobed	3	
2 (1).	Leaves variable from smooth to somewhat hairy on both surfaces; petioles smooth with prominent glands near leaf blade	*Viburnum trilobum*	94
2.	Leaves velvet-hairy beneath; petiole pubescent without glands, but with small, papery brown stipules near attachment	*Viburnum acerifolium*	80

3 (1).	Leaves serrated, without course teeth, veins curving and/or breaking up before reaching margin		4
3.	Leaves with coarse teeth; major lateral veins reach margin and extend into teeth, may fork once or twice before doing so		7
4 (3).	Margins with rounded teeth; petioles, midrib, and to some degree undersurface with yellow-brown scurf; flower/fruit heads with peduncles (½-1 inch long)	*Viburnum cassinoides*	96
4.	Leaf margins with fine, pointed serrations		5
5 (4).	Leaf petiole (along with terminal bud that sets in early summer) with dense, rusty pubescence; upper surface of leaf very shiny, undersurface with some rusty pubescence, especially on midrib	*Viburnum rufidulum*	92
5.	Leaves and buds without rusty tomentose; upper surface smooth but not shiny		6
6 (5).	Leaves with tips abruptly long pointed; marginal teeth pointing outward; petioled winged for at least part of its length	*Viburnum lentago*	84
6.	Leaves with tips somewhat pointed to rounded; marginal teeth pointing toward tip of leaf; petiole grooved but not winged	*Viburnum prunifolium*	88
7 (3).	Leaves with deeply heart-shaped base; margin with many (16-30 on each side) teeth; bark on older twigs peeling significantly; fruits linear and flattened	*Viburnum molle*	86
7.	Leaf bases not deeply heart-shaped; bark not peeling; fruits small (~ ¼ inch) and round		8
8 (7).	Leaves with stellate pubescence below; leaf margins with > 10 pointed or rounded teeth per side	*Viburnum dentatum*	82
8.	Leaves with lower surface soft and densely hairy; leaf margins with only 4-10 teeth, occasionally appearing to have 3 major veins from base of leaf	*Viburnum rafinesquianum*	90

Winter

1.	Buds scurfy, with scales as opposed to hairs		2
1.	Buds not scurfy, smooth or pubescent toward tip; bud scales may have hairs along edges		5
2 (1).	Buds densely scurfy with red-brown scales, blunt, less than 2x as long as wide; terminal flower bud about ½ inch long with swollen base	*Viburnum rufidulum*	92
2.	Buds not densely red-brown scurfy		3
3 (2).	Buds loosely scurfy brown; end of twig may be sparsely scurfy as well; flower bud about ¾ inches long and swollen about ½ way from base; lateral buds slender (> 2x as long as wide) and curved	*Viburnum cassinoides*	96

3.	Buds dull brown to gray-brown, appear smooth, but with very small scales		4

| 4 (3). | Terminal flower bud impressively large (to 1 inch) and swollen at base; lateral buds very slender, ½ inch or more long, and 4-5x as long as wide | *Viburnum lentago* | 84 |
| 4. | Terminal flower bud about ½ inch long, swollen toward middle with short tip; lateral buds about ¼ inch long and about 2x as long as wide; limbs diverge from main limbs at about 90 degrees | *Viburnum prunifolium* | 88 |

| 5 (1). | Outer scales of buds are grown together and as long as bud; buds roundish, green (or sometimes reddish), stalked; twigs (and buds) lustrous | *Viburnum trilobum* | 94 |
| 5. | Outer bud scales shorter than inner | | 6 |

| 6 (5). | Outer pair of bud scales blunt, forming a "collar" about ⅓ way up bud; inner scales not blunt | *Viburnum rafinesquianum* | 90 |
| 6. | Buds without "collar"; scales obviously paired | | 7 |

| 7 (6). | Bark is gray on second-year twigs and peels on these and older stems to reveal reddish brown inner bark | *Viburnum molle* | 86 |
| 7. | Bark not peeling | | 8 |

| 8 (7). | Twigs pubescent; outer bud scales very short, covering only about ¼ of the bud; frequently a shrub in the understory of deciduous woods | *Viburnum acerifolium* | 80 |
| 8. | Twigs smooth or with coarse hairs in groups; outer bud scales about ½ the length of the bud and strongly keeled | *Viburnum dentatum* | 82 |

Key to the genus *Vitis*

Summer

| 1. | Leaves broad with thick, felt-like tomentose, white to rusty-colored, which obscures the lower leaf surface; stem has tendrils or flower stalk opposite every leaf; large fruits, to 1 inch diameter | *Vitis labrusca* | 340 |
| 1. | Leaves glabrous beneath or with pubescence that does not obscure lower leaf surface; tendrils/flower stalks not opposite every leaf | | 2 |

| 2 (1). | Leaves often folded up on either side of center vein so that lighter underside is visible; a sprawling, trailing growth form, rarely climbing; tendrils quite variable, mostly absent except at uppermost leaves or tips of flowering stems (however, can be prominent in those locations) | *Vitis rupestris* | 346 |
| 2. | Leaves not typically folded, stem with tendrils or flower stalks absent at every third leaf | | 3 |

3 (2).	All leaves strongly lobed (typically 5) with deep, rounded sinuses, terminal lobe often triangular; new stems are bright purplish red	*Vitis palmata*	342
3.	Some leaves may be deeply lobed, but not all; new stem growth green to brown, occasionally with reddish blush	4	
4 (3).	Leaves green and glabrous beneath when mature except for straight hairs on veins and axes where veins meet, may be pubescent beneath when young but with short, straight hairs	5	
4.	Leaves pubescent or white glaucous at maturity; pubescence often with long, cobweb-like hairs	6	
5 (4).	Leaves not lobed, or with very short lobes that tend to point outward; teeth somewhat rounded in shape but with pointed tip; veins beneath generally have straight hairs; basal sinus (near petiole) narrow	*Vitis vulpina*	348
5.	Leaves generally with 2 short side lobes; teeth acute, pointed, and prominent, margins with short hairs; veins with some straight hairs; basal sinus wide	*Vitis riparia*	344
6 (4).	Young branches and stems round; leaf margins of some unlobed, other leaves may be shallowly or deeply 3- to 5-lobed; underneath of leaf with white cobweb-like pubescence; new leaves green on top but with cobweb-like hairs	*Vitis aestivalis*	336
6.	Young branches and stems angled; leaf margins unlobed or with 2 very small side-lobes; lower surface of older leaves with cobweb like pubescence and upper surface has some pubescence on veins; new leaves appear white, heavily covered with cobwebbing	*Vitis cinerea*	338

Winter

Note: Grapes are often very difficult to identify accurately even when in-leaf; identification in winter is tenuous. Readers are encouraged to mark plants in question and return several times during the year (as is necessary with *Salix* and several other genera) to confirm species using multiple characteristics.

1.	Stem with tendrils at each node and pubescent with red hairs; nodes that have (had) fruiting stem will not have tendril	*Vitis labrusca*	340
1.	Stems missing tendrils at some nodes	2	
2 (1).	Stems (red and smooth) with irregular tendrils, often with few or none; a running vine (does not generally climb), often dominating sandbars and other open areas near streams	*Vitis rupestris*	346
2.	Stems with tendril missing from every third node; vines climbing with peeling brown bark	3	
3 (2).	Stems slender and bright red, smooth, and finely striped; vine of very wet areas such as swamps and sloughs	*Vitis palmata*	342
3.	Stems not bright red	4	

4 (3).	Stems strongly angled, grayish, pubescent, with cobweb-like hairs	*Vitis cinerea*	338
4.	Stems not strongly angled, although all may be finely ridged	5	
5 (4).	Newest stems reddish brown, covered with red pubescence, which is lost later	*Vitis aestivalis*	336
5.	Newest stems usually retain some greenish coloration, smooth	6	
6 (5).	Very similar to *Vitis vulpina* in appearance and location, although generally occurs on wetter sites; woody partition at nodes (that divides areas of brown pith) thin, only about $1/_{16}$ inch thick	*Vitis riparia*	344
6.	Very similar to *Vitis riparia* in appearance and location, although this species may occupy drier sites as well; woody partition at nodes thicker, $1/_8$-$1/_4$ inch thick	*Vitis vulpina*	348

Plant Hardiness Zone Map

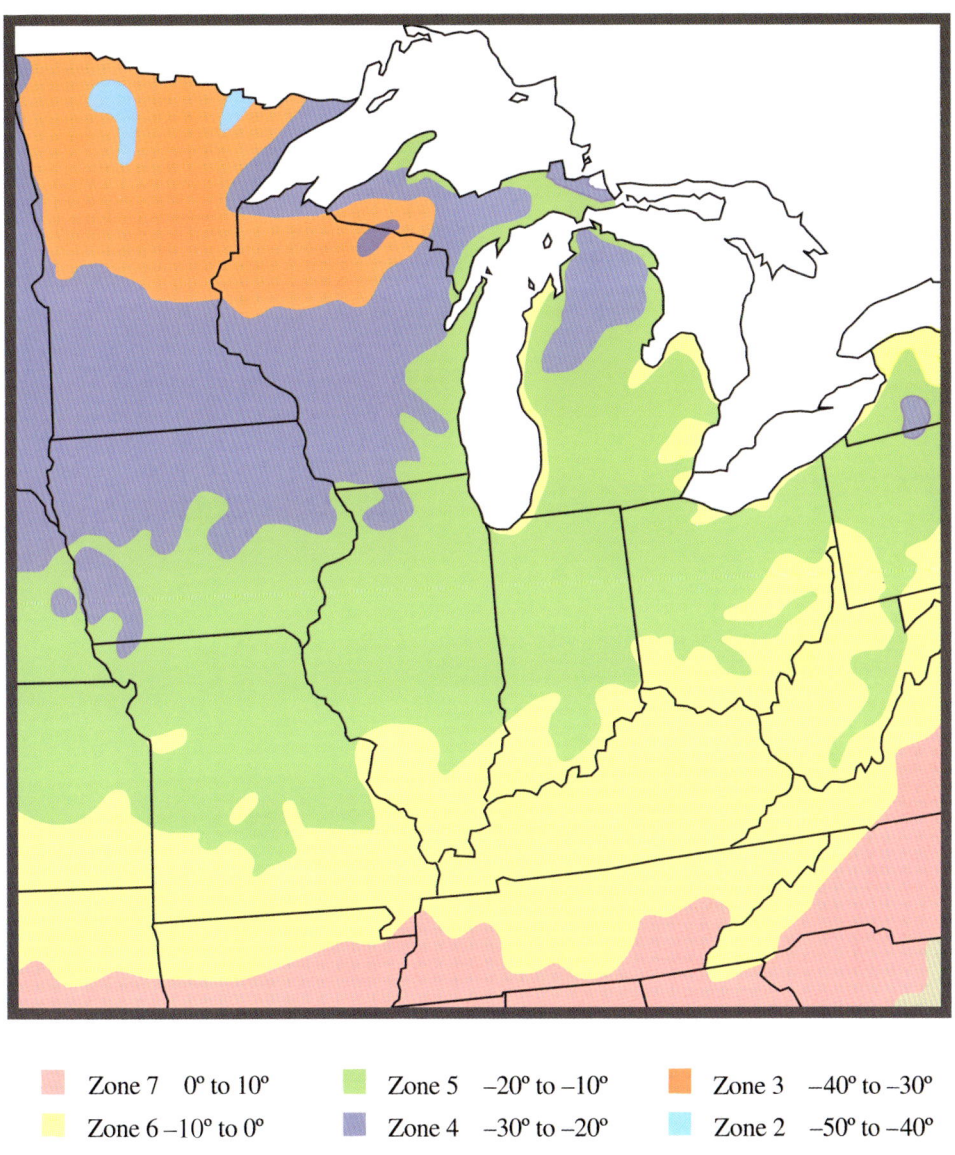

Zone 7	0° to 10°	Zone 5	−20° to −10°	Zone 3	−40° to −30°
Zone 6	−10° to 0°	Zone 4	−30° to −20°	Zone 2	−50° to −40°

Bibliography

Allen, C. M., D. A. Newman, and H. H. Winters. 2002. *Trees, Shrubs and Woody Vines of Louisiana*. Pitkin, Louisiana: Allen's Native Ventures, L.L.C.

Barnes, B. V., and W. H. Wagner, Jr. 1981. *Michigan Trees*. Ann Arbor: University of Michigan Press.

Billington, C. 1949. *Shrubs of Michigan*. Bloomfield Hills, Michigan: Cranbrook Institute of Science.

Braun, E. L. 1989. *The Woody Plants of Ohio*. Columbus: The Ohio State University Press.

Cullina, W. 2002. *Native Trees, Shrubs, and Vines*. New York: Houghton Mifflin Co.

Czarapata, E. J. 2005. *Invasive Plants of the Upper Midwest*. Madison, Wisconsin: University of Wisconsin Press.

Deam, C. C. 1932. *Shrubs of Indiana*. Indianapolis: Indiana Department of Conservation.

DeGraaf, R. M. 2002. *Trees, Shrubs, and Vines for Attracting Birds*. 2nd ed. Lebanon, New Hampshire: University Press of New England.

Dickinson, T. A. 2011. http:// labs.eeb.utoronto. ca/Dickinson/Crataegus/html.

Dirr, M. A. 1998. *Manual of Woody Landscape Plants*. Champaign, Illinois: Stipes Publ. L.L.C.

Dirr, M. A. 2007. *Viburnums*. Portland, Oregon: Timber Press.

Flora of North America Editorial Committee. 2009. *Flora of North America North of Mexico*. Vol. 8. New York: Oxford University Press, Inc.

Flora of North America Editorial Committee. 2010. *Flora of North America North of Mexico*. Vol. 7. New York: Oxford University Press, Inc.

Flora of North America Editorial Committee. In press. *Flora of North America North of Mexico*. Vol. 9. New York: Oxford University Press, Inc.

Foote, L. E., and S. B. Jones, Jr. 2001. *Native Shrubs and Woody Vines of the Southeast*. Portland, Oregon: Timber Press.

Gill, J. D., and W. M. Healy. 1974. *Shrubs and Vines for Northeastern Wildlife*. Forest Service General Technical Report NE–9.

Gleason, H. A., and A. Cronquist. 1998. *Manual of Vascular Plants of Northeastern United States and Adjacent Canada*. New York: The New York Botanical Garden.

Graves, A. H. 1992. *Illustrated Guide to Trees and Shrubs: A Handbook of the Woody Plants of the Northeastern United States and Adjacent Canada*. New York: Dover Publishing, Inc.

Harlow, W. H. 1942. *Trees of the Eastern United States and Canada: Their Woodcraft and Wildlife Uses*. New York: McGraw-Hill Book Co., Inc.

Hightshoe, G. L. 1988. *Native Trees, Shrubs and Vines for Urban and Rural America*. New York: John Wiley and Sons, Inc.

Hunter, C. G. 1989. *Trees, Shrubs and Vines of Arkansas*. 2nd ed. Little Rock: The Ozark Society Foundation.

Jaynes, R. A. 1997. *Kalmia*. Portland, Oregon: Timber Press.

Johnson, W. T., and H. H. Lyon. 1991. *Insects that Feed on Trees and Shrubs*. Ithaca, New York: Cornell University Press.

Kurz, D. 1997. *Shrubs and Woody Vines of Missouri*. Jefferson City: Missouri Department of Conservation.

Leopold, D. J., W. C. McComb, and R. N. Muller. 1998. *Trees of the Central Hardwood Forests of North America*. Portland, Oregon: Timber Press.

Martin, A. C., H. S. Zim, and A. L. Nelson. 1951. *American Wildlife and Plants: A Guide to Wildlife Food Habits*. New York: McGraw-Hill Book Co., Inc.

Miller, J. H., and K. V. Miller. 1999. *Forest Plants of the Southeast and Their Wildlife Uses*. Auburn, Alabama: Southern Weed Science Society, Craftmaster Printers.

Petrides, G. A. 1986. *A Field Guide to Trees and Shrubs*. Boston, Massachusetts: Houghton Mifflin Co.

Sinclair, W. A., H. H. Lyon, and W. T. Johnson. 1987. *Diseases of Trees and Shrubs*. Ithaca, New York: Cornell University Press.

Smith, W. R. 2008. *Trees and Shrubs of Minnesota*. Minneapolis: University of Minnesota Press.

Soper, J. H., and M. L. Heimburger. 1982. *Shrubs of Ontario*. Toronto: Royal Ontario Museum.

Swink, F., and G. Wilhelm. 1994. *Plants of the Chicago Region*. 4th ed. Indianapolis: Indiana Academy of Science.

Symonds, G. W. D. 1963. *The Shrub Identification Book*. New York: William Morrow & Co.

Taylor, P. A. 1996. *Easy Care Native Plants*. New York: Henry Holt and Co., Inc.

Vine, R. A. 1990. *Trees, Shrubs, and Woody Vines of the Southwest*. 7th ed. Austin: University of Texas Press.

Voigt, T. B., B. R. Hamilton, and F. A. Giles. 1983. *Ground Covers for the Midwest*. Champaign, Illinois: University of Illinois at Urbana-Champaign.

Voss, E. G. 1985. *Michigan Flora*. Part II. Ann Arbor, Michigan: Cranbrook Institute of Science.

Voss, E. G. 1996. *Michigan Flora*. Part III. Ann Arbor, Michigan: Cranbrook Institute of Science.

Week, S. S., H. P. Weeks, Jr., and G. R. Parker. 2010. *Native Trees of the Midwest*. 2nd ed. West Lafayette, Indiana: Purdue University Press.

Wharton, M. E., and R. W. Barbour. 1973. *Trees and Shrubs of Kentucky*. Lexington: University of Kentucky Press.

Species Index

Numbers in bold type indicate detailed species description.

453

About the Authors

Sally S. Weeks was born and grew up on a dairy farm near Winamac, Indiana. She received a BSF in Wildlife Management and an MS in Forestry from Purdue's Department of Forestry and Natural Resources, where she now teaches tree and shrub identification.

Harmon P. Weeks, Jr. is a professor of Wildlife Science in Forestry and Natural Resources at Purdue University and has taught habitat management for over thirty years.

Michael A. Homoya is State Botanist for the Indiana Division of Nature Preserves, holds two degrees in Botany from Southern Illinois University, and is renowned for his knowledge of midwestern flora.